Synthesis Lectures on Mathematics & Statistics

Series Editor

Steven G. Krantz, Department of Mathematics, Washington University, Saint Louis, USA

This series includes titles in applied mathematics and statistics for cross-disciplinary STEM professionals, educators, researchers, and students. The series focuses on new and traditional techniques to develop mathematical knowledge and skills, an understanding of core mathematical reasoning, and the ability to utilize data in specific applications.

Sam Efromovich

Survival Analysis

Efficient Nonparametric Curve Estimation
For Censored Data with R Examples

 Springer

Sam Efromovich
Department of Mathematical Sciences
The University of Texas at Dallas
Richardson, TX, USA

ISSN 1938-1743   ISSN 1938-1751   (electronic)
Synthesis Lectures on Mathematics & Statistics
ISBN 978-3-031-82813-3   ISBN 978-3-031-82814-0   (eBook)
https://doi.org/10.1007/978-3-031-82814-0

This Springer imprint is published by the registered company Springer Nature Switzerland AG
The registered company address is: Gewerbestrasse 11, 6330 Cham, Switzerland

If disposing of this product, please recycle the paper.

To My Family and My Students

Preface

Survival analysis studies duration of time, also called lifetime, until one or more events of interest occur, such as failure in mechanical system, death in biological organism, the waiting time until hospitalization, duration of a strike, the time for a startup company to its bankruptcy, the length of stay at a hospital, and the total amount paid by a health insurance. This is why survival analysis is a classical branch of statistics. Because the interest is in analysis of lifetimes, the primary interest of survival analysis is in estimation of the following three characteristics of the lifetime. (1) Survival function which is the probability that the lifetime will be longer than a specific time. (2) Probability density which is the negative derivative of the survival function and is the best characteristic for visualization of the distribution. (3) Hazard rate function which defines the instantaneous likelihood that the event occurs within a specific small interval given that it has not occurred earlier. The hazard rate quantifies the trajectory of imminent risk, and it may be referred to by other names in different sciences, for instance, the failure rate in reliability theory and the force of mortality in actuarial science and sociology. If a predictor (covariate of interest) is available, then the survival analysis is also interested in prediction of the lifetime of interest given the predictor.

A key characteristic of survival analysis, that distinguishes it from other areas of statistics, is that survival data are usually censored. This book considers two types of censoring. The former is the *right-censoring*, and the latter is the *case I interval censoring* also called the *current status censoring* and abbreviated as CSC. Under the right-censoring, the available observation is the smaller among the lifetime of interest and a censoring lifetime. Using formulas, if T is the lifetime of interest and C is the censoring lifetime, then under the right-censoring we observe a sample from the pair $(V, \Delta) := (\min(T, C), I(T \leq C))$ and Δ is called the indicator of censoring. Note that, under the right-censoring, we observe either the lifetime of interest T (the *uncensored observation V*) if $\Delta = 1$ or the censoring variable C (the *censored observation V*) if $\Delta = 0$. For instance, in a clinical trial of 5 years to study effects of a surgery, the censoring occurs if a participant leaves the study before the event of interest occurs or the event of interest does not occur during the study. Accordingly, the censoring lifetime C is the time of either leaving the study or its end. Under the current status censoring (CSC), the lifetime of interest is never observed, instead it is known that the event of interest already occurred or not at a known monitoring time Z. Therefore, we observe a sample from the pair $(Z, \Delta) := (Z, I(T \leq Z))$ and never observe the lifetime of interest T. For example, let the lifetime of interest be the time from a contact with the source of infection until becoming infected. A test at a monitoring time Z can show that a subject is either infected or not at the time of testing, but the time of interest T of becoming infected is still unknown. Hence we know Z and the *status* $I(T \leq Z)$ but not the T. The current status censoring is a dramatically more complicated data modification than the right-censoring because the lifetime of interest is never observed. Nonetheless, we will be able to propose optimal estimators. Furthermore, understanding of the CSC sheds light on optimal estimation for right-censored data, and this is why in the book the CSC is discussed first.

One of the important aims of the book is to explain what can be done if the rate of censoring is high. As it was above-explained, right-censoring divides observations into uncensored observations when the lifetimes of interest are observed, and censored observations when censoring times are observed. The two pillars of the survival analysis, the product-limit Kaplan–Meyer methodology of estimating survival functions and the partial likelihood Cox methodology of regression, are based on the paradigm of dominance of uncensored observations over censored ones. Moreover, censored observations are often referred to as "losses", the famous Kaplan–Meier estimator has jumps only at uncensored observations, and the classical Buckley–James method imputes censored observations (the losses) using uncensored ones. There are asymptotic theories supporting the dominance of uncensored observations, but what can be done for finite samples with high rate of censoring? Then it is prudent to understand how to aggregate censored and uncensored observations, and solution of this practically important problem is explored. Let us comment on it using analysis of a practical example.

The statistical approach, used in the book, is nonparametric meaning that no assumptions about shape of the underlying distribution of the lifetime of interest T or shape of nuisance distributions, like the censoring C or monitoring time Z, are made. Figure 1 sheds light on the nonparametric approach and the importance of the aggregation in analysis of right-censored data. Here we study the breast cancer data from the R package condSURV. The lifetime of interest T is the time in days from cancer treatment to the cancer recurrence. The available right-censored data are shown in the left-top diagram by the circles. Note that $\Delta = 1$ indicates an observed uncensored lifetime $T = V$, and $\Delta = 0$ indicates an observed censoring time $C = V$ when we only know that at time V the participant was cancer-free. There are $n = 686$ participants (observations, cases), and among them $N := \sum_{l=1}^{n} \Delta_l = 299$ participants experienced the cancer recurrence during the study. Accordingly, the rate of censoring is 66%, and by all means it is very high. The solid stepwise line is the famous Kaplan–Meier estimate of the survival function $S^T(v) := \mathbb{P}(T > v)$. The estimate jumps at uncensored observations, and to better realize that look at its right tail. Censored observations are used only to define the size of each step. For the data at hand, the Kaplan–Meier estimate provides us with little information about shape of the underlying distribution beyond 2000 days because there are only several large uncensored observations. At the same time, we have many large censored observations, and this explains the large steps in the right tail of the Kaplan–Meier estimate. From a pure theoretical point of view, the Kaplan–Meier survival function estimator is efficient (theoretically optimal), but that optimality is asymptotic as $n \to \infty$. In the considered breast cancer data the sample size n is relatively large but the high rate of censoring yields only $N = 299$ uncensored observations, and this number is not large enough for the onset of asymptotic theory. The issue of a more efficient using censored observations for small samples is addressed in the book. Further, it is difficult to make a conclusion about an underlying distribution of T via visualization of its survival function. The latter is easier to do via analysis of the probability density of T which is the negative derivative of the survival function. Due to the stepwise nature of the Kaplan–Meier estimator, it is a challenging problem to visualize the corresponding derivative, and special methods for constructing density estimators are presented in the book. Let us consider some of them.

One of the tempting approaches to deal with right-censored data is to ignore censored observations and consider uncensored observations as a sample from T. After all, if we are interested in the time to cancer recurrence, why not look at those in the study who experienced the recurrence? This approach is called *naive*, it is not scientific but often used by practitioners because it is easy for interpretation and also feasible when the rate of censoring is low. The naive approach is illustrated in the right-top diagram by the histogram of $N = 299$ uncensored lifetimes overlaid by the corresponding density E-estimate (the solid line) and its 95% confidence band (the dashed lines) explained in Chap. 1. Note how the density estimate smooths the histogram and exhibits several modes reflecting several categories of participants. It is clear that the distribution is far from being the traditionally assumed Normal distribution which is symmetric about its mean and whose famous bell-shaped density is defined by the mean and the standard deviation. What are the consequences of the shown asymmetric density? One of the typical questions, that a cancer patient explores before a treatment, is how long you may live cancer-free after the treatment. A typical online information is about average time and, sometimes, about standard deviation. This answer would be informative for a Normal distribution but not for the one seen in the right-top diagram. This is why the nonparametric approach is feasible in survival analysis. Now, returning to the naive approach, overall it is grossly unfair to efficiency of the underlying cancer treatment because it ignores "censored" patients who do not experience the cancer recurrence during the study. In other words, this approach ignores the better cases and skews the underlying distribution to the right.

Let us also add a formula that sheds light on the naive approach. Assume that both the lifetime of interest T and the censoring lifetime C are continuous and independent random variables. Denote by f^T and S^T the density and survival function of T, and by f^C and S^C the density and survival function of the censoring variable C, respectively. Then for the observed pair (V, Δ) we have the following formula for the joint mixed density

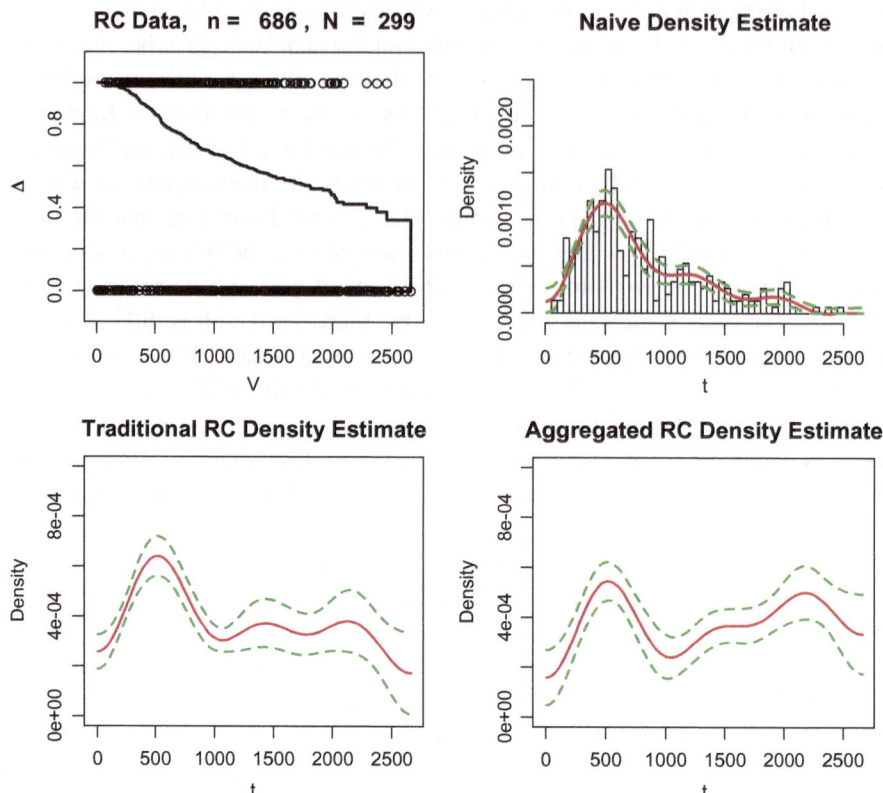

Fig. 1 Analysis of right-censored (RC) lifetimes T, in days, from breast cancer treatment to the cancer recurrence. The left-top diagram shows by the circle the RC data, namely the circles with $\Delta = 1$ show the uncensored observations when $V = T$, while the circles with $\Delta = 0$ show censored observations when $V = C$ with C being the censoring variable. The solid stepwise line is the famous Kaplan–Meier survival function estimate. The right-top diagram exhibits the histogram of uncensored observations. Here and in the other diagrams the solid line is the density estimate and the dashed lines show 95% confidence band. {Similarly to other figures in the book, the complementary R software allows the reader to repeat this figure with the same or different arguments of the used estimator, see Sect. 1.3. The arguments are shown in the square brackets.} [cJ0 $= 3$, cJ1 $= 0.8$, cTH $= 4$, alpha $= 0.05$]

(the likelihood):

$$f^{V,\Delta}(v, \delta) = [f^T(v)S^C(v)]^\delta [f^C(v)S^T(v)]^{1-\delta}. \tag{1}$$

Accordingly, if we restrict our attention to only uncensored observations with $\Delta = 1$, we get the following formula for the conditional density of uncensored observations:

$$f^{V|\Delta}(v|1) = \frac{S^C(v)}{\mathbb{P}(\Delta = 1)} f^T(v). \tag{2}$$

As we see, the uncensored observations are not from the underlying distribution of T, they are biased by the function S^C, and this bias should be taken into account.

A standard RC density estimate does exactly that and cures the above-explained bias. Such an estimate, explained in Chap. 3, is shown in the left-bottom diagram. It can be thought as the derivative of a smoothed Kaplan–Meier survival function estimate shown in the left-top diagram. This density estimate dramatically improves our opinion about the cancer-free living, and to see that look at its heavier right tail and compare scales of the y-axes.

Can that opinion about cancer-free living be improved even further by more rigorous aggregation of the censored observations? To answer the question, let us for a moment return to formula (1) and the case of a censored observation. Given $\Delta = 0$, we get the following conditional density of a censored observation,

$$f^{V|\Delta}(v|0) = \frac{f^C(v)}{\mathbb{P}(\Delta = 0)} S^T(v). \tag{3}$$

The formula presents both good and bad news. The good news is that we can estimate the survival S^T, and hence consistent estimation of the density f^T is possible. The bad news is: (i) Estimation of S^T is based on estimation of the density $f^{V|\Delta}$ which can be done with slower rates than estimation of S^T based on direct observations of T, and therefore the problem is inherently ill-posed; (ii) The censored observations are biased by the density f^C, which must be known or estimated; (iii) The density of interest f^T must be "restored" from an estimated S^T, and that restoration is a classical ill-posed problem on its own. In short, estimation based on censored observations is possible but it is ill-posed meaning asymptotically slower rates of risk convergence with respect to uncensored observations. Nonetheless, for small samples and high rates of censoring all that complications can be overcome, and the aggregated density estimate, explained in Chap. 3, is presented in the right-bottom diagram. Note how it improves our opinion about the cancer-free living due to more effective aggregation of censored observations.

To finish our brief introduction to the aggregation for right-censored data, let us note that it will be shown in Chap. 3 that the aggregation is always feasible for nonparametric regression where censored observations are no longer ill-posed with respect to uncensored ones. In other words, for some estimands censored observations are ill-posed and for others not, and we will learn about that in the book.

Now let us more formally explain the current status censoring (CSC). Here at a random monitoring time Z it is known that an event of interest already occurred or not occurred, and we observe the pair $(Z, \Delta) := (Z, I(T \leq Z))$. Accordingly, in a CSC sample, there are two subsamples with events of interest that already *occurred* (let us refer to these observations as CSCO) and *not occurred* (let us refer to them as CSCNO). Under a mild

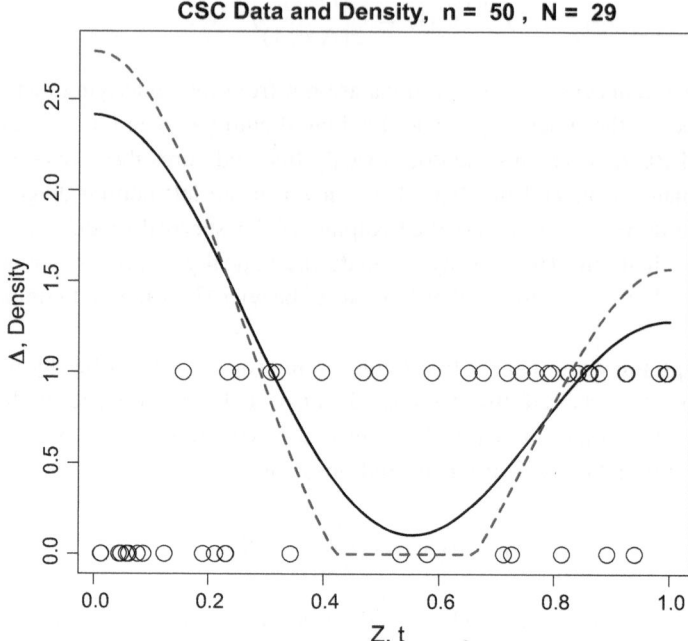

Fig. 2 Analysis of simulated CSC data. The pairs (Z_l, Δ_l), $l = 1, 2, ..., n$ are shown by the circles, $N := \sum_{l=1}^{n} \Delta_l$ is the size of CSCO subsample. The underlying density of T is the second corner function called the Bathtub, the distribution of independent monitoring time Z is Uniform$(0,1)$. The solid and dashed lines are the underlying density of T and its E-estimate, respectively. {The sample size n and the density T can be changed by the arguments n and corn, respectively, as well as parameters $cJ0$, $cJ1$ and cTH of the E-estimator.} [n = 50, corn = 2, cJ0 = 3, cJ1 = 0.8, cTH = 4]

assumption, the likelihood of a CSC observation, that is the joint mixed density of pair (Z, Δ), is defined by the formula

$$f^{Z,\Delta}(z, \delta) = f^Z(z)[1 - S^T(z)]^{\delta}[S^T(z)]^{1-\delta} I(z \geq 0)I(\delta \in \{0, 1\}). \tag{4}$$

Interestingly, this formula and (1) show that a CSCNO observation is identical to a censored observation whenever C and Z have the same distribution. Accordingly, as we already know from our discussion of censored observations, consistent estimation of S^T is possible but ill-posed. Similarly, a CSCO observation is also ill-posed, and accordingly, we will need to aggregate two ill-posed subsamples for efficient distribution estimation.

Figure 2 sheds light on CSC data and the problem of density estimation. The shown CSC observations are simulated and hence we know the underlying density of interest f^T, see details in the caption. This approach is informative because we can compare the proposed estimate (the dashed line) with the underlying density (the solid line), and simulations are often used in the book. The observations from $(Z, \Delta) := (Z, I(T \leq Z))$ are shown by the circles. The sample size is very small for the nonparametric density estimation, but it allows us to better visualize the data. There is no resemblance in the CSC data with the underlying density because we do not observe the lifetime of interest T. Nonetheless, the estimate nicely shows the bathtub shape. Again, it is difficult to believe that density estimation is possible for CSC data, and this is when simulations and the software may help us to gain first-hand experience in dealing with the CSC.

Finally, let us note that while CSC makes distribution estimation ill-posed with respect to direct observations of the lifetime of interest (it slows down the optimal rate of a risk convergence), nonparametric regression is not ill-posed with respect to the case of direct observations. We are discussing in the book different CSC problems where the rate of estimation remains the same or slows down.

This ends our brief introduction to the main censoring models studied in the book.

The book is self-contained, has a brief review of necessary facts from probability as well as from parametric and nonparametric statistics. It covers a wide range of nonparametric topics including estimation of survival functions, densities, hazard rate functions, regressions, conditional distributions, as well as adaptation, consistency, sharp minimax, and confidence bands among others.

Used as a text, it is appropriate for a one-semester course for diverse classes with master's-level and Ph.D.-level students in data science, statistics, and other sciences including engineering, business, social, medical, and biological among others. It is also suitable for practitioners and researchers who want to learn modern nonparametric survival analysis. A large number of exercises, with different levels of difficulty, will guide the reader and help an instructor to test understanding of the material.

There are three chapters in the book. Chapter 1 presents a brief review of random variables and their characteristics. Then the universal nonparametric E-estimator is introduced and explained for several classical settings of survival function, density, regression, and hazard rate estimation based on directly observed data. The complementary R software, that allows the reader to repeat and change figures in the book, is also explained. Current status censoring and aggregation of CSCO and CSCNO observations for distribution and regression estimation are discussed in Chap. 2. Because we are interested in optimal aggregation, each section contains the asymptotic theory, that explains what can and cannot be done for data at hand, and then the corresponding methodology and the recommended for small sample E-estimator are proposed. Numerical simulations and real-life examples will help the reader to get up to speed quickly on developed nonparametric methods. Chapter 3 is devoted to right-censored data. Here all classical topics, including estimation of survival function, hazard rate function, density, regression, conditional distribution, and mean residual life are considered.

Each chapter contains exercises that allow the reader to polish understanding of the topics, open problems for future research, and notes that contain a discussion and bibliographical references.

For the reader who would like to use this book for self-study and who is venturing for the first time into this area, the advice is as follows. Begin with Sect. 1.1 and review basic probability facts, and make yourself familiar with the complementary R software discussed in Sect. 1.3. Note that no knowledge of this freely available statistical software is required, and all steps in its installation and using are explained. Then review the classical models and nonparametric series E-estimator in Sect. 1.2. Please keep in mind that the same E-estimator is used for all studied problems. The rest of the reading depends on the interest in this or that topic. The reader, who would like to know the theory, will find it in the first subsection of a section. The reader who is more interested in methodology, may skip the theory and check the proposed E-estimator and its performance via simulations and analysis of real examples. A majority of figures are based on simulations, and hence you know the underlying model and can appreciate the available data and nonparametric estimates. The software also allows the reader to change parameters of the estimator. For instance, the reader will be able to change parameters of the density E-estimator used in Fig. 1.

The last but not the least remark is about the used terminology. It is traditional for the survival analysis, and the interested reader may "translate" its notions into those used in the science of interest. For instance, in the actuarial theory of loss models the lifetime is called the loss, the mean residual life is called the mean excess loss, etc.

To find updates to the book, use the link located at http://profiles.utdallas.edu/efrom in the section Publications, and the author may be contacted by electronic mail at efrom@ utdallas.edu.

Dallas, TX, USA   Sam Efromovich
2025

Acknowledgments

I thank everyone who, in various ways, has had an influence on this book. My biggest thanks go to my family for the constant support and understanding. My students, colleagues, and reviewers graciously read and gave comments on a draft of the book. Many thanks to colleagues and friends at the UT Dallas, UT Southwestern Medical School, the University of California at San Diego, the Moore Cancer Center, the BIFAR, and the Scripps Hospital for datasets and numerous discussions and consultations. Susanne Filler provided invaluable assistance through the publishing process. The support by NSF Grant DMS-1915845, NSA Grant H982301310212, and actuarial research Grants from the CAS, TAF, and CKER are greatly appreciated.

Dallas, TX, USA   Sam Efromovich
2025

Contents

Acronyms

:=	Equal by definition
$\otimes$	Tensor product
$I(\bullet)$	Indicator function
A	The availability: $A = 1$ if a variable is not missed and $A = 0$ otherwise
T	Lifetime of interest
C	Censoring lifetime
V	The minimal among lifetime of interest and censoring lifetime
Z	Monitoring time for current status censoring
Δ	The indicator of right-censoring $\Delta := I(T \leq C)$, or the status $\Delta := I(T \leq Z)$ for current status censoring
Δ'	$\Delta' := 1 - \Delta$
n	Sample size
r	The restriction, used for estimation of a function over interval [0, r]
ζ	Generic positive constant
$\lceil x \rceil$	The smallest integer not smaller than x
J_n	Sequence $J_n := \lceil c_{J0} + c_{J1} \ln(n) \rceil$, c_{J0} and c_{J1} are constants
φ_j	Elements of the cosine basis on [0, 1]
η_{rj}	Elements of the cosine basis on [0, r]
ψ_i	Elements of the sine basis on [0, 1]
θ_j	Fourier coefficient
$O_n(1)$	A bounded sequence in n
$o_n(1)$	A vanishing sequence in n
$\mathbb{E}\{T\}$	Expectation of the random variable T
$\mathbb{V}\{T\}$	Variance of the random variable T
$\mathbb{E}\{T|X\}$	Conditional expectation of the random variable T given X
$\mathbb{P}(\mathcal{B})$	Probability of the event $\mathcal{B}$
$\mathbb{P}(\mathcal{B}|X)$	Conditional probability of the event $\mathcal{B}$ given X
f^T	Probability density of continuous random variable T

$f^{T\|X}$	Conditional density of T given X
h^T	Hazard rate function
$h^{T\|X}$	Conditional hazard rate function of T given X
S^T	Survival function of T
$S^{T\|X}$	Conditional survival function of T given X
$w(x)$	The availability likelihood, $w(X) := \mathbb{P}(A = 1\|X = x)$
CSC	Current status censoring
RC	Right-censoring
MAR	Missing at random
MCAR	Missing completely at random
MNAR	Missing not at random
MISE	Mean Integrated Squared Error

Introduction

1

Survival analysis is interested in nonnegative random variables and their characteristics such as survival function, probability density, hazard rate and regression. Nonparametric curve estimation allows one to analyze data and to estimate the above-mentioned characteristics without assuming the shape of an estimated curve. This chapter reviews basics of probability, mathematical statistics and nonparametric curve estimation for directly observed random variables, and presents discussion of typical data modifications in survival analysis.

Section 1.1 is a brief probability review of random variables and their main characteristics. Section 1.2 is a review of nonparametric curve estimation. In particular, it presents results on orthogonal series approximation and the E-estimation algorithm used throughout the book. Installation of the book's R package and a short tutorial on its use can be found in Sect. 1.3. Exercises in Sect. 1.4 will allow the reader to review basics of probability and statistics. Section 1.5, called the Notes, presents historical and bibliographical remarks.

1.1 Random Variables and Their Characteristics

This section is devoted to a brief introduction to notions and basic results of probability, statistics, and survival analysis that are used in the book.

1.1.1 Traditional Random Variables

Many datasets, that are of theoretical and practical interest, are generated by a *random experiment* which is an act or process that leads to an outcome that cannot be predicted with certainty in advance. For instance, in a clinical study of the time to cancer relapse after

© The Author(s), under exclusive license to Springer Nature Switzerland AG 2026   1
S. Efromovich, *Survival Analysis*, Synthesis Lectures on Mathematics & Statistics,
https://doi.org/10.1007/978-3-031-82814-0_1

a surgery, outcomes are not known in advance, but we can always suppose that the set of possible outcomes is known. We shall refer to such a set as the *sample space* and denote it by Ω. A particular collection of possible outcomes is called an *event*. If D_1 and D_2 are two events, then: Their *union* is the event that occurs if either D_1 or D_2 or both occur on a single performance of the experiment, and it is denoted by the symbol $D_1 \cup D_2$; Their *intersection* is the event that occurs if both D_1 and D_2 occur on a single performance of the experiment, and it is denoted by the symbol $D_1 \cap D_2$; If the two events never occur simultaneously (they are *mutually exclusive*), then we define the *null* event $\emptyset := D_1 \cap D_2$, and if additionally $D_1 \cup D_2 = \Omega$, then D_2 is the *complementary* event of D_1 (and vise versa) and we use the notation $D_2 = D_1^c$, in particular $\emptyset = \Omega^c$.

The theory of probability assigns the *probability* (likelihood) to events, and we shall denote the probability of an event D by $\mathbb{P}(D)$. There are *three axioms of probability*: (1) The probability of any event should be between 0 and 1, that is $0 \leq \mathbb{P}(D) \leq 1$; (2) The probability that an outcome of a random experiment belongs to the sample space Ω is 1, that is $\mathbb{P}(\Omega) = 1$; (3) For any countable sequence of mutually exclusive events $D_1, D_2, \ldots$, that is $D_i \cap D_j = \emptyset$ whenever $i \neq j$ and the latter we can write down using the formula $\mathbb{P}(D_i \cap D_j) = \mathbb{P}(D_i)I(i = j)$, the following relation holds, $\mathbb{P}(\cup_{i=1}^{\infty} D_i) = \sum_{i=1}^{\infty} \mathbb{P}(D_i)$. Further, events D_1 and D_2 are called *independent* if $\mathbb{P}(D_1 \cap D_2) = \mathbb{P}(D_1)\mathbb{P}(D_2)$. It is important to stress that for events the notions of independence and mutual exclusivity are not the same. For instance, the event D, such that $0 < \mathbb{P}(D) < 1$, and the complementary event D^c are mutually exclusive but dependent because $\mathbb{P}(D \cap D^c) = 0$ while $\mathbb{P}(D)\mathbb{P}(D^c) = \mathbb{P}(D)[1 - \mathbb{P}(D)] > 0$.

Statistics is primarily interested in random quantities of interest that are numbers. These quantities of interest, or more formally, real-valued functions defined on a sample space, are known as *random variables*. A random variable that can take at most a countable number of possible values is said to be a *discrete random variable*. For instance, if we have 10 patients who were last week in contact with an infected individual, then the random number X of them who will be sick next week is the discrete random variable. Random variables are always denoted by upper case letters like X, Y, Z, T, A, C, Δ while their values by corresponding lower case letters, for instance we write $\mathbb{P}(X = x)$ for the probability that the discrete random variable X is equal to value x. Classical examples of discrete random variables, that are frequently used in the book, are: *Bernoulli*(p) random variable X which takes on values $\{0, 1\}$ and $\mathbb{P}(X = 1) = p$; *Binomial*(p, n) random variable $X := \sum_{i=1}^{n} X_i$ where X_i are independent and identically distributed (iid) Bernoulli(p). A random variable X is called *continuous* if there exists a nonnegative function $f^X(x)$ such that $\int_{-\infty}^{\infty} f^X(x)dx = 1$ and $\mathbb{P}(X \in (a, b]) := \int_a^b f^X(x)dx$. Example of a continuous random variable is the time until event of interest. The function f^X is called the *probability density*, and it characterizes (uniquely defines) the continuous random variable X. Its analog for a discrete variable is the *probability mass function* $p^X(x) := \mathbb{P}(X = x)$ which is calculated for all x such that $p^X(x) > 0$. In general, the set $\mathcal{X}^X$ is called *support* of the random variable X if it is the minimal set such that $\mathbb{P}(X \in \mathcal{X}^X) = 1$. For instance, for Bernoulli random variable X its support $\mathcal{X}^X = \{0, 1\}$ is the set of two numbers 0 and 1.

A general characteristic, used to define a random variable X that can be either continuous or discrete, is the *cumulative distribution function* $F^X(x) := \mathbb{P}(X \leq x)$.

If F^X is known, then the *mean*   of a function $g(X)$ is defined as $\mathbb{E}\{g(X)\} := \int_{\mathcal{X}^X} g(x) d F^X(x)$, and its *variance*   as $\mathbb{V}\{g(X)\} := \mathbb{E}\{(g(X) - \mathbb{E}\{g(X)\})^2\}$. In the book we are dealing only with random variables having finite mean and variance, and note that

$$\mathbb{V}\{X\} = \mathbb{E}\{X^2\} - [\mathbb{E}\{X\}]^2. \tag{1.1.1}$$

We can also consider two random variables X and Y together as a pair (X, Y), then their joint distribution is completely defined (characterized) by the *joint cumulative distribution function* $F^{X,Y}(x, y) := \mathbb{P}(\{X \leq x\} \cap \{Y \leq y\}) =: \mathbb{P}(X \leq x, Y \leq y)$. Here the right side of the formula introduces a convenient notation, via the comma, for the intersection of two events. The joint distribution allows us to introduce the *marginal* cumulative distribution functions $F^X(x) := F^{X,Y}(x, \infty)$ and $F^Y(y) := F^{X,Y}(\infty, y)$. Random variables X and Y are *independent* if $F^{X,Y}(x, y) = F^X(x) F^Y(y)$ for all x and y. All these notions are straightforwardly extended to the case of n variables $X_1, \ldots, X_n$, for instance $F^{X_1, \ldots, X_n}(x_1, \ldots, x_n) := \mathbb{P}(X_1 \leq x_1, \ldots, X_n \leq x_n)$ and these n variables are independent if $F^{X_1, \ldots, X_n}(x_1, \ldots, x_n) = \prod_{i=1}^{n} F^{X_i}(x_i)$.

Two random variables X and Y are *jointly continuous* if there exists a bivariate nonnegative function $f^{X,Y}(x, y)$ on a plane $(-\infty, \infty) \times (-\infty, \infty)$, referred to as *two-dimensional or bivariate probability density*, such that $F^{X,Y}(x, y) =: \int_{-\infty}^{x} \int_{-\infty}^{y} f^{X,Y}(u, v) dv du$. Then we may define the corresponding marginal densities, say $f^X(x) := \int_{-\infty}^{\infty} f^{X,Y}(x, y) dy$. These definitions are straightforwardly extended to n-dimensional vector $(X_1, \ldots, X_n)$ of continuous random variables, and they are called *independent* if $f^{X_1, \ldots, X_n}(x_1, \ldots, x_n) = \prod_{i=1}^{n} f^{X_i}(x_i)$ and otherwise they are dependent. Absolutely the same approach is used for a vector of discrete random variables only the probability density is replaced by the probability mass function.

For a pair of continuous random variables (X, Y) we can write $f^{X,Y}(x, y) =: f^X(x) f^{Y|X}(y|x)$. Here $f^{Y|X}(y|x)$ is the *conditional density* of Y given $X = x$, and it is a valid density in y. In particular, we can introduce the *conditional cumulative distribution function* $F^{Y|X}(y|x) := \int_0^y f^{Y|X}(v|x) dv$ and the *conditional mean*

$$\mathbb{E}\{Y|X = x\} := \int_{-\infty}^{\infty} y f^{Y|X}(y|x) dy. \tag{1.1.2}$$

Let us note that in statistics the conditional expectation $\mathbb{E}\{Y|X = x\}$ may be called the *regression* of the *response* Y on the *predictor* X. Further, the *conditional variance* is defined as

$$\mathbb{V}\{Y|X = x\} := \mathbb{E}\{(Y - \mathbb{E}\{Y|X = x\})^2 | X = x\}. \tag{1.1.3}$$

There are two useful formulas to know,

$$\mathbb{E}\{Y\} = \mathbb{E}\{\mathbb{E}\{Y|X\}\}, \tag{1.1.4}$$

and

$$\mathbb{V}\{Y\} = \mathbb{V}\{\mathbb{E}\{Y|X\}\} + \mathbb{E}\{\mathbb{V}\{Y|X\}\}. \tag{1.1.5}$$

1.1.2   Random Variables in Survival Analysis

Survival analysis traditionally focuses on the analysis of time duration until one or more events happen and, more generally, positive-valued random variables. Classical examples are the time to death in biological organisms, the time from diagnosis of a disease until death, the time to develop immunity after vaccination, the time from the start of treatment of a symptomatic disease and the suppression of symptoms, the time to failure in mechanical systems, the length of stay in a hospital, duration of a strike, the total amount paid by a health insurance, the time to getting a high school diploma. This topic may be called reliability theory or reliability analysis in engineering, duration analysis or duration modeling in economics, and event history analysis in sociology. Survival analysis attempts to answer questions such as: what is the proportion of a population which will survive past a certain time? Of those that survive, at what rate will they die or fail? Can multiple causes of death or failure be taken into account? How do particular circumstances or characteristics increase or decrease the probability of survival?

To answer such questions, it is necessary to define the notion of "lifetime". In the case of biological survival, death is unambiguous, but for mechanical reliability, failure may not be well defined, for there may well be mechanical systems in which failure is partial, a matter of degree, or not otherwise localized in time. Even in biological problems, some events (for example, heart attack or other organ failure) may have the same ambiguity. In the book it is assumed that all events are well-defined at specific times.

In what follows we denote the random continuous time of interest by T. It still may be characterized by the cumulative distribution function $F^T(t) := \mathbb{P}(T \leq t)$, but more frequently used characteristic is the *survival function* $S^T(t) := \mathbb{P}(T > t) = \mathbb{P}(T \geq t) = 1 - F^T(t)$, which is the probability that the event of interest occurs after time t. Note that for a continuous lifetime T we have $S^T(t) = \mathbb{P}(T \geq t)$ because $\mathbb{P}(T = t) = 0$. The survival function is also called the survivor function or survivorship function in problems of biological survival, and the reliability function in mechanical survival problems. Usually one assumes $S(0) = 1$, although it could be less than 1 if there is the possibility of immediate death or failure.

Survival analysis is often interested in studying a lifetime given that the underlying event of interest has not occurred at a specific time yet. For instance, it may be of interest to know how long a patient, who survived one year after cancer surgery, will live. Another example is the time to becoming sick for a patient being in contact with an infected person, given that the patient is not sick after two days from the contact. Let us present several probability characteristics motivated by that conditional setting. We begin with the classical conditional characteristic of a continuous random lifetime T called the *hazard rate function*

$$h^T(t) := \lim_{v \to 0} \frac{\mathbb{P}(t < T < t + v | T > t)}{v} = \frac{f^T(t)}{S^T(t)}, \quad S^T(t) > 0, \ t \ge 0. \tag{1.1.6}$$

The hazard rate represents the instantaneous likelihood that the event occurs within the interval $(t, t + dt)$ given that the event has not occurred at time t. The hazard rate quantifies the trajectory of imminent risk, and it may be referred to by other names in different sciences, for instance as the *failure rate* in reliability theory and the *force of mortality* in actuarial science and sociology.

Similarly to the probability density and the survival function, the hazard rate h^T characterizes the random lifetime T. Indeed, if the hazard rate is known, then the corresponding probability density is

$$f^X(t) = h^T(t)e^{-\int_0^t h^T(v)dv} =: h^T(t)e^{-H^T(t)}. \tag{1.1.7}$$

Here

$$H^T(t) := \int_0^t h^T(v)dv \tag{1.1.8}$$

is another classical conditional characteristic of the lifetime T called the *cumulative hazard function*. It is important to know the following formula that relates the survival function and the cumulative hazard,

$$S^T(t) = e^{-\int_0^t h^T(v)dv} = e^{-H^T(t)}. \tag{1.1.9}$$

The preceding identity follows from integrating both sides of the equality

$$h^T(t) = -\frac{dS^T(t)/dt}{S^T(t)}, \tag{1.1.10}$$

and then using $S^T(0) = 1$.

One important corollary of (1.1.9) is that if T is supported on $[0, b]$, $b < \infty$, then we get

$$\lim_{t \to b} h^T(t) = \infty. \tag{1.1.11}$$

Note that this property of the hazard rate precludes us from reliable estimation of its right tail. Another important property of the hazard rate is that if T and C are independent, then for $V := \min(T, C)$ the formula

$$h^V(t) = h^T(t) + h^C(t) \tag{1.1.12}$$

holds. Formula (1.1.12) follows from

$$S^V(t) = \mathbb{P}(\min(T, C) > t) = \mathbb{P}(T > t)\mathbb{P}(C > c) = S^T(t)S^C(t). \tag{1.1.13}$$

Further, given that an event of interest has not occurred at time t yet, it is natural to introduce the *mean residual life* (MRL)

$$\mu^T(t) := \mathbb{E}\{T - t | T > t\} = \int_t^\infty (v - t) f^{T|T>t}(v) dv, \qquad (1.1.14)$$

where $f^{T|T>t}(v) := f^T(v)/S^T(t)$, $v > t$ is the *conditional probability density* of T given $T > t$. In survival analysis it is often difficult or impossible to study tails of the lifetime of interest, and then the *restricted mean residual life* (RMRL)

$$\mu_r^T(t) := \mathbb{E}\{(T - t)I(T \le r) | T > t\} = \int_t^r (v - t) f^{T|T>t}(v) dv, \; t \le r \qquad (1.1.15)$$

may be studied instead of the mean residual life.

In applications it is often the case that survival data are not fully and/or directly observed, but rather censored. Two typical censoring mechanisms are the right-censoring (RC) and the current status censoring (CSC), and we consider them in turn.

Let us begin with several examples of right-censoring. Suppose that after a cancer surgery patients are followed for 12 weeks to study the cancer relapse. A patient who does not experience the event of interest for the duration of the study is said to be right-censored. The survival time for this person is considered to be at least as long as the duration of the study. Another example of right-censoring is when a person drops out of the study before the end of the study observation time and did not experience the event. This person's survival time is said to be right-censored, since we know that the event of interest did not happen while this person was under observation. Censoring is an important issue in survival analysis. In what follows we always denote by C the censoring lifetime. The right-censoring implies that instead of a sample from T we observe a sample from $(V, \Delta) := (\min(T, C), I(T \le C))$ where Δ is called the *indicator of censoring*, and recall that $I(D)$ is the *indicator function* of event D which is equal to 1 if the event D occurs and 0 otherwise.

Current status censoring (CSC), also known as *case I interval censoring*, implies that there is no direct access to the lifetime of interest. Instead it is known if the event already occurred or not at a random monitoring time. CSC is a simple sampling procedure and in many cases the only possibility to assess the lifetime of interest. For instance, serial-sacrifice carcinogenicity experiments of an occult non-lethal tumor provide simple example of CSC because at the time of sacrifice the cancer may or may not be present. Under a current status censoring, the underlying time T to an event of interest is never observed. Instead, at a monitoring time Z it is known if the event of interest already occurred or not. In other words, under CSC the pair $(Z, \Delta) := (Z, I(T \le Z))$ is observed. Despite of not observing the lifetime of interest T, we will learn how to estimate all its classical characteristics.

There are two other data modification mechanisms that are not studied in the book but are worthwhile to be mentioned and be familiar with. They are the *missing* data and *biased* data. We are considering them in turn.

Missingness means that some cases in data (you may think about rows in a matrix) are incomplete and instead of numbers some elements in cases are missed (empty). In R language missed elements are denoted by a logical flag "NA" which stands for "Not Available", and this is why we are saying that some elements in a case are available (not missed) and others not available (missed); we may similarly say that a case is complete (all elements are available) or incomplete (when some elements are not available). Using the probability approach, the mechanism of missing lifetime T may be described by the Bernoulli random variable A, called the *availability*, when in place of T we observe the pair (AT, A).

For a continuous T we have $\mathbb{P}(T = 0) = 0$, and then knowing that pair is equivalent to knowing AT. In other words, not availability of T is equivalent to $A = 0$ or $AT = 0$. As an example, applying this approach to right-censored data with missing lifetimes, instead of (V, Δ) we are observing (AV, Δ), and the missing is characterized by the *availability likelihood* $w(v, \delta) := \mathbb{P}(A = 1|V = v, \Delta = \delta)$. Depending on the availability likelihood, there are three main missing mechanisms. The simplest one is the *missing completely at random* (MCAR) when $w(v, \delta) = w$, that is the missing does not depend on the underlying variables. *Missing at random* (MAR) mechanism means $w(v, \delta) = w(\delta)$, that is the missing depends only on always available variables. All other mechanisms are referred to as *missing not at random* (MNAR).

Biased data is another data modification that often occurs in applications. Let us begin with an example. Suppose that we would like to know the distribution of the ratio of alcohol X^* in the blood of liquor-intoxicated drivers traveling along a particular highway. The data analyst has a sample $X_1, \ldots, X_n$ of the ratios of interest from routine police reports on arrested drivers charged with driving under the influence. Because a drunker driver has a larger likelihood to be stopped and arrested, the sample is biased. Using the probability approach, we can describe the sample $X_1, \ldots, X_n$ as follows. Let A be the availability random variable equal to 1 if a driver is stopped and arrested and $A = 0$ otherwise. Then we have $Y_i := A_i X_i^*$, $i = 1, 2, \ldots$ and $B(x) := \mathbb{P}(A = 1|X^* = x)$ is the probability that a driver with the ratio of alcolol x is stopped and arrested. If $A_1 = 1$ then $X_1 = Y_1$, and otherwise the Y_1 is skipped. We continue until a sample of size n is collected, that is we are dealing with the negative binomial experiment with n successes. Then

$$f^X(x) = \frac{f^{X^*}(x)B(x)}{\int_0^\infty f^{X^*}(v)B(v)dv}. \tag{1.1.16}$$

The function $B(x)$ is called the *biasing* function.

As we will see throughout the book, censoring may yield missingness and/or biased data. This is why these two modifications and corresponding remedies are to be known.

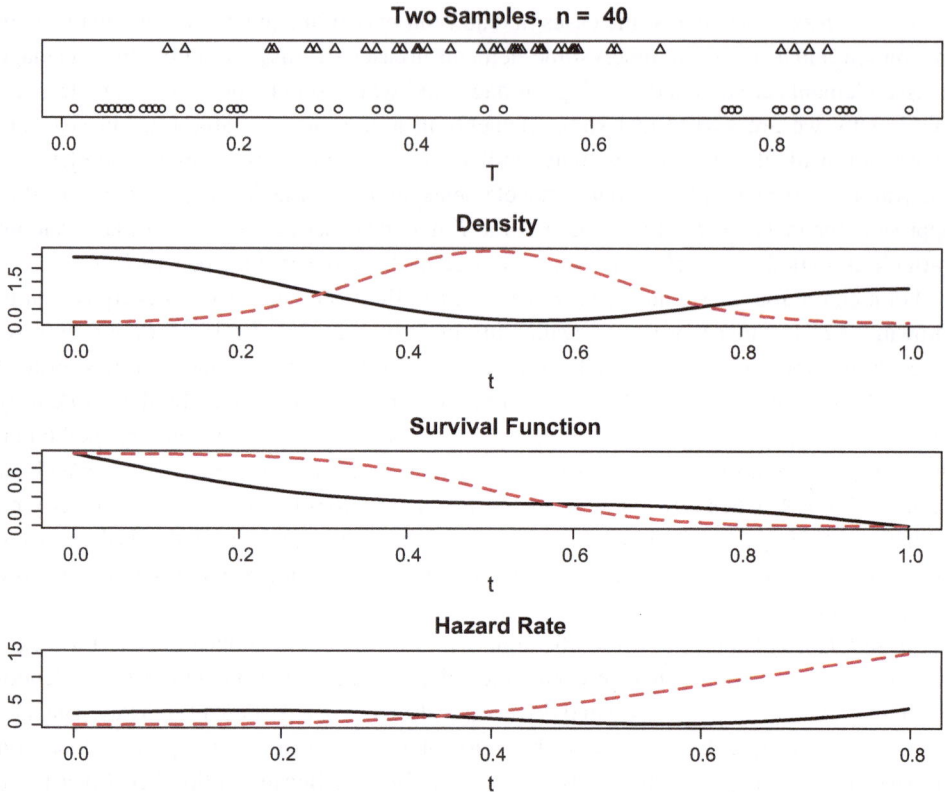

Fig. 1.1 Simulated samples from two lifetimes and the corresponding distribution characteristics. The circles and the triangles show samples from distributions shown by the solid and dashed lines, respectively. {Here and in all other captions information in the curly brackets is devoted to explanation of arguments of the figure that can be changed. For Fig. 1.1 the simulation can be repeated and results visualized by using the book's R package. See how to install and use it in Sect. 1.3. When the R package is installed, start R and simply enter (after the R prompt >) **ch1(Fig = 1)**. Note that using **ch1(fi = 1)** or **ch1(f=1)** yields the same outcome. Default values of arguments, used by the package for Fig. 1.1, are shown below in the square brackets. All indicated default values may be changed. For instance, after entering **ch1(Fig = 1, n = 100, corn = c(3, 5))** samples of size $n = 100$ from corner densities 3 and 5 will be generated and exhibited together with their characteristics. There are 5 fixed corner densities (distributions, functions): 1. Uniform; 2. Bathtub; 3. Normal; 4. Bimodal; 5. Strata. These densities (functions) are shown in Fig. 1.8. The optional argument *CFUN* allows one to substitute a corner function by a custom-made corner function. For instance, the choice $CFUN = list(3, "2 * x - 3 * \cos(x)")$ would imply that the third corner function (the Normal) is substituted by the positive part of $2x - 3\cos(x)$ divided by its integral over [0, 1], i.e., the third corner function will be $(2x - 3\cos(x))_+ I(x \in [0, 1]) / \int_0^1 (2u - 3\cos(u))_+ du$, where $x_+ := \max(0, x)$. Any valid R formula in x (use the lowercase x) may be used to define a custom-made corner function. This option is available for all figures where corner functions are used. See an example in Fig. 1.3.} [$n = 40$, corn $= c(2, 3)$]

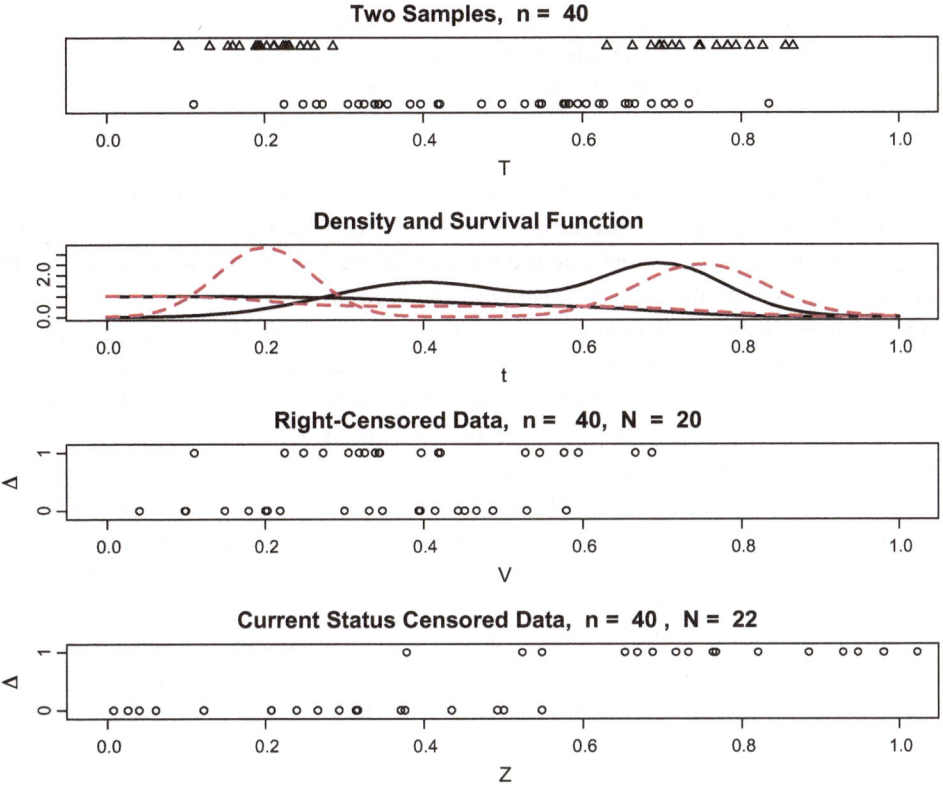

Fig. 1.2 Simulated right-censored and current status censored data. The censoring and monitoring distributions are the same Uniform(0, b), $N := \sum_{i=1}^{n} \Delta_i$. [n = 40, corn = c(4, 5), b = 1.2]

1.1.3   Simulated Examples

The companion R software package allows the reader to gain practical experience and verify theoretical propositions, as well as to test and modify proposed estimators. The manual on how to install and use it can be found Sect. 1.3. In this subsection the software is used to shed extra light on notions introduced in Sects. 1.1.1 and 1.1.2.

Figure 1.1 shows us two simulated samples of size $n = 40$ from two lifetimes with different distributions. The two samples, correspondingly shown in the top diagram by the circles and the triangles, clearly indicate different underlying distributions. The circles point upon heavy tails while the triangles point upon light tails and a pronounced mode. The corresponding densities, shown in the second from the top diagram, support the visual impression. Overall, among all descriptive characteristics of a random variable, visual analysis of the density presents the best opportunity to appreciate the variable. The corresponding survival functions are shown in the third diagram. It is a dramatically more difficult task to get a right impression about a random variable from its survival function. At the same time, the survival function tells us about the probability to survive beyond time t. In the bottom diagram we see the hazard rates that exhibit the likelihood of imminent risk for event of interest at moment t. Please look at how large the hazard rate may be, and recall that it is not integrable over the support.

Figure 1.2 complements Fig. 1.1 by exhibiting samples from two different distributions as well as the corresponding censored datasets. The top two diagrams are similar to Fig. 1.1. Note that while the densities indicate pronounced differences in the distributions, the differences are muted by the survival functions. The third diagram shows us how one of the samples, shown in the top diagram by the circles, is modified by the Uniform$(0, 1.2)$ censoring variable C. Here we observe pairs from $(V, \Delta) = (\min(T, C), I(T \leq C))$. Further, only $N := \sum_{l=1}^{n} \Delta_l = 20$ of the $n = 40$ underlying observations of T are uncensored implying the rate of censoring $(n - N)/n = 0.5$. Further, the uncensored observations $(\Delta = 1)$ are clearly biased. Further, there are no observations in the right tail. These are main challenges of dealing with right-censored data. Another interesting remark is that if someone considers censored observations $(\Delta = 0)$ as missed ones, then the missingness is MNAR. The bottom diagram shows us the current status censoring (CSC) modification of the same underlying data shown by the circles in the top diagram. Here the monitoring time Z has the same Uniform$(0, 1.2)$ distribution as the censoring C. Note that in the CSC data we do not observe the underlying lifetimes, and the modified observations do not resemble the underlying ones. Nonetheless, we will learn in Chap. 2 how to extract information about the underlying lifetime of interest from CSC observations.

1.2   Nonparametric Estimation for Direct Data

The aim of this section is to present main relevant results for the classical case of directly observed sample. We begin with estimation of the survival function, and then consider the topics of density, hazard rate and regression estimation. The main result is the universal nonparametric series estimator, called the E-estimator. This estimator will be also used for censored data.

1.2.1   Survival Function Estimation

The aim is to estimate the survival function $S^T(t) := \mathbb{P}(T > t)$ of the continuous lifetime of interest T based on a sample $T_1, \ldots, T_n$ of size n from T. Recall that in a sample all observations are independent and identically distributed. For the continuous lifetime T we have $\mathbb{P}(T > t) = \mathbb{E}\{I(T > t)\} = \mathbb{E}\{I(T \geq t)\}$, where $I(\cdot)$ is the indicator function and the last equality holds because T is a continuous random variable. Accordingly, S^T can be estimated by the method of moments estimator

$$\hat{S}^T(t) := n^{-1} \sum_{l=1}^{n} I(T_l \geq t).$$ (1.2.1)

Note that this is a classical sample mean estimator, and it has a number of excellent statistical properties. First of all, it is *unbiased* estimator meaning that $\mathbb{E}_{S^T}\{\hat{S}^T(t)\} = S^T(t)$. Second, its variance is easily calculated due to identically distributed and independent observations in the sample,

$$\mathbb{V}\{\hat{S}^T(t)\} = n^{-1} S^T(t)[1 - S^T(t)].$$ (1.2.2)

The latter is a remarkable result because the nonparametric function S^T can be estimated with the parametric rate n^{-1}. As we will see shortly, a typical rate for nonparametric functions is slower than n^{-1}. Third, recall the Hoeffding inequality for a sample of size n from a random variable X satisfying $\mathbb{P}(a \leq X \leq b) = 1$,

$$\mathbb{P}(|n^{-1} \sum_{l=1}^{n} X_l - \mathbb{E}\{X\}| > \epsilon) \leq 2e^{-2n\epsilon^2/(b-a)^2}, \quad \epsilon > 0.$$ (1.2.3)

This inequality yields

$$\mathbb{P}(|\hat{S}^T(t) - S^T(t)| > \epsilon) \leq 2e^{-2n\epsilon^2}, \quad \epsilon > 0.$$ (1.2.4)

The last but not the least, $\min_{l=1,\ldots,n} \hat{S}^T(T_l) \geq n^{-1}$, and hence $\hat{S}^T(T_l)$ can be used in a denominator.

These are the results to match for all studied cases of indirect observations.

1.2.2 Series Approximation

Suppose that we want to estimate a finite and integrable function $f(x)$ over interval $[0, 1]$ based on some data. We can write

$$f(x) = \sum_{j=0}^{\infty} \theta_j \varphi_j(x), \; x \in [0, 1] \quad \text{where} \quad \theta_j := \int_0^1 f(x)\varphi_j(x)dx.$$ (1.2.5)

Here the functions $\varphi_j(x)$ are known, fixed, and referred to as the *orthonormal functions* or *elements* of the *orthonormal basis* (or simply *basis*) $\{\varphi_0, \varphi_1, \ldots\}$, and the θ_j are called the *Fourier* coefficients of $f(x)$, $x \in [0, 1]$. A system of functions is called *orthonormal* if $\int_0^1 \varphi_s(x)\varphi_j(x)dx = I(s = j)$.

To describe a function via an infinite series expansion (1.2.5) one needs to know the infinite number of Fourier coefficients. Instead, a *truncated (finite) series* (or so-called partial sum)

$$f_J(x) := \sum_{j=0}^{J} \theta_j \varphi_j(x), \quad x \in [0, 1]$$ (1.2.6)

is used to approximate f, and then the *integrated squared error* (ISE) $\int_0^1 [f(x) - f_J(x)]^2 dx$ $\to 0$ as $J \to \infty$. The integer parameter J is called the *cutoff*. The advantage of the truncated series approach is the possibility to compress the data and describe a function using just several Fourier coefficients. Roughly speaking, the main statistical issue will be how to estimate Fourier coefficients θ_j and choose a cutoff J. In what follows, the *cosine orthonormal basis* on [0, 1]

$$\varphi_0(x) := 1 \quad \text{and} \quad \varphi_j(x) := \sqrt{2} \cos(\pi j x) \quad \text{for} \quad j = 1, 2, \ldots. \quad (1.2.7)$$

is used.

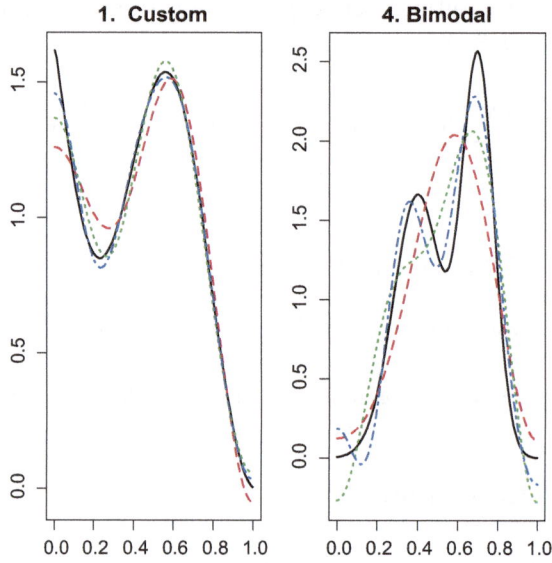

Fig. 1.3 Approximations of the custom-made and corner functions by cosine series with different cutoffs J. In each diagram the solid line is the underlying density, while the short-dashed, dotted and dot-dashed lines are approximations with cutoffs from Jset. The first function is custom- made, and the explanation of how to construct it is given in the curly brackets. {The optional argument CFUN allows one to analyze a custom-made corner density. For instance, the choice $CFUN = list(1, "4 * x - 2 * \cos(x)")$ would imply that the first density (the Uniform) is substituted by the positive part of $4x - 2\cos(x)$ divided by its integral over [0, 1], i.e., by $(4x - 2\cos(x))_+ I(0 \leq x \leq 1)/\int_0^1 (4u - 2\cos(u))_+ du$. Any valid R formula in x (use only the lower case x) may be used to define a custom-made corner function. This option is available for all figures where corner functions are used.} [Jset = c(3, 4, 6), corn = c(1, 5), $CFUN = list(1, "2 - 2 * x - \sin(8 * x)")$]

Figure 1.3 allows the reader to check how well different functions can be approximated by cosine series. As we can see, even functions with rather complicated shape can be well approximated using just a few Fourier coefficients. Also note that while cosine approximations are not perfect for some corner functions, understanding how these partial sums perform may help us to "read" messages of these approximations and guess about underlying functions. Overall, the cosine system does an impressive job in both representing a function and data compression.

Now let us present several useful theoretical results about the Fourier coefficients. The beauty of a theoretical approach is that it allows us to analyze simultaneously large classes of functions; in particular we are interested in functions f that are square integrable on $[0, 1]$, i.e., $\int_0^1 f^2(x)dx < \infty$. For square-integrable functions the famous *Parseval identity* states that

$$\int_0^1 f^2(x)dx = \sum_{j=0}^{\infty} \theta_j^2 \quad \text{and} \quad \text{ISE}(f, f_J) := \int_0^1 (f(x) - f_J(x))^2 dx = \sum_{j>J} \theta_j^2, \quad (1.2.8)$$

where f_J is the partial sum (1.2.6) and the ISE is the integrated squared error of approximation of f by f_J. Thus, the faster Fourier coefficients decrease, the smaller cutoff J is needed to get a good approximation of f by a partial sum f_J in terms of the ISE.

Let us explain the main characteristics of a function f that influence the rate at which its Fourier coefficients decrease. Namely, we would like to understand what determines the rate at which Fourier coefficients of an integrable function f decrease as $j \to \infty$. Suppose that function f is differentiable and denote its first derivative by $f^{(1)}(x) := df(x)/dx$. Using integration by parts, it is straightforward to establish that

$$|\theta_j| \le 2^{1/2}(\pi j)^{-1} \int_0^1 |f^{(1)}(x)|dx, j \ge 1. \quad (1.2.9)$$

If the function f is twice differentiable, set $f^{(2)}(x) := d^2 f(x)/dx^2$ and again via integration by parts get

$$|\theta_j| \le j^{-2}[|f^{(1)}(1)(-1)^j - f^{(1)}(0)| + \int_0^1 |f^{(2)}(x)|dx], \; j \ge 1. \quad (1.2.10)$$

Assume that $f^{(2)}$ is integrable on $[0, 1]$. Then the Fourier coefficients θ_j of a large class of twice differentiable functions decrease with rate not slower than j^{-2}. This is the good news, and note that *boundary conditions* (that is values of $f(x)$ at boundaries of the unit interval $[0, 1]$) do not affect the rate. The bad news is that if the function is smoother and has three derivatives, the boundary conditions may not allow the rate to decrease faster than j^{-2}. To avoid this classical boundary problem, a polynomial-trigonometric basis may be used, see the Notes.

In approximation theory (harmonic analysis) and in asymptotic statistics, it is a tradition to work with Sobolev classes of functions defined by Fourier coefficients, often also referred to as *Sobolev ellipsoids*. Let us define function classes that will be used in the book. Introduce a positive and continuous on $[0, 1]$ *anchor* function $f_0(x)$, $x \in [0, \infty)$. Note that the latter yields $\min_{x \in [0,1]} f_0(x) > 0$. Define a sequence, in the sample size n, of shrinking toward the anchor f_0 *local Sobolev classes*

$$\mathcal{S}(f_0, \alpha, Q, n) := \Big\{ f : \ f(x) = f_0(x) + g(x)I(x \in [0, 1]), \ x \in [0, \infty),$$

$$\min_{x \in [0,1]} f_0(x) > 0, \quad \max_{x \in [0,1]} |g(x)| \le \frac{\min_{x \in [0,1]} f_0(x)}{\ln(\ln(n + 20))}, \ g \in \mathcal{S}_1(\alpha, Q) \Big\}. \tag{1.2.11}$$

Note that $\ln(\ln(n + 20)) > 1$, and for $k = 0, 1$ set

$$\mathcal{S}_k(\alpha, Q) := \Big\{ g : g(x) = \sum_{j=k}^{\infty} \theta_j \varphi_j(x), \ x \in [0, 1],$$

$$\sum_{j=k}^{\infty} [1 + (\pi j)^{2\alpha}] \theta_j^2 \le Q < \infty \Big\}. \tag{1.2.12}$$

In (1.2.12) the class $\mathcal{S}_0$ is called the *global Sobolev class* of α-fold differentiable functions whenever α is a positive integer, and this is the main case of interest.

Using integration by parts, similarly to (1.2.10) it is possible to show that if functions g are periodic and satisfy some mild boundary conditions, then $\mathcal{S}_0(\alpha, Q) = \{g : \int_0^1 [g^2(x) + (g^{(\alpha)}(x))^2] dx \le Q\}$, $g^{(\alpha)}(x) := d^{\alpha} g(x)/dx^{\alpha}$. The latter sheds light on parameters α and Q of the global Sobolev class. For the case of aperiodic functions see the Notes.

The Sobolev class $\mathcal{S}_1(\alpha, Q)$ does not include the constant component $\varphi_0(x) = 1$ of the cosine basis, and accordingly $\int_0^1 g(x) dx = 0$ for $g \in \mathcal{S}_1(\alpha, Q)$. The latter is convenient because for a function $f \in \mathcal{S}(f_0, \alpha, Q, n)$ we have $\int_0^1 f(x) dx = \int_0^1 f_0(x) dx$. In particular, if f_0 is a positive density supported on $[0, 1]$, then $\mathcal{S}(f_0, \alpha, Q, n)$ is the sequence of Sobolev classes of densities supported on $[0, 1]$ and shrinking toward the anchor density f_0 as $n \to \infty$. Let us stress that in general the anchor f_0 and an underlying function of interest (the *estimand*) f are two different functions. The main role of the anchor is to point upon shrinking toward the anchor local classes of estimated functions because, as we will see shortly, the sharp constant of the mean integrated squared error (MISE) may depend on the anchor.

Sobolev classes are convenient for theoretical analysis of square integrable functions due to the Parseval identity (1.2.8). In particular, it yields the following upper bound for the ISE,

$$\sup_{f \in \mathcal{S}_0(\alpha, Q)} \int_0^1 (f(x) - f_J(x))^2 dx < Q(\pi J)^{-2\alpha}. \tag{1.2.13}$$

As we will see shortly, this upper bound is used to evaluate the integrated squared bias of nonparametric series estimators.

1.2.3 Density Estimation, E-Estimator and Confidence Band

This section is important because it presents the E-estimation algorithm (E-estimator). As we will see shortly, the E-estimator is a universal series estimator that can be used for any nonparametric curve estimation problem as soon as a sample mean Fourier coefficient estimator is proposed. Here the E-estimator is presented for the classical probability density estimation problem, and it is complemented by the nonparametric confidence band.

Density estimation is one of the most fundamental and practically important problems in statistics. We are considering the classical model of probability density estimation when all n independent and identically distributed observations $T_1, T_2, \ldots, T_n$ (the *sample* of size n) of a continuous random lifetime T are available. It is supposed that T is distributed according to an unknown probability density $f^T(t)$ supported on $[0, 1]$. Because no assumption about shape of the density is made (we do not assume that the density has a particular parametric distribution), the estimation procedure and the corresponding estimators are called *nonparametric*.

A note on notation. In what follows we use diacritics (e.g., "hat," "tilde," or "bar") above a parameter or a function to indicate that this is an estimator (statistic) of the corresponding parameter or function, for instance $\tilde{f}$ is an estimator of f. Also recall that $I(D)$ is the *indicator function* of the event D, that is, the indicator is equal to 1 if D occurs and is 0 otherwise, $\mathbb{E}_{f^T}\{\cdot\}$ denotes the expectation (population mean) given density f^T, and in the expectation the subscript f^T may be skipped whenever the underlying distribution is well understood. The used criterion for the quality of nonparametric estimation of the density $f^T(t)$, $t \in [0, 1]$ by an estimator $\tilde{f}$ is the *Mean Integrated Squared Error (MISE)* defined as

$$\text{MISE}(\tilde{f}, f^T) := \mathbb{E}\left\{ \int_0^1 [\tilde{f}(t) - f^T(t)]^2 dx \right\}. \qquad (1.2.14)$$

Note that the considered interval $[0, 1]$ of estimation is a simple convenience that allows us to use the same chosen basis. Any interval $[a, b]$ can be considered instead by using the *scale-location transformation* $T' := (T - a)/(b - a)$. Then the formula

$$f^T(t) = (b - a)^{-1} f^{T'}((t - a)/(b - a)), \quad t \in [a, b], \qquad (1.2.15)$$

relates the two densities.

In what follows and for any considered setting, we first present a brief review of the asymptotic theory. The asymptotic theory, presented here and for all other studied models, is based on a sharp lower bound for oracle-estimators and a data-driven adaptive estimator that attains the lower bounds and matches performance of the oracle. In what follows, the *oracle* knows data, all nuisance functions, and both global and local Sobolev classes that define smoothness of an estimated function. The underlying idea of the theory is to develop a *sharp lower bound* (the fastest rate in n and the minimal constant) for the MISE of oracle-estimators, and then propose a data-driven estimator that matches the best oracle-estimator. Finally, the E-estimator is developed that mimics the sharp asymptotic estimator for practically important small samples.

1.2.3.1 Asymptotic Theory

Recall the defined in (1.2.11) and (1.2.12) local and global Sobolev classes of α-fold differentiable densities supported on $[0, 1]$. Consider the oracle who knows a sample of size n from the lifetime of interest T and the Sobolev classes. Here and in what follows $o_n(1)$ denote generic sequences in n such that $o_n(1) \to 0$ as $n \to \infty$, $\lceil x \rceil$ is the smallest integer not smaller than x.

Also, c_{J0}, c_{J1} and c_{TH} are some positive constants used by the software.

In what follows we also use notation d for the *coefficient of difficulty* which is specific for a particular considered problem.

The following proposition is written in such a way that it can be used later for hazard rate and regression settings with the main difference being in a particular coefficient of difficulty d that may depend on the setting. Recall that the anchor f_0 is assumed to be continuous and positive on $[0, 1]$.

Theorem 1.1 *Consider a sample of size n from T and the oracle who knows the data and local shrinking Sobolev function classes $\mathcal{S}(f_0, \alpha, Q, n)$ with anchor f_0 being a continuous and positive density supported on $[0, 1]$. The following lower bound holds for all oracle-estimators $\check{f}^*$,*

$$\inf_{\check{f}^*} \sup_{f^T \in \mathcal{S}(f_0, \alpha, Q, n)} (n/d)^{2\alpha/(2\alpha+1)} \, \mathbb{E}\{ \int_0^1 (\check{f}^*(t) - f^T(t))^2 dt \}$$

$$\geq P(\alpha, Q)(1 + o_n(1)). \tag{1.2.16}$$

Here the coefficient of difficulty $d = \int_0^1 f^X(x)dx = 1$ and the Pinsker constant

$$P(\alpha, Q) := Q^{1/(2\alpha+1)}(2\alpha + 1)^{1/(2\alpha+1)} \left[\frac{\alpha}{\pi(\alpha + 1)} \right]^{2\alpha/(2\alpha+1)}. \tag{1.2.17}$$

The lower bound is sharp. In particular, if the anchor $f_0 \in \mathcal{S}_0(\alpha + 1, Q')$, $Q' < \infty$, then the lower bound is attainable by the following oracle-estimator not using the anchor f_0,

$$\check{f}^*(t) := \sum_{j=0}^{J_n} \hat{\theta}_j I(\hat{\theta}_j^2 > c_{TH} \mathbb{V}\{\hat{\theta}_j\}) \varphi_j(t) + \sum_{j=J_n+1}^{J_n^*} [1 - (j/J_n^*)^\alpha] \hat{\theta}_j \varphi_j(t). \tag{1.2.18}$$

Here

$$\hat{\theta}_j := n^{-1} \sum_{l=1}^n \varphi_j(T_l) \tag{1.2.19}$$

is the sample mean estimator of Fourier coefficient θ_j,

$$J_n := \lceil c_{J0} + c_{J1} \ln(n) \rceil \tag{1.2.20}$$

with c_{J0} and c_{J1} being nonnegative constants, and

$$J_n^* := J_n + \lceil [(n/d)Q\pi^{-2\alpha}(\alpha+1)(2\alpha+1)/\alpha]^{1/(2\alpha+1)} \rceil. \tag{1.2.21}$$

Further,

$$\sup_{f^T \in \{S(f_0,\alpha,Q,n) \cup S_0(\alpha,Q)\}} (n/d)^{2\alpha/(2\alpha+1)} \mathbb{E}\{ \int_0^1 (\tilde{f}^*(t) - f^T(t))^2 dt \}$$

$$\leq P(\alpha, Q)(1 + o_n(1)). \tag{1.2.22}$$

We conclude that the lower bound (1.2.16) is sharp and holds for densities shrinking toward an anchor density f_0. The upper bound (1.2.22) holds for both local and global Sobolev classes. Let us comment about the coefficient of difficulty. For the considered density estimation problem $d = \int_0^1 f^X(x)dx = 1$ because an underlying density is supported on $[0, 1]$. The parameter (in general it is a function of the estimand and nuisance functions) is called the coefficient of difficulty because, according to (1.2.16), it describes complexity of a particular problem. For instance, if for another problem the coefficient of difficulty $d = d_1 > 1$, then a larger sample size nd_1 is needed for attaining the same MISE as for the density estimation. It is of interest to note that for the considered density estimation problem the coefficient of difficulty is equal to 1 for all Sobolev classes. This is a nice and unique feature of density estimation, and it creates a simple benchmark for other statistical models. Further, let us stress that the oracle-estimator (1.2.18) uses only data and nuisance parameters (α, Q) but not the anchor. The latter is a helpful hint of the oracle.

The oracle-estimator (1.2.18) has the *low-frequency component* which is the first sum with indexes $0 \leq j \leq J_n$ where J_n is defined in (1.2.20), and the *high-frequency component* with indexes $J_n < j \leq J_n^*$ where J_n^* is defined in (1.2.21). The low-frequency component uses the so-called *hard thresholding* of Fourier estimates $\hat{\theta}_j$ defined by the factor (weight) $I(\hat{\theta}_j^2 > c_{TH}\mathbb{V}\{\hat{\theta}_j\})$. Further, note that the low-frequency component does not depend on an underlying Sobolev class and the nuisance parameters (α, Q) that define smoothness of f^T. Asymptotically the more important part is the high-frequency component which utilizes very special *smoothing coefficients* (weights) $[1 - (j/J_n^*)^\alpha]$ and the special *cutoff* $J_n^*(d)$ to satisfy the upper bound (1.2.22).

As we will see throughout the book, roles of the low-frequency and high-frequency components are different. The former is used for developing the recommended E-estimator for small samples, the latter is needed for attaining the sharp lower bound and it shortly will be modified into a blockwise-shrinkage estimator that adapts to unknown smoothness of the estimand, that is to unknown parameters α and Q, see Sect. 1.2.3.3.

Theorem 1.1 has inspired many different data-driven density estimators mimicking performance of the oracle-estimator and adapting to the nuisance parameters. Note that it is a tradition in nonparametric curve estimation literature to refer to adaptation to nuisance parameters (α, Q) as *adaptation to smoothness* of an underlying estimand. Further, with some abuse of the notion, we may refer to a nonparametric curve estimator that attains the corresponding sharp lower bound, as well as to the used Fourier coefficient estimator, as *efficient*. In particular, we can say that the density oracle-estimator (1.2.18) is efficient, and the Fourier coefficient estimator (1.2.19) is efficient.

1.2.3.2 Algorithm of E-Estimator

Let us explain how the low-frequency component of the oracle-estimator (1.2.18) can be used for construction of a data-driven density estimator for small samples. The proposed methodology is applicable for all statistical problems considered in the book (density estimation, hazard rate estimation, regression, etc.) and it is called the *E-estimator*. The E-estimator is based on the following three steps.

E-estimator of a function $f(x)$, $x \in [0, 1]$:
1. Consider a series expansion $f(x) = \sum_{j=0}^{\infty} \theta_j \varphi_j(x)$. Suggest a sample mean estimator $\hat{\theta}_j$ of Fourier coefficients $\theta_j := \int_0^1 f(x)\varphi_j(x)dx$. Then calculate a corresponding sample variance estimator $\hat{\sigma}_{jn}^2$ of variance $\mathbb{V}\{\hat{\theta}_j\}$ of the sample mean estimator.
2. The E-estimator uses three nonnegative constants (parameters) c_{J0}, c_{J1}, c_{TH}, and it is defined as

$$\hat{f}(x) := \sum_{j=0}^{\hat{j}} I(\hat{\theta}_j^2 > c_{TH}\hat{\sigma}_{jn}^2)\hat{\theta}_j \varphi_j(x). \tag{1.2.23}$$

Here the empirical cutoff is

$$\hat{J} := \text{argmin}_{0 \le J \le c_{J0}+c_{J1}\ln(n)}\{\sum_{j=0}^{J}[2\hat{\sigma}_{jn}^2 - \hat{\theta}_j^2]\}. \tag{1.2.24}$$

3. If there are bona fide restrictions on $f(x)$ (for instance, the probability density is nonnegative and integrated to one, or it is known that the function is monotonic) then a projection of $\hat{f}(x)$ on the bona fide function class is performed as explained in Efromovich (1999a).

Let us recall that the function argmin $_{0 \le k \le s}\{a_k\}$, used in (1.2.24), returns the value k^* that is the index of the smallest element among $\{a_0, a_1, \ldots, a_s\}$.

Steps 2 and 3 in construction of the E-estimator are the same for all nonparametric statistical problems. As a result, as soon as a sample mean estimator of Fourier coefficients is proposed, this Fourier estimator yields the corresponding nonparametric E-estimator. Further, the choice of parameters of E-estimator may depend on a priori information about smoothness of f. In this chapter the default values for the parameters are $c_{J0} = 3, c_{J1} = 0.8$

and $c_{TH} = 4$. It will be left as an exercise to propose better values for each considered problem. Further, in what follows we are considering only settings where the variance $\mathbb{V}\{\hat{\theta}_j\}$ is of order n^{-1}. This remark sheds useful light on the second step. More about the algorithm can be found in the Notes.

Now we can explain how to use the E-estimator for density estimation. Because

$$\theta_j := \int_0^1 f^T(t)\varphi_j(t)dt = \mathbb{E}\{\varphi_j(T)\}, \tag{1.2.25}$$

we can use the sample mean (method of moments) Fourier estimator and its sample variance

$$\hat{\theta}_j := n^{-1}\sum_{l=1}^n \varphi_j(T_l), \quad \hat{\sigma}_{jn}^2 := (n-1)^2 \sum_{l=1}^n (\varphi_j(T_l) - \hat{\theta}_j)^2. \tag{1.2.26}$$

To evaluate performance of the E-estimator for small samples, we use Monte Carlo simulations where samples are generated according to the corner densities. Figure 1.4 presents simulations and E-estimates, the caption explains the diagrams, and the confidence band is explained in Sect. 1.2.3.4. Note that the estimates are based on simulations, hence another simulation may yield different estimates. The particular shown outcomes in Fig. 1.4, as well as outcomes in all other figures, are chosen with the objective of their discussion. For the shown simulations, very interesting outcome is for the underlying Uniform density. The E-estimate is clearly poor, but it follows the histogram and correctly describes the data at hand. This simulation presents a teachable moment that even for a large sample the data not necessarily follow the underlying distribution. For the two other simulations the generated samples are fair and the E-estimator performs very well.

Two practical conclusions may be drawn from the analysis of these particular simulations. First, visualization of a particular estimate is useful and sheds light on the estimator. Second, a conclusion about an estimator may not be robust if it is based only on analysis of several simulations. The reason is that for any estimator one can find a dataset where an estimate is perfect or, conversely, very bad. This is why it is important to complement your reading of the book by repeating figures using the R software and analyzing the results. Every experiment (every figure) should be repeated many times; the rule of thumb, based on the author's experience, is that results of at least 20 simulations should be analyzed before making a conclusion. Also, it is useful to try different values of the parameters and understand their role in the E-estimation.

Remark 1.1 If the density should be estimated over an interval $[a, b]$ (or data are given only for this interval), do the following. First, rescale the data and compute $Y_l := (T_l - a)/(b - a)$. Second, calculate the sample mean Fourier coefficient estimates

$$\check{\theta}_j := n^{-1}\sum_{l=1}^n I(Y_l \in [0, 1])\varphi_j(Y_l). \tag{1.2.27}$$

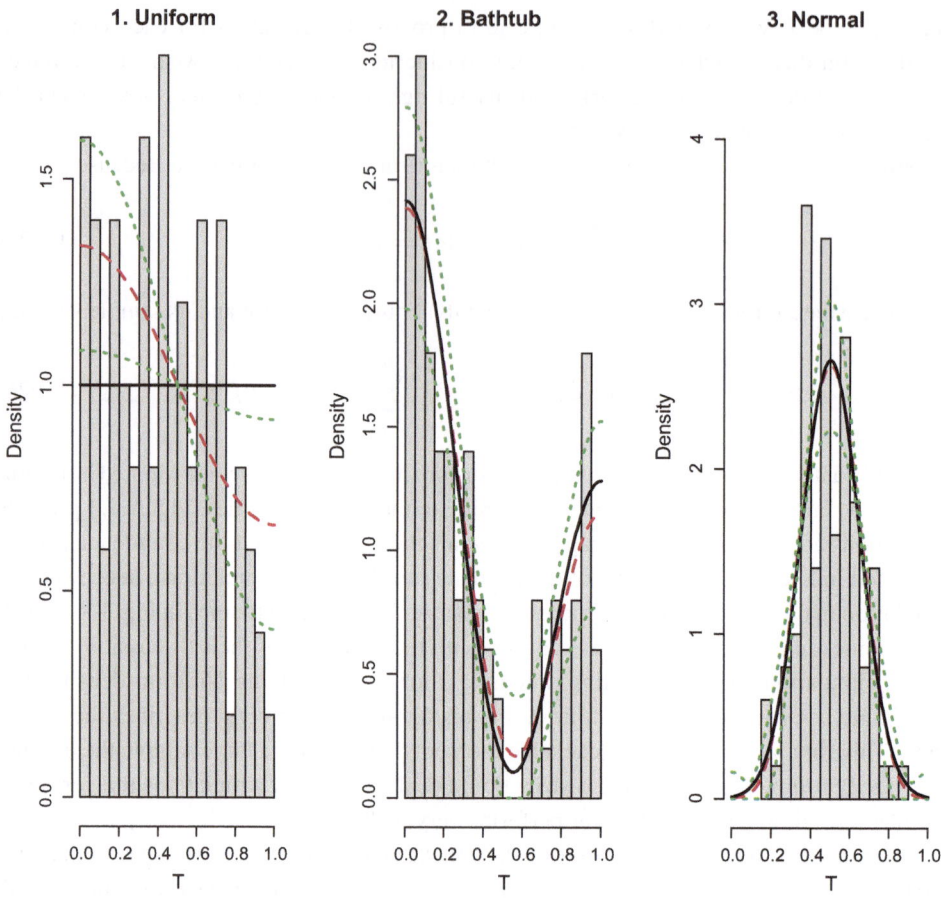

Fig. 1.4 Performance of the density E-estimator. A sample is shown by the histogram overlaid by the underlying density (the solid line), the E-estimate (the dashed line), and the $1 - \alpha$ coefficient confidence band (the dotted lines), $\alpha = 0.05$. Note that a solid line may "hide" other lines. {Recall that this figure may be repeated (with other simulated datasets) by calling after the R-prompt > the R-function > **ch1(f = 4)**. All the arguments, shown below in square brackets, may be changed. Let us review these arguments. The argument n allows one to choose the sample size n. The argument *corn* controls the three underlying corner densities defined in Fig. 1.8. For instance, to look at three simulations from the Uniform distribution, make the call > **ch1(f = 4, corn = c(1, 1, 1))**. The arguments *cJ0*, *cJ1*, and *cTH* control the parameters c_{J0}, c_{J1}, and c_{TH} used by the E-estimator. Note that R does not recognize subscripts, so *cJ0* is used instead of c_{J0}, etc. The confidence coefficient is controlled by *alpha*. Also recall that below in the square brackets the default values for these arguments are given. Thus, after the call > **ch2(f = 4)** the estimates will be calculated with these default values of the arguments. If one would like to change them, for instance to use a different threshold level, say $c_{TH} = 3$, make the call > **ch1(f = 4, cTH = 3)**.} $[n = 100, corn = c(1, 2, 3), cJ0 = 3, cJ1 = 0.8, cTH = 4, alpha = 0.05]$

Third, use these sample mean Fourier coefficient estimates in the E-algorithm and calculate the density estimate $\check{f}^Y(t)$, $t \in [0, 1]$. Finally, the E-estimator of $f^T(t)$ for $t \in [a, b]$ is $\check{f}^T(t) := (b-a)^{-1}\check{f}^Y((t-a)/(b-a))$, $t \in [a, b]$.

Remark 1.2 It is a well-established methodology in statistics to use the empirical cdf or the empirical survival function for estimating the density. Is there a merit in the unconventional methodology of estimating the cdf or the survival function using a density estimate? After all, $F^T(t) = \int_0^t f^T(v)dv$, and if a density estimate $\tilde{f}^T(t)$ is bona fide (nonnegative and integrated to 1 over the support), then the cdf estimate $\tilde{F}^T(t) := \int_0^t \tilde{f}^T(v)dv$ is also bona fide. Now note that a series density estimate implies a series cdf estimate. Accordingly, is it a feasible methodology to estimate the cdf directly using a series estimator? The following theoretical result gives a positive answer. Consider an *analytic class* of distributions supported on $[0, 1]$,

$$\mathcal{A}(\gamma, \Gamma) := \{F : dF(t)/dt = 1 + \sum_{j=1}^{\infty} \theta_j \varphi_j(t), t \in [0, 1], |\theta_j| \le \Gamma e^{-\gamma j}, \gamma > 0, \Gamma < \infty\}.$$

Note how fast, namely exponentially, decrease the Fourier coefficients of analytic cumulative distribution functions. Introduce the series cdf estimator

$$\hat{F}^T(t) = t + \sum_{j=1}^{J_n''} \hat{\theta}_j (\pi j)^{-1} 2^{1/2} \sin(\pi j t).$$

The used sample mean Fourier coefficient estimator $\hat{\theta}_j$ and the cutoff J_n'' are defined as follows,

$$\hat{\theta}_j := n^{-1} \sum_{l=1}^{n} \varphi_j(T_l), \text{ and } J_n'' := \lceil \ln(n)(2\gamma)^{-1}(1 + 1/\ln(\ln(n + 20))) \rceil.$$

The proposed cdf estimator has excellent statistical properties. Not only the $\hat{F}^T$ matches all nice properties of the classical empirical cdf, but it is better asymptotically in terms of the second order accuracy and it is a continuous function. At the same time, the series estimate $\hat{F}^T$ may be not bona fide, in particular may be not monotone. Is it a bad news? In general the answer is "no". Indeed, the non-monotonicity may point upon violating specific underlying assumptions, and at the same time the monotonic projection solves the issue. In other words, in nonparametric curve estimation the main step is a good estimator in terms of a chosen criterion, then a projection on a bona fide class may be performed.

Now let us comment about the low-frequency component (the first sum) on the right side of (1.2.18) and the second step of the E-estimation algorithm. We begin with the following lemma. Recall that ζ, ζ_i, ζ_*, ζ^* denote generic positive constants.

Lemma 1.1 *Consider a square integrable on* $[0, 1]$ *function* $f(x) = \sum_{j=0}^{\infty} \theta_j \varphi_j(x)$, $x \in$ $[0, 1]$ *where* $\theta_j := \int_0^1 f(x) \varphi_j(x) dx$ *are Fourier coefficients of* f. *Consider a sequence* J'_n *of positive cutoffs and the approximations* $f_{J'_n}(x) := \sum_{j=0}^{J'_n} \theta_j \varphi_j(x)$ *of* $f(x)$, $x \in [0, 1]$. *Let* $\breve{\theta}_j$ *be an unbiased estimator of* θ_j *based on a sample of size* n *such that* $\mathbb{V}\{\breve{\theta}_j\} \leq \zeta^* n^{-1}$. *Consider a nonnegative sequence* c'_n *and the following estimator of* $f_{J'_n}$,

$$\breve{f}_{J'_n}(x) := \sum_{j=0}^{J'_n} \breve{\theta}_j I(\breve{\theta}_j^2 > c'_n \mathbb{V}\{\breve{\theta}_j\}) \varphi_j(x).$$

Then the following upper bound for the mean integrated squared error holds,

$$\mathbb{E}\{\int_0^1 (\breve{f}_{J'_n}(x) - f_{J'_n}(x))^2 dx\} \leq 2\zeta^* (1 + c'_n) J'_n n^{-1}.$$

This theoretical result, together with the fact that the optimal rate of the MISE convergence of an α-fold differentiable function is $n^{-2\alpha/(2\alpha+1)}$, implies that any sequences c'_n and J'_n such that $(1 + c'_n) J'_n = o_n(1) n^{1/(2\alpha+1)}$ can be used in the low-frequency component of an efficient estimator. Accordingly, there is a wide spectrum of coefficients c_{TH} and the upper bounds for $\hat{J}$ in (1.2.24). The particular logarithmic upper bound $c_{J0} + c_{J1} \ln(n)$ satisfies the above-presented restriction, and it is supported by numerical simulations.

It is instructive to prove Lemma 1.1. Write using the *Cauchy inequality* $(a + b)^2 \leq 2(a^2 + b^2)$,

$$\mathbb{E}\{[\breve{\theta}_j I(\breve{\theta}_j^2 > c'_n \mathbb{V}\{\breve{\theta}_j\}) - \theta_j]^2\} = \mathbb{E}\{[(\breve{\theta}_j - \theta_j) - \breve{\theta}_j I(\breve{\theta}_j^2 \leq c'_n \mathbb{V}\{\breve{\theta}_j\})]^2\}$$

$$\leq 2[\mathbb{V}\{\breve{\theta}_j\} + c'_n \mathbb{V}\{\breve{\theta}_j\}] \leq 2\zeta^* [1 + c'_n] n^{-1}.$$

This inequality and the Parseval identity verify Lemma 1.1.

1.2.3.3 Efficient Adaptation

The oracle-estimator (1.2.18) is sharp, or we may say *efficient*, meaning that it attains the lower bound (1.2.16). It does not use the anchor f_0, and its low-frequency component becomes data-driven (based only on data) if $\mathbb{V}(\hat{\theta}_j)$ is replaced by the sample variance as explained in the previous subsection. At the same time, the high-frequency component of the oracle-estimator for indexes $j > J_n$ depends on parameters (α, Q) of the underlying Sobolev class. In other words, the oracle-estimator depends on smoothness of an underlying density, and it is paramount to understand if the lower bound is sharp (attainable) by data-driven estimators and how to construct an estimator that adapts to the smoothness of an underlying density. This subsection is devoted to solving this problem. A *blockwise-shrinkage* (we may also say *blockwise*) methodology of adaptation to unknown smoothness is presented that proposes a data-driven estimator matching performance of the sharp oracle-estimator.

Further, the blockwise-shrinkage adaptation is universal and can be applied to all studied problems like density, regression or hazard rate estimation.

We begin with heuristic of the blockwise-shrinkage methodology of adaptive estimation, and explain it via several steps. Note that we would like to present a universal blockwise-shrinkage methodology, and accordingly in what follows we are considering a general setting of estimating a function $f(x) = \sum_{j=0}^{\infty} \theta_j \varphi_j(x)$, $x \in [0, 1]$. The studied density estimation problem is a particular example of that general setting.

The first step is to consider estimation of a particular Fourier coefficient θ_j of the function f under the mean squared error (MSE) criterion. Let $\bar{\theta}_j$ be an unbiased estimator of parameter θ_j. A particular example is the sample mean estimator (1.2.19) for the density model. To decrease the mean squared error $\mathbb{E}\{(\bar{\theta}_j - \theta_j)^2\}$ of $\bar{\theta}_j$, it may be beneficial to look at the *shrinking estimator* $\lambda_j \bar{\theta}_j$, $\lambda_j \in [0, 1]$ which minimizes the *mean squared error* $\mathbb{E}\{(\lambda_j \bar{\theta}_j - \theta_j)^2\}$. The oracle's solution, known as the *Wiener filter*, is to use the shrinking weight

$$\lambda_j^* = \frac{\theta_j^2}{\theta_j^2 + \sigma_{jn}^2}, \qquad \sigma_{jn}^2 := \mathbb{E}\{(\bar{\theta}_j - \theta_j)^2\}. \tag{1.2.28}$$

To verify (1.2.28), using unbiasedness of $\bar{\theta}_j$ and (1.1.1) we note that

$$\mathbb{E}\{(\lambda_j \bar{\theta}_j - \theta_j)^2\} = \lambda_j^2 (\sigma_{jn}^2 + \theta_j^2) - 2\lambda_j \theta_j^2 + \theta_j^2,$$

and the result follows. Further, for all traditional parametric problems we have σ_{jn}^2 proportional to n^{-1}. The latter is the classical parametric rate of the MSE convergence. We may conclude that the Wiener shrinkage can be beneficial if θ_j^2 is compared to or smaller in order than n^{-1}, otherwise no shrinkage is necessary. That is the heuristic of the famous Wiener filter.

It may be tempting to plug in appropriate estimates of θ_j^2 and σ_{jn}^2 in (1.2.28) and replace λ_j^* by the corresponding estimate. Unfortunately, this idea is not feasible because θ_j^2 is estimable with the parametric rate n^{-1}, and this accuracy is not enough for mimicking the Wiener filter. Nonetheless, the idea of mimicking the Winner filter is inspiring as it will be shown shortly.

The second step in the heuristic of blockwise-shrinkage adaptation is as follows. Let us, for a moment, return to the high-frequency component of the oracle-estimator (1.2.18). Note that the shrinking weights $1 - (j/J_n^*)^\alpha$ are close to each other for adjacent indexes j. In other words, the oracle hints that it is not necessary to use individual Wiener filters for each θ_j and instead it is possible to use a single Wiener filter for a block of adjacent indexes j. Namely, let $B := \{i + 1, \ldots, i + L\}$ be a block of length L of positive integers. Then the shrinking coefficient Λ^*, which minimizes $\mathbb{E}\{\sum_{j \in B} (\Lambda \bar{\theta}_j - \theta_j)^2\}$, is

$$\Lambda^* = \frac{L^{-1} \sum_{j \in B} \theta_j^2}{L^{-1} \sum_{j \in B} \theta_j^2 + [L^{-1} \sum_{j \in B} \sigma_{jn}^2]} =: \frac{\Theta}{\Theta + \sigma_n^2}. \tag{1.2.29}$$

In (1.2.29)

$$\Theta := L^{-1} \sum_{j \in B} \theta_j^2 \qquad (1.2.30)$$

is the classical *Sobolev functional* which is the focal point of the blockwise nonparametric adaptation. The theory of estimating Sobolev functionals is well developed, and while θ_j^2 may be estimated with the classical parametric rate n^{-1}, the Sobolev functional is estimable with the same rate but the constant decreases as L (length of the block) increases. This is what creates the opportunity for estimating Λ^* with sufficient accuracy for matching oracle-estimators.

The third step is to propose a simple estimator of the ratio (1.2.29). This step is explained via the density model and the particular sample mean Fourier estimator (1.2.26), namely $\hat{\theta}_j = n^{-1} \sum_{l=1}^{n} \varphi_j(T_l)$. Set

$$\hat{\Theta} := \frac{2}{n(n-1)} \sum_{1 \le l < s \le n} [\sum_{j \in B} \varphi_j(T_l)\varphi_j(T_s)]. \qquad (1.2.31)$$

Relation (1.2.25) yields $\mathbb{E}\{\hat{\Theta}\} = \Theta$, and accordingly $\hat{\Theta}$ is unbiased estimator of Θ. Further, the reader familiar with U-statistics may notice that $\hat{\Theta}$ is the *U-statistic* with the *kernel* $\sum_{j \in B} \varphi_j(T_1)\varphi_j(T_2)$. There is a well developed theory of U-statistics, and in particular it implies the following upper bound for the MSE,

$$\mathbb{E}\{(\hat{\Theta} - \Theta)^2\} \le \zeta L^{-1} n^{-1}(\Theta + n^{-1}). \qquad (1.2.32)$$

Here ζ is a finite positive constant. The factor L^{-1} in the MSE, together with the factor $\Theta + n^{-1}$, are the key for successful mimicking the oracle's blockwise shrinking (1.2.29). The recommended estimator of Λ^* is

$$\hat{\Lambda} := \frac{\hat{\Theta}}{L^{-1} \sum_{j \in B} \hat{\theta}_j^2} I(\hat{\Theta}_j > \rho n^{-1}). \qquad (1.2.33)$$

Here ρ is a positive constant.

Note that the same shrinkage $\hat{\Lambda}\hat{\theta}_j$ is used for all Fourier coefficient estimators $\hat{\theta}_j$ from the block B. This is why the procedure of adaptation to unknown smoothness of an estimand is called the blockwise-shrinkage adaptation.

The final step is to explain how to choose blocks B and constants ρ. Recall that we need to mimic only the high-frequency component of the oracle-estimator (1.2.18). The blocks are consecutive in the indexes, begin with $J_n + 1$ and are defined by their length L. Accordingly, consider a sequence of monotonically increasing to infinity integers $L_1, L_2, \ldots$ and monotonically vanishing positive $\rho_1, \rho_2, \ldots$ There are many sequences to choose from, see the Notes. For instance, one may set $L_k = k^2$ or $L_k = L_{nk} = \lceil (1 + 1/\ln(n+3))^k \rceil$, and $\rho_k = 1/\ln(k+3)$ or $\rho_k = \rho_{nk} = 1/\ln(n+3)$. Further, it is also possible to restrict the number of considered blocks. Returning to the oracle-estimator (1.2.18), note that no more than J_n^* of Fourier coefficients are estimated. Accordingly, no more than $n^{1/(2\alpha+1)} \ln(\ln(n + 20))$ Fourier coefficients are estimated by the oracle for large n. This allows us to restrict the number of considered blocks to k_n defined as the minimal integer such that $\sum_{k=1}^{k_n} L_k > n^{1/(2\alpha+1)} \ln(\ln(n + 20))$. Note that $k_n = k_n(\alpha)$ decreases in α. We do not know α, but may assume from our understanding of the problem that an underlying function f has some minimal smoothness. It is a tradition in the nonparametric statistical practice to assume that $\alpha \geq 2$ or at least $\alpha \geq 1$, that is the estimated function is at least differentiable. This assumption defines the sequence k_n for a chosen system of blocks.

The blockwise-shrinkage adaptation is defined, and it may be used for all considered in the book settings. More about adaptation can be found in the Notes.

1.2.3.4 Confidence Band for E-Estimator

We begin with recalling the classical notion of the $(1 - \alpha)$ coefficient (equivalently $100(1 - \alpha)\%$ level) *confidence interval* for the population mean $\mu = \mathbb{E}\{g(T)\}$ of the random variable $g(T)$. Here g is a known bounded function defined on the support of the lifetime T. Suppose that a sample of size n from T is available, then the sample mean estimator $\bar{g}_n := n^{-1} \sum_{l=1}^{n} g(T_l)$ is a traditional and feasible estimator of μ with a nice bouquet of statistical properties including the formula $\mathbb{V}\{\bar{g}_n\} = n^{-1}\sigma^2$, $\sigma^2 := \mathbb{V}\{g(T)\}$ and the *Central Limit Theorem* which states that asymptotically in n (the rule of thumb is $n \geq 30$) the distribution of $(n^{1/2}/\sigma)[\bar{g}_n - \mu]$ is standard Normal. Set $S^\xi(z_\alpha) := \alpha$ where ξ is the standard Normal random variable. Then we can propose the classical $1 - \alpha$ coefficient confidence interval $[\bar{g}_n - M_n, \bar{g}_n + M_n]$ where $M_n := z_{\alpha/2}\sigma/n^{1/2}$ is the *margin of error*, and the interval covers the unknown population mean μ with the probability asymptotically close to $1 - \alpha$, that is

$$\mathbb{P}(\mu \in [\bar{g}_n - M_n, \bar{g}_n + M_n]) = 1 - \alpha + o_n(1), \tag{1.2.34}$$

and $o_n(1)$ is negligibly small for the above-mentioned large n.

For a nonparametric estimator $\hat{f}(t)$ of a function $f(t)$, $t \in [0, 1]$, we would like to construct a similar confidence interval for each t, implying that we would get a confidence band around $\hat{f}(t)$ for all $t \in [0, 1]$. Unfortunately, there is no direct generalization of the parametric confidence interval due to the bias of any nonparametric density estimator. Nonetheless, the following approach can be used for the E-estimator of a function f which is not necessarily the density. The underlying idea is to assume that $f(t) = \sum_{j\in N} \theta_j \varphi_j(t)$. Consider the estimator $\hat{f}_N(t) = \sum_{j\in N} \hat{\theta}_j \varphi_j(t)$ where $\hat{\theta}_j$ is unbiased Fourier coefficient estimator of θ_j. This yields the formula

$$\sigma_N^2(t) := \mathbb{V}\{\sum_{j\in N}(\hat{\theta}_j - \theta_j)\varphi_j(t)\} = \sum_{i,j\in N} \mathrm{Cov}(\hat{\theta}_i, \hat{\theta}_j)\varphi_i(t)\varphi_j(t), \tag{1.2.35}$$

where $\mathrm{Cov}(\hat{\theta}_i, \hat{\theta}_j) := \mathbb{E}\{(\hat{\theta}_i - \mathbb{E}\{\hat{\theta}_i\})(\hat{\theta}_j - \mathbb{E}\{\hat{\theta}_j\})\}$ is the *covariance*. The covariance can be estimated by the empirical covariance, and this yields the estimate $\hat{\sigma}_N(t)$ of $\sigma_N(t)$. This and the Central Limit Theorem allow us to propose the $1 - \alpha$ coefficient *confidence band* satisfying

$$\mathbb{P}\left(|f_N(t) - \hat{f}_N(t)| \leq z_{\alpha/2}\hat{\sigma}_N(t)\,\big|\,f_N\right) = 1 - \alpha + o_n(1), \quad t \in [0, 1]. \qquad (1.2.36)$$

This band is the nonparametric analog of the parametric confidence band (1.2.34). Note how simple the band is and how it fits the E-estimator.

The dotted lines in Fig. 1.4 show bands for the density E-estimates based on the three simulations. As we have already discussed it, the left diagram exhibits the abnormal simulation. The confidence band "feels" that and exhibits dramatic left and right tails that should be compared with tails for the two other E-estimates. Let us stress that the E-estimator and its confidence band know only data, and the data points upon the decreasing density with large variability in the tails. The confidence bands for the Bathtub and the Normal underlying densities are reasonable and reflect the data variability. The final remark is that if $\hat{f}^T(t) = 1, t \in [0, 1]$, then by the definition the confidence band has zero width.

More about confidence bands can be found in the Notes.

1.2.4   Hazard Rate Estimation

The *hazard rate* of a continuous lifetime T (also referred to as the *failure rate* or the *force of mortality*) is defined as $h^T(t) := f^T(t)/S^T(t)$ when $S^T(t) > 0$. On first glance, it is natural to use the previous subsections and to estimate the hazard rate via ratio of the above-estimated density and the above-estimated survival function. This approach is feasible but has its drawbacks. The first one is that it is not optimal and does not minimizes the MISE, and as we will see shortly, the oracle recommends to use a direct series estimation of the hazard rate. The second drawback is that, as we will see shortly in Chap. 3, for right-censored observations it is simpler to estimate the hazard rate rather than the density. The latter is one of the interesting specifics of censored data.

As we already know from Sect. 1.1, hazard rate is not integrable over its support. Accordingly, to find an estimator with minimal MISE we restrict estimation to a subset of the support. To condense presentation of results and be able to refer to Theorem 1.1, it is assumed that $S^T(1) = e^{-\int_0^1 h^T(v)dv} > 0$ and h^T is estimated over the unit interval $[0, 1]$. To estimate h^T over an interval $[a, a + b]$, we can use the scale-location transformation $Z = (T - a)/b$ of the lifetime T, and then note that $h^T(t) = b^{-1}h^Z((t - a)/b), t \in [a, a + b]$.

1.2.4.1 Asymptotic Theory

Given that $h^T(t)$ is square-integrable on $[0, 1]$, we can write $h^T(t) = \sum_{j=0}^{\infty} \theta_j \varphi_j(t), t \in [0, 1]$ where the Fourier coefficients can be written as the expectation,

$$\theta_j = \int_0^1 h^T(t)\varphi_j(t)dt = \mathbb{E}\{\varphi_j(T)I(T \in [0, 1])/S^T(T)\}. \tag{1.2.37}$$

The expectation on the right side of (1.2.37) shows that the survival function S^T is the *natural nuisance function* for the hazard rate estimation. This is an interesting outcome because theoretically knowing S^T is equivalent to knowing the estimand h^T. Further, (1.2.37) yields the sample mean Fourier coefficient oracle-estimator

$$\hat{\theta}_j^* := n^{-1}\sum_{l=1}^{n}\frac{\varphi_j(T_l)I(T_l \in [0, 1])}{S^T(T_l)}. \tag{1.2.38}$$

In its turn, by plugging in (1.2.38) the empirical survival estimator (1.2.1), we get the data-driven (empirical) Fourier coefficient estimator

$$\hat{\theta}_j := n^{-1}\sum_{l=1}^{n}\frac{\varphi_j(T_l)I(T_l \in [0, 1])}{\hat{S}^T(T_l)}. \tag{1.2.39}$$

Theorem 1.2 *Consider estimation of the hazard rate $h^T(t)$ over interval $[0, 1]$ under the MISE criterion. Let $h_0(t), t \in [0, \infty)$ be the anchor hazard rate such that $h_0(t)$ is continuous, bounded and positive on $[0, 1]$. Then the assertion of Theorem 1.1, formulated for the density, also holds for the hazard rate if we replace density f^T by hazard rate h^T, replace density anchor f_0 by hazard rate anchor h_0, use $\hat{\theta}_j$ defined in (1.2.39), and use the hazard rate coefficient of difficulty*

$$d := d(h^T, S^T) := \int_0^1 \frac{h^T(t)}{S^T(t)}dt. \tag{1.2.40}$$

Let us comment on the proposition. First of all, the coefficient of difficulty indicates that hazard rate estimation may be dramatically more complicated than density estimation. Indeed, the hazard rate is not integrable over the support of T, and $S^T(t)$ is a monotonically decreasing function. Second, let us one more time repeat that the oracle's Fourier coefficient estimator (1.2.38) tells us that the oracle treats the survival function S^T as the natural nuisance function which should be estimated and then plugged in. This is an interesting example of nuisance function because knowledge of the hazard rate (the estimand) and knowledge of the survival function (nuisance function) are equivalent. At the same time, apart of the natural nuisance function, there is a remarkable similarity between the density and the hazard rate estimation.

1.2.4.2 E-Estimator for Hazard Rate
Several preliminary remarks are due. First of all, Theorem 1.2 tells us that we cannot estimate the hazard rate over its support because the corresponding coefficient of difficulty (1.2.40)

is infinity. Accordingly, we need to estimate the hazard rate over a subinterval $[0, r]$ of the support. Recall that r is called the *restriction*.

Then the corresponding formula for coefficient of difficulty is

$$d(r) := r^{-1} \int_0^r \frac{f^T(t)}{[S^T(t)]^2} dt = r^{-1} \mathbb{E}\left\{\frac{I(T \in [0, r])}{[S^T(T)]^2}\right\}. \tag{1.2.41}$$

This formula yields the method of moments estimator of the coefficient of difficulty,

$$\hat{d}(r) := \frac{1}{nr} \sum_{l=1}^n \frac{I(T_l \in [0, r])}{[\hat{S}^T(T_l)]^2}. \tag{1.2.42}$$

This estimator allows us to visualize the function $\hat{d}(r)$ and choose a feasible interval of estimation. Another remark is a practical one. For direct observations estimation of the density is simpler than of the hazard rate, and hence it may be a good idea to first look at simulated data using the histogram and/or the density E-estimate. Further, the E-estimator is a natural hazard rate estimation methodology due to the proposed Fourier coefficient estimator (1.2.39).

Figure 1.5 helps us to understand how to deal with estimation of the hazard rate. It presents two simulations of the same lifetime T in its two columns of diagrams. The two histograms show us the samples. The density E-estimates (the dashed lines) exhibit well the underlying Bimodal density (the solid line). The shown integrated squared error (ISE) allows us to quantify the quality of estimation. Estimated coefficients of difficulty $\hat{d}(t)$, $t \in [0, r]$, $r = 0.7$ are shown in the middle diagrams. Please note that the choice of $r = 0.7$ yields the coefficient of difficulty close to 5. According to Theorems 1.1 and 1.2, the latter means that for estimating the hazard rate over interval $[0, 0.7]$ one needs approximately five times larger sample size to get the same MISE as for estimation of the underlying Bimodal density over its support $[0, 1]$. This theoretical result allows us to appreciate the relatively good performance of the hazard rate E-estimator for the data at hand, see the bottom diagrams. Further, note how the estimated coefficient of difficulty allows us to choose a feasible interval of estimation. Finally, recall that if it is known that an underlying hazard rate is monotone, then the monotonic projection should be used.

It is highly recommended to repeat this figure several times with different intervals of estimation, sample sizes and underlying distributions. Hazard rate is an important characteristic of a lifetime, but it is less intuitively understandable than the density. Accordingly, Fig. 1.5 may help the reader to better understand the hazard rate and its estimation. Further, as we will see shortly in Chap. 3, for right-censored lifetimes estimation of the hazard rate is dramatically simpler than of the density.

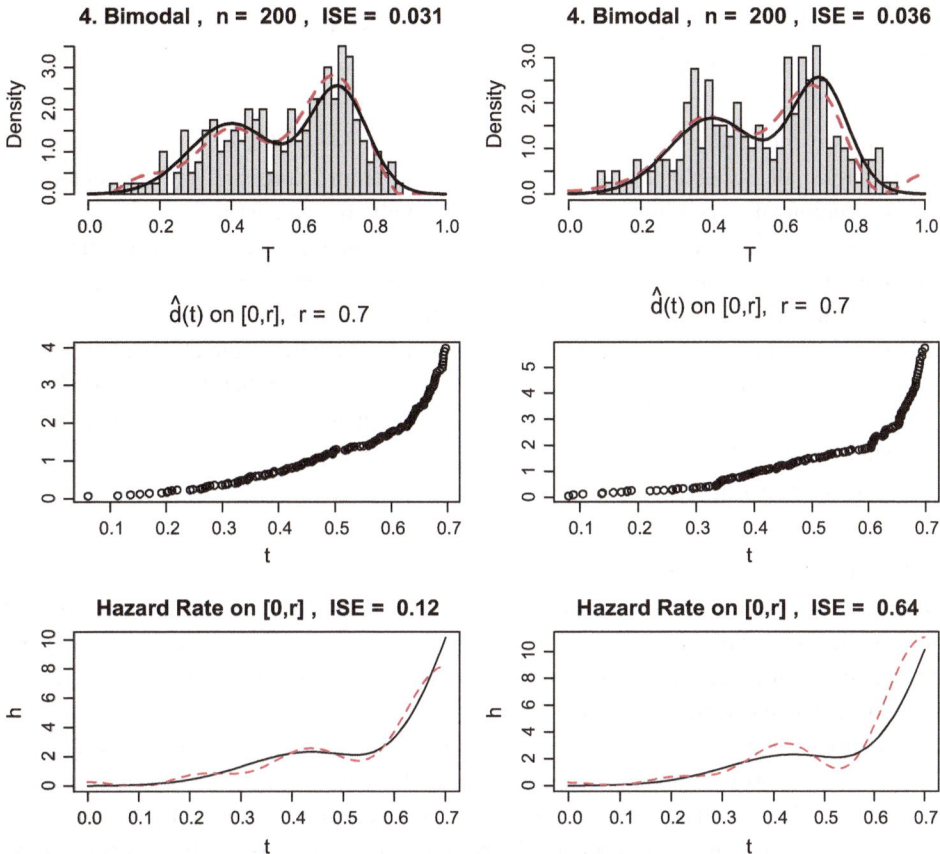

Fig. 1.5 Hazard rate estimation. Two independent and identical simulations are shown in the two columns of diagrams. A top diagram shows the histogram overlaid by the underlying density (the solid line) and its E-estimate (the dashed line). The circles in a middle diagram show estimated values $\hat{d}(T_l)$, $l = 1, 2, \ldots, n$ of the coefficient of difficulty. The solid line and the dashed line in a bottom diagram are the underlying hazard rate and its E-estimate, respectively. Hazard rate is estimated over interval $[0, r]$ with r being the controlled parameter. [$n = 200$, $corn = 4$, $r = 0.7$, $cJ0 = 3$, $cJ1 = 0.8$, $cTH = 4$]

1.2.5 Nonparametric Regression

The aim of nonparametric regression is to predict the lifetime of interest T using an observed random variable X. It is a tradition to refer to X as the *predictor* or independent variable and to T as the *response* or dependent variable. Accordingly, we are dealing with a sample $(X_1, T_1), \ldots, (X_n, T_n)$ of size n from the pair (X, T). Suppose that given $X = x$ we would like to find $m(x)$ which minimizes the mean squared error $\mathbb{E}\{(T - m(x))^2 | X = x\}$. Then the classical inequality $\mathbb{V}(Z) \leq \mathbb{E}\{(Z - \zeta)^2\}$, which becomes the equality if and only if the constant $\zeta = \mathbb{E}\{Z\}$, yields

$$m(x) := \mathbb{E}\{T|X = x\}. \qquad (1.2.43)$$

Furthermore, the random variable $m(X)$ minimizes the MSE, that is it minimizes $\mathbb{E}\{(T - \mu(X))^2\}$ over all possible $\mu(X)$. The latter follows from expressing the expectation via the expectation of a conditional expectation, namely we can write $\mathbb{E}\{(T - \mu(X))^2\} = \mathbb{E}\{\mathbb{E}\{(T - \mu(X))^2|X\}\}$ as explained in Sect. 1.1. The function (1.2.43) is called the *regression* function, and it represents the best prediction of T given $X = x$ in terms of the MSE criterion. The problem is to estimate $m(x)$ based on a sample from the pair (X, T) and without any assumption about shape of the regression. The latter is the classical problem of *nonparametric regression*. In what follows, it is always assumed that X is supported on $[0, 1]$, otherwise X should be rescaled on $[0, 1]$ as was explained in the previous section.

A simpler (than the above-discussed) regression model, which is also often considered in the literature, is when

$$T =: m(X) + v(X)\xi. \qquad (1.2.44)$$

In (1.2.44) $m(x)$ is the nonparametric regression of interest, $v(x) > 0$ is called the *scale* or the *spread*, ξ is the zero mean and unit variance random variable which is independent of X and may be referred to as the regression error, and then $v^2(x) = \mathbb{V}\{v(X)\xi|X = x\}$. Typical example of ξ is the standard normal variable, and then $v(x)\xi$ is the normal random variable with zero mean and variance $v^2(x)$.

Predictor X can be either deterministic (equidistant regression with $X_l = l/n, l = 1, 2, \ldots, n$ is a particular example) or random (in this case $X_1, \ldots, X_n$ is a sample from a random variable X with density f^X), and then the regression is referred to as fixed-design or random-design regression, respectively. Note that both designs can be described by the design density f^X defining the predictor. In the book we are considering the more complicated case of the random-design regression when X is a continuous random variable with density f^X that may be unknown. Further, regression model (1.2.44) is called *homoscedastic* if $v(x)$ is constant (does not depend on the predictor), otherwise it is *heteroscedastic*.

It is easy to propose a sample mean oracle-estimator of the Fourier coefficient $\theta_j = \int_0^1 m(x)\varphi_j(x)$. Indeed, for both models (1.2.43) and (1.2.44) we can write

$$\theta_j = \mathbb{E}\{\frac{T\varphi_j(X)}{f^X(X)}\}. \qquad (1.2.45)$$

This formula yields the sample mean Fourier coefficient oracle-estimator

$$\tilde{\theta}_j^* := n^{-1} \sum_{l=1}^{n} \frac{T_l\varphi_j(X_l)}{f^X(X_l)}. \qquad (1.2.46)$$

Note that the oracle uses the design density f^X of the predictor, which is the *natural nuisance function* for the regression. At the same time, neither the scale $v(x)$ nor the distribution of ξ are used by the oracle. Accordingly, it is sufficient to estimate the design density

f^X and plug it in (1.2.46). The latter can be done by the density E-estimator $\hat{f}^X$ developed in the previous subsection. Let us assume that $f^X(x)$ is differentiable and bounded below from zero on $[0, 1]$. Then the Fourier coefficient estimator for nonparametric regression is

$$\tilde{\theta}_j := n^{-1} \sum_{l=1}^{n} \frac{T_l \varphi_j(X_l)}{\max(\hat{f}^X(X_l), c_L/\ln(n+20))}. \tag{1.2.47}$$

Here c_L is a positive parameter of the used R software.

Remark 1.3 The Fourier coefficient oracle-estimator $\bar{\theta}_j^*$ is efficient for the fixed-design regression with the classical example being equidistant predictors. For the random-design regression, when X is a random variable, the oracle-estimator (1.2.46) is not efficient. There are several methods to remedy this issue discussed in the Notes. Here we restrict our attention to a remedy used in a number of settings considered in the book. Set $m_{-j}(x) := m(x) - \theta_j \varphi_j(x)$, that is we remove the jth Fourier component from the regression function. Then $\theta_j := \int_0^1 m(x)\varphi_j(x)dx = \int_0^1 [m(x) - m_{-j}(x)]\varphi_j(x)dx$. This is the underlying idea of the remedy. Namely, set $\tilde{m}_{-j}(x) := \sum_{s \in \{\{0,1,...,s_n\}\setminus\{j\}\}} \bar{\theta}_s \varphi_s(x)$, here s_n increases to infinity as slowly as desired as n increases, for instance one may set $s_n := \lceil \ln(\ln(n+20)) \rceil$. Then

$$\hat{\theta}_j := n^{-1} \sum_{l=1}^{n} \frac{[T_l - \tilde{m}_{-j}(X_l)]\varphi_j(X_l)}{\max(\hat{f}^X(X_l), c_L/\ln(n+20))} \tag{1.2.48}$$

is efficient Fourier coefficient estimator. As we will see shortly, the proposed remedy is useful in construction efficient Fourier coefficient estimators for a large number of statistical problems.

Theorem 1.3 *Consider estimation of the regression $m(x)$ over interval $[0, 1]$ for model (1.2.44) under the MISE criterion. Let $m_0(x)$, $x \in [0, 1]$ be a continuous on $[0, 1]$ anchor regression. Suppose that $v^2(x)$ is continuous and bounded on $[0, 1]$, and $f^X(x)$ is continuous and positive on $[0, 1]$. Then the assertion of Theorem 1.1, formulated for the density, also holds for the nonparametric regression if we replace density f^T by regression m, replace density anchor f_0 by the anchor regression m_0, in the lower bound additionally consider supremum over all distributions of the regression error ξ with zero mean and variance bounded by 1, utilize $\hat{\theta}_j$ defined in (1.2.48), and use the regression coefficient of difficulty*

$$d := d(v, f^X) := \int_0^1 \frac{v^2(x)}{f^X(x)} dx. \tag{1.2.49}$$

Similarly to the previously considered nonparametric statistical problems, the regression coefficient of difficulty (1.2.49) is of interest because it explains how the pair of nuisance functions (v, f^X) affects the nonparametric regression estimation. As an interesting corol-

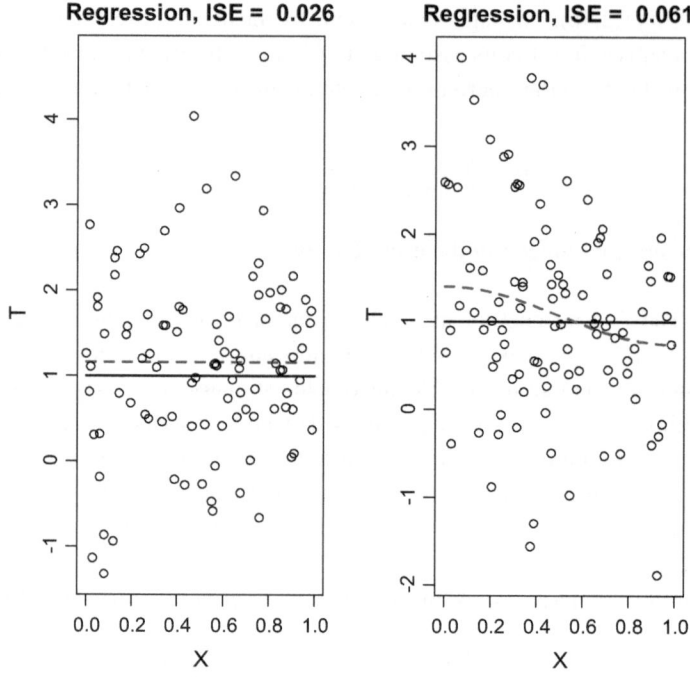

Fig. 1.6 Two simulations of a heteroscedastic nonparametric regression $T = m(X) + \sigma s(X)\xi$. Here $m(x)$ is the Uniform corner function, ξ is standard normal and independent of X, and $\sigma s(x)$ is the scale function. A scattergram of observations from (X, T) is shown by the circles. The solid and dashed lines are the underlying regression and its estimate, respectively. ISE is the integrated squared error $\int_0^1 (m(x) - \hat{m}(x))^2 dx$. The E-estimator knows that an underlying regression function is nonnegative. {The arguments are: n controls the sample size, *corn* controls the regression function which can be one of the corner functions, *sigma* controls σ, the string *scalefun* defines the shape of a custom function $s(x)$ which is truncated from below by the value *dscale* and then rescaled into a bona fide density supported on [0, 1], the string *desden* controls shape of the design density which is then truncated from below by the value *dden* and then rescaled into a bona fide density. Argument cL controls parameter c_L in (1.2.48). Arguments $cJ0$, $cJ1$ and cTH are parameters of the E-estimator.} [$n = 100, corn = 1, desden = "1 + 0.6 * x", scalefun = "3 - (x - 0.5) \wedge 2",$ $sigma = 1, dscale = 0, dden = 0.2, cL = 1, cJ0 = 3, cJ1 = 0.8, cTH = 4$]

lary, a direct calculation shows that the design density, minimizing the coefficient of difficulty and hence improving the estimation, is $f_*^T(x) := v(x)/\int_0^1 v(v)dv$.

The proposed sample mean Fourier coefficient estimate (1.2.48) allows us to use the nonparametric series E-estimator for the regression. Figure 1.6 illustrates its performance and the caption explains the two simulations. Here the underlying regression is number 1 corner function Uniform. Let us look at the left diagram. The scattergram exhibits the pronounced heteroscedasticity, and it is not easy to visualize a curve that goes through the middle of the scattergram. Nonetheless the E-estimator does a good job in showing the flat shape of the Uniform. The right diagram shows us a repeated simulation, but the outcome is really far from what could be expected. The E-estimate shows the decreasing regression (the dashed line), and this shape is supported by the data at hand. The latter is the important lesson to learn. Indeed, the E-estimator knows only data, and then it describes the data. To get used to nonparametric regression and gain experience in "reading" scattergrams, it is highly recommended to repeat this figure a number of times using different underlying models.

1.2.6   Multivariate Function

As we will see shortly, the above-presented series approach and the E-estimation are effortlessly extended to multivariate models. This is the good news. The bad news is that for the same sample size the quality of estimation, measured by the mean integrated squared error, is dramatically worse. To state this differently, to get the accuracy of estimation similar to a univariate case, dramatically larger sample sizes are needed. This fact is often referred to as the *curse of dimensionality*, and it will be explained shortly.

We begin with discussion of bivariate functions and the corresponding E-estimation, a general case will follow shortly. Consider a square integrable on $[0, 1]^2$ bivariate function $f(x_1, x_2)$, that is a function satisfying the inequality $\int_{[0,1]^2} f^2(x_1, x_2)dx_1dx_2 < \infty$. Introduce two univariate bases $\{\phi_j(x), j = 0, 1, \ldots\}$ and $\{\eta_j(x), j = 0, 1, \ldots\}$ on the unit interval $[0, 1]$. Then all possible products of elements from these two bases,

$$\{\varphi_{j_1 j_2}(x_1, x_2) := \phi_{j_1}(x_1)\eta_{j_2}(x_2), \quad j_1, j_2 = 0, 1, \ldots\}, \tag{1.2.50}$$

create a basis on $[0, 1]^2$ which is called a *tensor-product* basis. This useful mathematical fact implies a great flexibility in creating convenient bivariate bases. In the previous section the cosine basis on $[0, 1]$ was used, for bivariate functions we will use the cosine tensor-product basis with elements $\varphi_{j_1 j_2}(x_1, x_2) := \varphi_{j_1}(x_1)\varphi_{j_2}(x_2)$. Recall that $\varphi_0(x) := 1$ and $\varphi_j(x) := 2^{1/2} \cos(\pi jx), j = 1, 2, \ldots$

If a function $f(x_1, x_2)$ is square integrable on $[0, 1]^2$, then its partial sum approximation with cutoffs J_1 and J_2 is

$$f_{J_1 J_2}(x_1, x_2) = \sum_{j_1=0}^{J_1} \sum_{j_2=0}^{J_2} \theta_{j_1 j_2} \varphi_{j_1 j_2}(x_1, x_2), \tag{1.2.51}$$

and the Fourier coefficients $\theta_{j_1 j_2}$ are defined by the formula

$$\theta_{j_1 j_2} := \int_{[0,1]^2} f(x_1, x_2)\varphi_{j_1 j_2}(x_1, x_2)dx_1dx_2. \tag{1.2.52}$$

This formula immediately yields the sample mean estimator of Fourier coefficients for the case of a bivariate density $f^{X_1X_2}(x_1, x_2)$ supported on $[0, 1]^2$. Denote by $(X_{11}, X_{21}), \ldots,$ (X_{1n}, X_{2n}) a sample of size n from (X_1, X_2), and define the sample mean Fourier coefficient estimator

$$\hat{\theta}_{j_1 j_2} := n^{-1} \sum_{l=1}^{n} \varphi_{j_1 j_2}(X_{1l}, X_{2l}). \tag{1.2.53}$$

The Fourier coefficient estimator allows us to use the E-estimation methodology. The only new element to mention is that now a pair of cutoffs $(\hat{J}_1, \hat{J}_2)$, which minimizes the empirical risk of the projection estimator, should be calculated. Namely, we use the cutoffs

$$(\hat{J}_1, \hat{J}_2) := \operatorname{argmin}_{\{(J_1, J_2) \in \{0, 1, \ldots, c_{J0} + c_{J1} \ln(n)\}^2\}} \sum_{j_1=0}^{J_1} \sum_{j_2=0}^{J_2} [2\hat{\sigma}^2_{j_1 j_2 n} - \hat{\theta}^2_{j_1 j_2}]. \tag{1.2.54}$$

Here $\hat{\sigma}^2_{j_1 j_2 n}$ is the sample variance estimator of the variance $\mathbb{V}\{\hat{\theta}_{j_1 j_2}\}$.

Figure 1.7 illustrates performance of the bivariate density E-estimator. As it is explained in the caption, the bivariate density is the product of two corner functions. Now let us look at the left-top diagram which shows us a simulated sample from (X_1, X_2). First of all, it is difficult to believe that we have $n = 100$ observations, but we do. This is a serious issue with analysis of multivariate data because the observations are sparse. The data is generated by two independent variables X_1 and X_2 with different corner distributions. Knowing this fact, can you guess their distributions? This may not be a simple task because you need to project observations on the axes and then guess the densities. You may find the correct answer in the caption. A training in reading data is needed, the interested reader may repeat this figure several times to get a better "feeling" of data. For instance, to realize the independence, you may move a horizontal or vertical line along the range of a corresponding variable and then convince yourself that the univariate shape of the density along the moving line remains the same.

Another complication with multivariate curves is how to visualize them. The left-middle diagram shows us the surface of the underlying bivariate density. It helps to get a feeling of the density, but many different views of the surface are necessary for its understanding. Nonetheless, this surface helps us to see the Bathtub shape in x_1 and the Normal shape in x_2. The left-bottom diagram shows us the corresponding E-estimate. The estimate looks good. At the same time, it is difficult to appreciate how close it is to the underlying bivariate density.

The right column of diagrams helps us to understand how well the E-estimator performs. Here three slices of $f^{X_1, X_2}(x_1, x_2)$ and its E-estimates, for the specific values of x_2 shown in the titles, are exhibited. We note that near the boundary, namely for $x_2 = 0.1$, the estimate is zero and does not exhibit the underlying Bathtub. At the same time, note the scale and that the underlying density is less than 0.08. The estimate is symmetric about 0.5 for $x_2 = 0.5$, it correctly shows the bathtub nature of the density and its magnitude. The same comment can be made about the estimate for $x_2 = 0.8$. Note how well the E-estimator evaluates magnitudes of the slices.

While the bivariate E-estimator is not perfect, it is one of the best that can be proposed for such a small sample size. The reader is advised to repeat this figure with different parameters and get used to the quality of bivariate estimation. Moreover, because for this bivariate nonparametric problem the coefficient of difficulty is 1, this simulation may serve as a benchmark for other more complicated bivariate problems discussed in the next chapters.

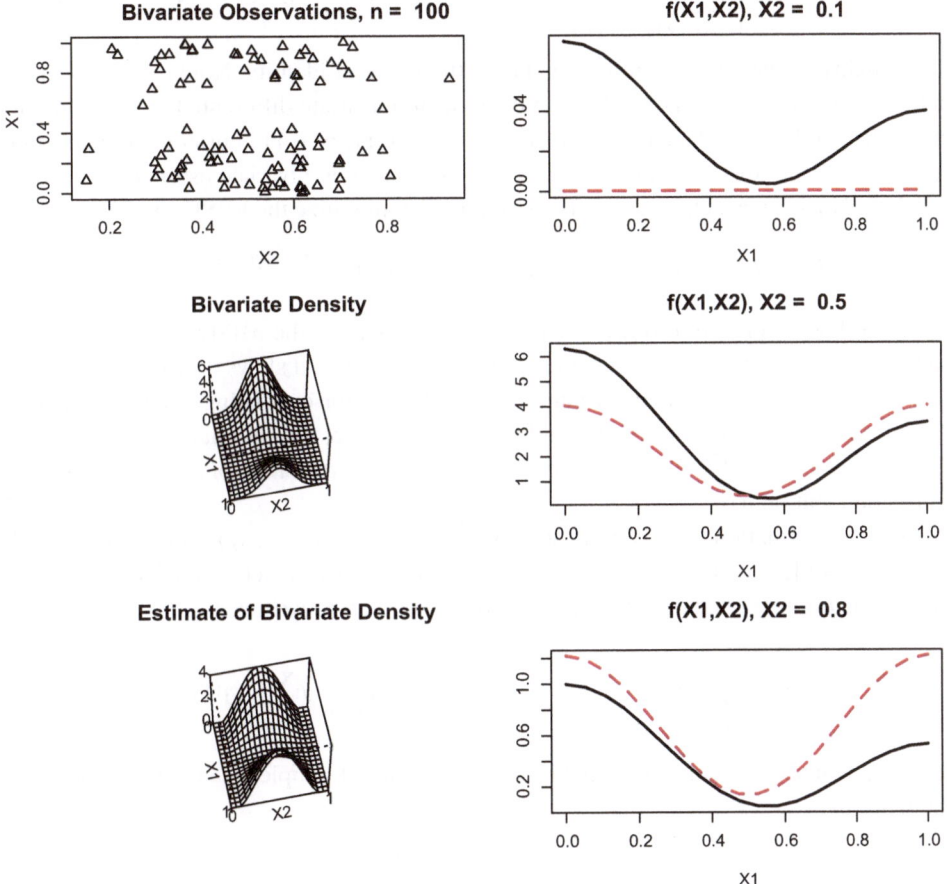

Fig. 1.7 E-estimation of bivariate density. The bivariate density $f^{X_1,X_2} = f^{X_1} f^{X_2}$ where f^{X_1} and f^{X_2} are the Bathtub and the Normal, respectively. Diagrams in the left column show the data, the underlying bivariate density and its E-estimate. Diagrams in the right column show slices of the bivariate density (the solid line) and slices of the bivariate estimate (the dashed line). {An underlying bivariate density is the product of two corner densities defined by the arguments $c1$ and $c2$.} [$n = 100$, $c1 = 2$, $c2 = 3$, $cJ0 = 3$, $cJ1 = 0.8$, $cTH = 4$]

Now let us comment on the asymptotic theory of the bivariate projection estimation. If a function $f(x_1, x_2)$ has β bounded partial derivatives in x_1 and x_2, then under a mild additional assumption it is known that the *integrated squared bias* (ISB) can be bounded from above as follows,

$$\text{ISB}(J_1, J_2) := \int_{[0,1]^2} [f(x_1, x_2) - \sum_{j_1=0}^{J_1} \sum_{j_2=0}^{J_2} \theta_{j_1 j_2} \varphi_{j_1 j_2}(x_1, x_2)]^2 dx_1 dx_2$$

$$= \sum_{(j_1, j_2) \notin \{0,1,\dots,J_1\} \times \{0,1,\dots,J_2\}} \theta_{j_1 j_2}^2 \leq \zeta^* [J_1^{-2\beta} + J_2^{-2\beta}], \quad \zeta^* < \infty. \tag{1.2.55}$$

The equality in the last bottom line is due to the Parseval identity. Note that the decrease in the ISB is similar to what we have in the case of univariate differentiable functions, and this is the good news. The bad news is that now $(J_1 + 1)(J_2 + 1)$ Fourier coefficients must be estimated, and this yields that variance of the projection bivariate estimator is of order $n^{-1} J_1 J_2$. Now we can choose cutoffs J_1 and J_2 that minimize the MISE and get

$$J_1^* = J_2^* = O_n(1) n^{1/(2\beta+2)}, \quad \text{MISE} = O_n(1) n^{-2\beta/(2\beta+2)}. \tag{1.2.56}$$

As we know, for β-fold differentiable univariate density the MISE decreases with the rate $n^{-2\beta/(2\beta+1)}$, while for the considered bivariate densities it slows down to $n^{-2\beta/(2\beta+2)}$. To shed an extra light on that formula, let us note that for estimating a k-variate density $f^{X_1,\dots,X_k}$, which is β-fold differentiable in each variable, the optimal rate of the MISE convergence is $n^{-2\beta/(2\beta+k)}$. This is the result underlying the *curse of dimensionality*, and let us comment on it.

Consider estimation of a k-variate density $f^X(\mathbf{x}), \mathbf{x} := (x_1, \dots, x_k)$ supported on $[0, 1]^k$. Let $\varphi_{\mathbf{j}}(\mathbf{x}) := \prod_{i=1}^k \varphi_{j_i}(x_{j_i}), \mathbf{j} := (j_1, \dots, j_k)$ be the tensor-product cosine basis on $[0, 1]^k$. Then a square-integrated density supported on $[0, 1]^k$ can be written as

$$f^X(\mathbf{x}) = \sum_{j_1,\dots,j_k=0}^{\infty} \theta_{\mathbf{j}} \varphi_{\mathbf{j}}(\mathbf{x}), \quad \mathbf{x} \in [0, 1]^k, \quad \theta_{\mathbf{j}} := \int_{[0,1]^k} f^X(\mathbf{x}) \varphi_{\mathbf{j}}(\mathbf{x}) d\mathbf{x}. \tag{1.2.57}$$

If a sample of size n from $\mathbf{X}$ is available, then the unbiased sample mean Fourier coefficient estimator is

$$\hat{\theta}_{\mathbf{j}} := n^{-1} \sum_{l=1}^n \varphi_{\mathbf{j}}(\mathbf{X}_l). \tag{1.2.58}$$

Note that variance of this estimator is of order n^{-1}, that is of the same order as for univariate and bivariate densities.

Set

$$\hat{f}(\mathbf{x}) := \sum_{i=1}^k \sum_{j_i=1}^{J_i} \hat{\theta}_{\mathbf{j}} \varphi_{\mathbf{j}}(\mathbf{x}). \tag{1.2.59}$$

Using the Parseval identity, we can repeat the above-presented analysis of the MISE of the bivariate density estimator and conclude that choosing cutoffs

$$J_1^* = \ldots = J_k^* = J_* := O_n(1)n^{1/(2\beta+k)} \tag{1.2.60}$$

yields the optimal rate of the MISE convergence

$$MISE = O_n(1)[J_*^k n^{-1} + J_*^{-2\beta}] = O_n(1)n^{-2\beta/(2\beta+k)}. \tag{1.2.61}$$

This formula quantifies the effect of dimensionality k on the MISE convergence.

1.3   Software

The provided complementary R package allows the reader to repeat and modify all figures in the book. R is a commonly used free statistical software. No experience in dealing with R is required, and below its brief installation and usage are explained. Download the book's R-package called *book3.r* and the data-files using the link located at http://profiles.utdallas.edu/ efrom in the section Publications. The link also leads to a file containing relevant information about the book.

Use Installation and Administration manual on CRAN (www.r-project.org, http://cran. r-project.org). After installing R (or if it was already installed), you need to choose your working directory for the book's software package. For instance, for MAC you can use "/Users/Me/Book3". Download to this directory the file *book3.r* and the data-files. By downloading the package, the user agrees to consider it as a "black-box" and employ it for educational purposes only.

Now you need to start the R on your computer. This should bring up a new window, which is the R console. In the R console you will see:

>

This is the R prompt. Now, only once while you are working with the book, you need to install several standard R packages. Type the following command in the R console to install required packages

> install.packages("MASS")
> install.packages("survival")
> install.packages("monotone")
> install.packages("scatterplot3d")

These packages are installed just once, you do not need to repeat this step when you start your next R session.

Next you need to source (install) the book's package, and you do this with R operating in the chosen working directory. To do this, first type

> getwd()

This R command will allow you to see a working directory in which R is currently operating. If it is not your chosen working directory "/Users/Me/Book3", type

> setwd("/Users/Me/Book3") and then R will operate in the chosen directory for the book. Then type

> source("book3.r") and the book's R software will be installed in the chosen directory and you are ready to use it. This sourcing should be done every time when you start a new R session.

For the novice, it is important to stress that no knowledge of R is needed to use the package and repeat/modify figures. Nonetheless, a bit of information may be useful to understand the semantics of calling a figure. What you are typing is the name of an R function and then in its parentheses you assign values to its arguments. For instance,

> ch1(Fig = 1, n = 50) is the call to R function ch1 that will be run with arguments Fig = 1 and n = 50, and all other arguments of the function will be equal to their default values indicated in the caption of the corresponding Fig. 1.1. Note that ch1 indicates that the figure is from this chapter while its argument Fig = 1 indicates that it is the first figure in this chapter. In other words, ch1(Fig = 1, n = 50) runs Fig. 1.1 with the sample size $n = 50$. In R, arguments of a function may be either scalars (for instance, a=3 implies that a function will use value 3 for argument a), or vectors (for instance, vec = c(1,3) implies that a function will use vector vec with the first element equal to 1 and the second element equal to 3, and note that c() is a special R function called "combine" that creates vectors), or a string (say den= "normal" implies that a function will use name *normal* for its argument den). R functions are smart in terms of using shorter names for its arguments, for instance both Fig = 5, fi = 5 and f = 5 will be correctly recognized and imply the same value 5 for the argument fig unless there is a confusion with other arguments.

In general, to repeat Figure k.j, which is jth figure in Chapter k, and to use the default values of its arguments outlined in square brackets of the caption of Figure k.j, type

> chk(f = j)

If you want to use new values for two arguments, say n = 300 and b = 5, type

> chk(f = j, n = 300, b = 5)

Note that values of other arguments will be equal to default ones. Further, recall that typing f = j and fig = j implies the same outcome.

To finish the R session, type

> q()

Figure 1.8 shows the five corner functions supported by the software. The argument corn is always used to choose specific corner functions used in simulations, and Fig. 1.3 explains how to create a custom-made corner function. Also recall that Fig. 1.1 explains how to use the R package, and Fig. 1.4 explains parameters of the E-estimator and how to change them. It is possible to make all curves of the same black color by setting argument $COL = 0$.

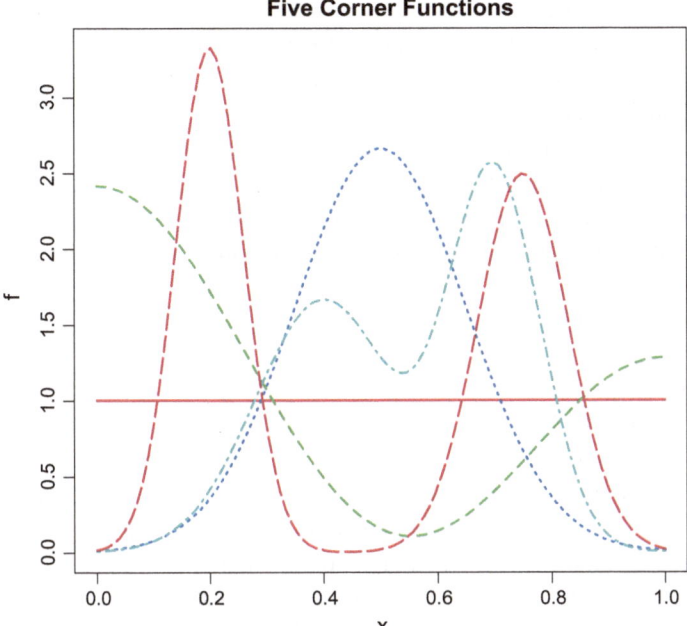

Fig. 1.8 Corner functions $f(x)$, $x \in [0, 1]$ supported by the software. The functions have both the assigned numbers from 1 to 5 and the following corresponding names: The Uniform (the solid line); The Bathtub (the short-dashed line); The Normal (the dotted line); The Bimodal (the dot-dashed line); the Strata (the long-dashed line). Note that the name describes the shape of a corner function. In the text the functions may be referred to as f_i, $i = 1, 2, 3, 4, 5$. Each corner function is a density supported on $[0, 1]$. [corn = c(1, 2, 3, 4, 5)]

1.4 Exercises

Here and in all other chapters, the asterisk denotes a more difficult exercise.

1.1.1 Let D_1 and D_2 be two events. Describe the events $D_1 \cup D_2$, $D_1 \cap D_2$, D_1^c, $D_1 \cup D_1^c$.

1.1.2 Formulate the three axioms of probability.

1.1.3 Present examples of discrete random variables. What are the main probability characteristics of these variables?

1.1.4 Present examples of continuous random variables. What are the main probability characteristics of these variables?

1.1.5 Definition of the mean and variance of a random variable.

1.1.6* Consider a pair (X, Y) of continuous random variables. Define the joint distribution function, the joint density, the conditional distribution function, the conditional density, the marginal density.

1.1.7 Prove formula (1.1.1).

1.1.8 Show that $\mathbb{E}\{(X - \zeta)^2\} \geq \mathbb{V}\{X\}$ with the equality iff $\zeta = \mathbb{E}\{X\}$.

1.1.9* Show that $\mathbb{E}\{|X - \zeta|\} \geq \mathbb{E}\{X - \nu\}$ with the equality iff ν is the median.

1.1.10 Let events D_1 and D_2 be independent. Are the events D_1 and D_2^c independent?

1.1.11 Let X and Y be independent continuous random variables. Show that $S^{X,Y}(x, y) = S^X(x)S^Y(y)$ and $f^{X,Y}(x, y) = f^X(x)f^Y(y)$.

1.1.12 What random variables are studied in survival analysis? Present several examples.

1.1.13 Definition of the survival function.

1.1.14 Let T be a continuous lifetime. Explain why $S^T(t) = \mathbb{P}(T \geq t)$.

1.1.15 Explain the definition (1.1.2) of the hazard rate function, and then verify the equality in (1.1.2).

1.1.16 What is the limit of $H^T(t)$ as $t \to \infty$?

1.1.17 Prove (1.1.4).

1.1.18 Prove (1.1.5)

1.1.19 Verify formula (1.1.9).

1.1.20 Prove formula (1.1.12).

1.1.21 Explain formula (1.1.13). Hint: Use the assumption that T and C are independent.

1.1.22* How may formula (1.1.13) look if T and C are dependent?

1.1.23 Is the mean residual life $\mu^T(t)$ a monotonic function in t?

1.1.24 Is the restricted mean residual life $\mu_r^T(t)$ a monotonic function in t? Is it a monotonic function in r?

1.1.25 Describe the model of right-censoring. Hint: Begin with the lifetime of interest T and the censoring lifetime C.

1.1.26 Definition of the hazard rate and cumulative hazard rate.

1.1.27 Consider a right-censored sample of size n from (V, Δ). Set $N := \sum_{l=1}^n \Delta_l$. What is the distribution of N? Find the mean and variance of N.

1.1.28 What is the model of current status censoring? Is it also referred to as the case I interval censoring?

1.1.29* Consider the CSC model with the lifetime of interest T and the monitoring time Z being independent. What is the probability mass function of $\Delta := I(T \leq Z)$? Can you name the distribution of Δ?

1.1.30 Describe the MCAR, MAR and MNAR missing mechanisms. Hint: Begin with the Bernoulli random variable A, called the availability, and then introduce the availability likelihood.

1.1.31 When may a data be called biased? Present an example of biased data.

1.1.32 Present an example of missingness that creates biased data.

1.1.33 Explain formula (1.1.16). What is the biasing function?

1.1.34 Repeat Fig. 1.1 ten times and describe the observed simulations. Then change the underlying corner functions and compare the results.

1.1.35 Explain all diagrams in Fig. 1.2. Then repeat it with different corner functions. Compare the outcomes.

1.1.36 Repeat Fig. 1.2 five times with $b = 1$ and then with $b = 2$. Compare the outcomes and explain them.

1.2.1 Explain formula (1.2.1) for the empirical survival function and prove the relation (1.2.2).

1.2.2* Verify inequality (1.2.4).

1.2.3 Explain why in formula (1.2.1) the indicator $I(T_l \geq t)$ is used in place of $I(T_l > t)$.

1.2.4 Definition of an orthonormal basis.

1.2.5 Definition of a Fourier coefficient.

1.2.6 Why is notion of the truncated orthonormal series important for nonparametric curve estimation?

1.2.7 Definition of the integrated squared error.

1.2.8 Define elements of the cosine basis on $[0, 1]$.

1.2.9 What is the Parseval identity?

1.2.10 Verify the inequality (1.2.9).

1.2.11* Verify the inequality (1.2.10).

1.2.12* Propose an upper bound, similar to (1.2.10), given that f has three derivatives. Comment on the outcome for a series approximation.

1.2.13* Use Fig. 1.3 and for each of the five corner functions suggest a reasonably small cutoff which, in your opinion, yields a fair visual impression of the corner function. Then explain why there is a difference or no difference between the recommended cutoffs.

1.2.14* Present definitions and then explain the local shrinking and global Sobolev function classes. Hint: Think about bias of a series approximation.

1.2.15 Verify inequality (1.2.13).

1.2.16 Describe the problem of density estimation.

1.2.17 What is the mean integrated squared error (MISE) of a density estimator?

1.2.18* Suppose that f^T is the density of a random variable T. Consider the scale-location transformation $Z := b(T - a)$ of T, $b > 0$ What is the density f^Z of Z? Hint: Begin with finding the cumulative distribution function $F^Z(z) := \mathbb{P}(Z \leq z)$ and then take its derivative using the definition $f^Z(z) := dF^Z(z)/dz$.

1.2.19 Explain the assertion of Theorem 1.1. Hint: The assertion includes the lower bound and the upper bound that asymptotically (as $n \to \infty$) coincide. Also explain the oracle's role in the assertion.

1.2.20 Explain the notion of the coefficient of difficulty and its role in the theory of density estimation.

1.2.21* Explain how the smoothness of an underlying density, defined by the parameter α, affects the rate of MISE convergence. Then explain how the smoothness affects the cutoff J_n^*.

1.2.22 What are the statistical techniques used by the oracle for low- and high-frequency components of the proposed density oracle-estimator?

1.2.23 Describe the 3 steps of the E-estimation algorithm.

1.2.24 Explain the estimates shown in Fig. 1.4. Then repeat the simulation ten times and describe the results.

1.2.25* Use Fig. 1.4 and choose optimal parameters of the E-estimator for estimation of the Bathtub and the Bivariate corner functions.

1.2.26* Use Fig. 1.4 and choose optimal parameters of the E-estimator for estimation of the Bathtub and the Bivariate corner functions for $n = 50$ and $n = 200$. Explain your findings.

1.2.27 Suppose that a density should be estimated over an interval $[a, b]$. Explain how the E-estimator can be used for solving that problem. Hint: Read Remark 1.1.

1.2.28* Explain the methodology of the series estimation of the survival function for analytic distributions. Hint: Read Remark 1.2.

1.2.29 Explain the notion of efficient adaptation.

1.2.30* Explain heuristic of the blockwise-shrinkage adaptation.

1.2.31 Prove optimality of the Wiener filter.

1.2.32 Prove optimality of the oracle's blockwise-shrinkage (1.2.29).

1.2.33 What is the Sobolev functional?

1.2.34 Prove that the estimator (1.2.31) of the Sobolev functional Θ is unbiased.

1.2.35* Prove (1.2.32). Hint: Either use the U-statistic technique or a direct calculation.

1.2.36 Explain the underlying idea of the estimator (1.2.33).

1.2.37 Consider a normal random variable X with mean μ and variance σ^2. Let a sample of size n from X be given. Propose unbiased estimator of the mean and the confidence interval.

1.2.38 Explain the underlying idea of the pointwise confidence band defined in (1.2.36) and the relation (1.2.36).

1.2.39 Definition of the hazard rate.

1.2.40 Why is there a problem with estimation of the right tail of the hazard rate?

1.2.41 Show that the Fourier coefficient oracle-estimator (1.2.38) is unbiased.

1.2.42 What is the natural nuisance function for the hazard rate estimation?

1.2.43 What is the coefficient of difficulty for estimation of the hazard rate?

1.2.44 How can an estimate of the coefficient of difficulty help in choosing a feasible interval of estimation of the hazard rate?

1.2.45 Compare assertions of Theorems 1.1 and 1.2.

1.2.46 Describe the hazard rate E-estimator.

1.2.47 Use Fig. 1.5 and describe the effect of interval of estimation, controlled by parameter r, on quality of hazard rate estimation.

1.2.48* Use Fig. 1.5 and propose optimal parameters of the E-estimator for the Bathtub distribution and sample sizes 100 and 300.

1.2.49 What is the problem of nonparametric regression?

1.2.50 For nonparametric regression, define the predictor, the response, the scale function, the design density.

1.2.51 Is regression model (1.2.44) more general than (1.2.43)? Explain your answer.

1.2.52 Define homoscedastic and heteroscedastic regressions.

1.2.53 Let (1.2.44) hold. Prove that $m(x) = \mathbb{E}\{T|X = x\}$.

1.2.54 Assume that the design density f^X is given. Prove that the method of moments estimator (1.2.46) is unbiased.

1.2.55 What is the natural nuisance function for the nonparametric regression?

1.2.56 Explain the underlying idea of the Fourier coefficient estimator (1.2.47).

1.2.57* Prove that the estimator (1.2.47) is asymptotically unbiased.

1.2.58* Verify the assertion of Remark 1.3.

1.2.59 Explain the assertion of Theorem 1.3.

1.2.60 Compare coefficients of difficulty for regression and density estimation.

1.2.61 Repeat Fig. 1.6 and explain the outcomes.

1.2.62 Use Fig. 1.6 and explain the effect of the scale function on quality of estimation.

1.2.63* Use Fig. 1.6 and find optimal parameters of the E-estimator for several corner regressions and sample sizes 100 and 300.

1.2.64 Consider the tensor-product basis (1.2.50) on the unit squarer $[0, 1]^2$. Verify that $\int_{[0,1]^2} \varphi_{j_1 j_2}(x_1, x_2) \varphi_{i_1 i_2}(x_1, x_2) dx_1 dx_2 = I(j_1 = i_1) I(j_2 = i_2)$.

1.2.65 Explain why the projection (1.2.51) can approximate a square integrable function $f(x_1, x_2)$ as J_1 and J_2 increase. Hint: Recall the Parseval identity.

1.2.66 Explain formula (1.2.52).

1.2.67 Find the mean and the variance of the Fourier coefficient estimator (1.2.53).

1.2.68* Write down the E-estimation algorithm for bivariate density estimation. Hint: Follow the steps outlined in Sect. 1.2.3.

1.2.69 Explain the six diagrams in Fig. 1.7.

1.2.70 Analyze the left-top diagram in Fig. 1.7 and try to explain why the data correspond to the underlying bivariate density. Hint: Begin with the independence, and then look at the marginal densities.

1.2.71* Repeat Fig. 1.7 several times for different underlying corner functions. Rank them from the better to the worse in quality of estimation. Then compare your ranking with smoothness of the corner functions.

1.2.72 Figure 1.7 exhibits two possibilities to exhibit a bivariate function. Which one do you prefer? Do they complement each other?

1.2.73 Verify (1.2.55).

1.2.74 Explain and then verify relations (1.2.56). How do they shed light on the curse of dimensionality?

1.2.75* Find the mean and variance of the Fourier coefficient estimator (1.2.58).

1.2.76* Show that cutoffs (1.2.60) yield the optimal rate $n^{-2\beta/(2\beta+k)}$ of the MISE convergence.

1.2.77* Describe E-estimator for estimation of a k-variate density supported on $[0, 1]^k$.

1.2.78 Verify formula (1.2.61).

1.2.79 Explain why the curse of dimensionality precludes us from estimating high-frequency components.

1.3.1 What are the five corner functions supported by the software? What is so special in their shapes?

1.3.2 Consider a corner function f. What is $\int_0^1 f(x)dx$?

1.3.3 Suppose that you are interested in the corner density $f(x) = (1 + 3x)I(x \in [0, 1])/\int_0^1 [1 + 3v)dv$. How can you explore it using the software?

1.5 Notes

In what follows the numeration corresponds to the sections and subsections in the chapter.

1.1.1 Textbooks by Bickel and Doksum (2007), Ross (2023) and more advanced Casella and Berger (2002) may be recommended for the reader who would like to master basics of probability and statistics. A short review can be found in Efromovich (1999a).

1.1.2 Books by Miller (1981), Cox and Oakes (1984), Klein and Moeschberger (2003), Allison (2014), Klein et al. (2014), Efromovich (2018), and Collett (2023) contain useful information about survival analysis and many interesting examples. Biased data is discussed in the books Efromovich (1999a, 2018). The theory and methodology of efficient nonparametric estimation for biased data can be found in Efromovich (2004a), Efromovich (2004b). There are many excellent books devoted to the missing data, for instance see Rubin (1987), Little and Rubin (2002), Enders (2010), van Buuren (2012), Molenberghs et al. (2014), Sullivan et al. (2017). The theory and methodology of efficient nonparametric estimation for missing data is presented in Efromovich (2011a), Efromovich (2011b), Efromovich (2012a), Efromovich (2012b), Efromovich (2013c), Efromovich (2014b), Efromovich (2014c), Efromovich (2014d), Efromovich (2014e), Efromovich (2015), Efromovich (2016c), Efromovich (2017), Efromovich (2018), Efromovich (2019a), Efromovich (2020a), Efromovich (2021a), Efromovich (2021b), Jiang (2022), and Efromovich and Fuksman (2024a), Efromovich and Fuksman (2024b), Efromovich and Fuksman (2024c).

1.2 A relatively simple, brief and introductory level discussion of nonparametric curve estimation can be found in Wasserman (2006). Mathematically more rigorous statistical theory of series estimation can be found in Efromovich (1999a), Tsybakov (2009) and Johnstone (2024). A number of interesting statistical models, efficient estimators and applications can be found in Efromovich (1980a–2024e), Efromovich and Pinsker (1981), Efromovich and Pinsker (1982), Efromovich and Pinsker (1984), Efromovich and Pinsker (1986), Efromovich and Pinsker (1989), Efromovich and Pinsker (1996), Efromovich and Low (1994), Efromovich and Low (1996a), Efromovich and Low (1996b), Efromovich and Samarov (1996), Efromovich and Samarov (2000), Efromovich and Thomas (1996), Efromovich and Ganzburg (1999), Efromovich and Koltchinskii (2001), Efromovich et al. (2004), Efromovich et al. (2008), Efromovich and Salter-Kubatko (2008), Efromovich and Valdez-Jasso (2010), Efromovich and Smirnova (2014a), Efromovich and Smirnova (2014b), Efromovich and Wu (2017), Efromovich and Wu (2018a), Efromovich and Wu (2018b), Efromovich and Wu (2023a), Efromovich and Wu (2023b), Efromovich and Chu (2018a), Efromovich and Chu (2018b).

1.2.1 Properties of the empirical survival function may be found in Surfling (1980). Useful probability inequalities are presented in Efromovich (2018). One of the best books devoted to analysis of sums of independent random variables is Petrov (1975).

1.2.2 There are several excellent books to read about Fourier series. A relatively brief discussion of the history and main theoretical results can be found in the text Dym and McKean (1972). The books by Bary (1964) and Kolmogorov and Fomin (1957) give a relatively simple discussion (with rigorous proofs) of Fourier series. Butzer and Nessel (1971) and DeVore and Lorentz (1993) are good references on approximation theory. Useful discussion of different orthogonal systems (bases) can be found in Walter (1994) and Efromovich (1999a). Sobolev classes are discussed in the classical book Nikolskii (1975). A nice discussion of Sobolev classes and Sobolev ellipsoids can be found in Tsybakov (2009). Polynomial-cosine bases, that can be used for Sobolev classes of aperiodic functions, are discussed in Efromovich (2019c), Efromovich (2021b). Nice discussion of Bathtub shapes can be found in Jankowski and Wellner (2009).

1.2.3 Density estimation is a classical topic in nonparametric curve estimation, see a nice discussion in the books Silverman (1986) and Scott (2015), as well as an overview in Chap. 3 of Efromovich (1999a). The first sharp minimax oracle's result for Sobolev classes is due to Efromovich and Pinsker (1982). The idea of blockwise shrinkage, that leads to efficient nonparametric estimation for the problem of filtering a signal from gaussian noise, is due to Efromovich and Pinsker (1984). This idea was developed further for density estimation in Efromovich (1985) and for sequential density estimation in Efromovich (1989). A comprehensive discussion of hard thresholding can be found in the monograph Johnstone (2024). A review of many useful methods of adaptation yielding rate optimal density estimation can be found in Efromovich (1999a). The E-estimation algorithm was first proposed in Efromovich (1999a) where its discussion and interesting modifications can be found. One of them will be used in Sect. 3.6. Projections on bona fide classes are discussed in Efromovich (1999a), Glad et al. (2003). A discussion of analytic distributions and the efficient estimation can be found in Efromovich (2021b). Nonparametric confidence bands are discussed in Cai and Low (2004), Giné and Nickl (2011), Hoffmann and Nickl (2011), Efromovich and Chu (2018a, b), and Efromovich (2018). Nonparametric hypotheses testing is discussed in the book Ingster and Suslina (2003). Kernel estimators are also popular, see Efromovich (1999a), Wasserman (2006), Sakhanenko (2015), Sakhanenko (2017). Aggregation of density estimators is discussed in Samarov and Tsybakov (2007). R software is discussed in Helsel (2011).

1.2.4 Hazard rate function and its estimation is discussed in a number of books including Prakasa Rao (1983), Cox and Oakes (1984), Gill (2006), Fleming and Harrington (2011). The theory and methodology of efficient estimation for Sobolev classes is due to Efromovich (2016a), Efromovich (2017), Efromovich (2018).

1.2.5 Nonparametric regression is definitely the most popular topic in nonparametric curve estimation. The interested reader can find a lot of information in books Carroll and Ruppert (1988), Eubank (1988), Härdle (1990), Green and Silverman (1994), Nemirovskii

(1999), Li and Racine (2009). The E-estimator is discussed in Chap. 4 of Efromovich (1999a). The theory and methodology of sharp-minimax estimation is due to Efromovich (1986), Efromovich (1992), Efromovich (1994b), Efromovich (1996a), Efromovich (2001c), Efromovich (2002), Efromovich (2005a), Efromovich (2007d), Efromovich (2007e). The heteroscedastic regression was studied in Efromovich (1992), Efromovich (2013a) and Efromovich and Pinsker (1996). Nonparametric estimation under different types of shape restrictions is discussed in Efromovich (1999a), Efromovich (2001a), Horowitz and Lee (2017), and Groeneboom and Jongbloed (2014).

1.2.6 A book-length discussion of multivariate density estimation can be found in Scott (2015). The first sharp-minimax results for multivariate Sobolev classes, with applications to density and regression, are due to Efromovich (2000b), Efromovich (2002), Efromovich (2010c), Efromovich (2010d). Anisotropic Sobolev classes, with application to rate optimal estimation of regression, were first considered in Hoffmann and Lepski (2002). The term "curse of dimensionality" is due to Bellman (1961), see also discussions in Silverman (1986), Hastie and Tibshirani (1990), and Izenman (2008).

1.3 The corner functions are defined in Sect. 2.1 of Efromovich (1999a).

Current Status Censoring

2

Under a current status censoring (CSC), the underlying time T to an event of interest is not observed directly. Instead, at a monitoring time Z it is known if the event of interest already occurred or not. In other words, under the CSC pair $(Z, \Delta) := (Z, I(T \leq Z))$ is observed. Then the aim is to estimate distribution of the lifetime T or predict it based on available covariates. The CSC divides a sample of observations from (Z, Δ) into two subsamples according to the status Δ. The first one is the subsample when the events of interest already occurred, that is $\Delta = 1$. In what follows we are referring to a CSC observation with $\Delta = 1$ as the CSC with the *occurred* event of interest (CSCO). The second subsample is supplementary and consists of CSC cases with *not occurred* events of interest (CSCNO). For CSCNO we have $\Delta = 0$ or equivalently $\Delta' := I(T > Z) = 1$. It is of interest to understand how CSCO and CSCNO observations should be aggregated for optimal distribution and regression estimation.

As we will see shortly, for CSC there is a dramatic difference between estimation of distribution and regression estimation. Namely, with respect to the case of direct observations of T, the CSC slows down the rate of MISE convergence for distribution estimation but not for regression estimation. This is an important specific of the CSC to know. Another important remark is that the CSCNO subsample is the same as the subsample of censored observations for right-censored data whenever the monitoring time Z has the same distribution as the censoring lifetime C. The latter will help us to develop optimal estimators for right-censored data discussed in the next chapter.

Section 2.1 presents introduction to the CSC. CSC may be not familiar to some readers, and a number of interesting examples sheds light on this censoring model. Distribution estimation is considered in Sect. 2.2. Univariate and multivariate regressions are discussed in Sects. 2.3 and 2.4, respectively. Estimation of conditional survival function and general

© The Author(s), under exclusive license to Springer Nature Switzerland AG 2026
S. Efromovich, *Survival Analysis*, Synthesis Lectures on Mathematics & Statistics,
https://doi.org/10.1007/978-3-031-82814-0_2

conditional linear functionals can be found in Sect. 2.5. In that section several related to the CSC topics are considered as well. Literature review and conclusions can be found in the Notes.

2.1   The Problem of Current Status Censoring

Let T be the lifetime of interest, that is the time to an event of interest. In a classical setting we observe a sample $T_1, \ldots, T_n$ from T, and then use it to estimate the distribution of T or predict T using available covariates. In some applications it is impossible to observe the lifetime T directly due to different mechanisms of censoring. In this chapter we are considering a *current status censoring* (CSC). The CSC occurs when it is not feasible to constantly monitor the time. Instead, there is a possibility to check status of the event of interest at some random moment of time Z, called the *monitoring* time. Accordingly, the available observation is the pair (Z, Δ) where Z is the monitoring time and $\Delta := I(X \leq Z)$ is the *status* of the event of interest. The status (the indicator) is equal to 1 if the event of interest already occurred at moment Z, and $\Delta = 0$ or equivalently $\Delta' := I(X > Z) = 1$ otherwise. We will refer to the corresponding observations as CSCO (CSC with *occurred* event when $\Delta = 1$) and CSCNO (CSC with *not occurred* event when $\Delta' = 1$). This terminology simplifies future references to these two subsamples of observations that we would like to aggregate for optimal estimation. In particular, we may say that CSC sample is the union of CSCO and CSCNO subsamples. Moreover, in some practical examples we are dealing with missing CSC when either CSCO or CSCNO is available. For instance, under CSCNO we observe only cases when the event of interest has not occurred at the monitoring time and skip (miss) cases when the event of interest already occurred at the monitoring time.

Note that under the CSC the underlying lifetime of interest T is never observed, only status of the event of interest at a given monitoring time Z is known. Nonetheless we will be able to estimate the distribution of T and to predict T for given covariates, that is will be able to solve distribution estimation and regression problems. Let us one more time stress that for distribution estimation the available CSC observations are $(Z_1, \Delta_1), \ldots, (Z_n, \Delta_n)$, while for regression the available CSC observations are $(X_1, Z_1, \Delta_1), \ldots, (X_n, Z_n, \Delta_n)$ and the task is to estimate the regression $\mathbb{E}\{T | X = x\}$.

The CSC arises in different applications ranging from biostatistics and engineering to econometrics. Let us present several practical examples of CSC.

Example 1 (Clinical Study) The lifetime of interest T is the time from cancer surgery to cancer reoccurrence. The follow-up examination at time Z after the surgery determines whether or not the cancer is present. We do not observe T and instead observe the monitoring time Z and the indicator of cancer $\Delta = I(T \leq Z)$. Note that $\Delta' = 1$ means that at the time of examination the patient is cancer free. A number of covariates, including age and size of tumor, may be of interest.

Example 2 (Bioassay) In rodent bioassay experiments, when the time from inducing a chemical to developing a disease is the lifetime of interest T, sacrifices are often used to detect the disease. Then the available information is the time Z of the sacrifice, the indicator of disease, and some characteristics of rodents and the chemical.

Example 3 (Engineering Experiment) Destructive tests are often used to find whether a system has failed. Then at the time Z of destruction we know status of the system (it is still working or failed).

Example 4 (Econometrics) A large cluster of CSC applications is in econometrics when the interest is in developing choice models for individual and household behavior (Nobel Prize in Economic Sciences in 2000). The following example sheds light on the binary choice model. The interest is in price T that an individual is ready to pay for an item whose asking price is Z, and the available observations are the asking price Z and the indicator of sale $\Delta' = I(T \geq Z)$. Note that in the econometrics the indicator Δ' is called the observable binary outcome.

Another important issue to mention is the independence of T and Z. This is a traditional assumption that implies consistent estimation of the distribution of T based on a sample from (Z, Δ). In what follows we refer to this setting as *independent* CSC. If T and Z are dependent, then the problem becomes dramatically more complicated and we will refer to it as the *dependent* CSC.

Example 5 (Incidence of an occult non-fatal disease) This example clarifies notions of the CSCO, CSCNO, independent CSC and dependent CSC. Suppose that we are interested in a population of single men ages 23–55 and working in the transportation industry. For this population we would like to estimate the distribution of age at incidence of an occult non-fatal disease for which an accurate diagnostic test is available. Recall that in medicine the term occult means that the disease may exist but there are no readily discernible signs. A doctor may conduct such a test for a cross-sectional sample of size n and then, if required, submit files with information about sick individuals to agency "S" and about healthy individuals to agency "H". In the example T is the (unknown, hidden and of primary interest) age of an individual at the disease incidence, Z is the individual's age at the time of medical testing, the status $\Delta = 1$ if the disease is present and $\Delta = 0$ otherwise. If the statistician has access to the doctor's files then the available data are CSC of size n. If the statistician knows the sample size n and has access only to files of agency "S", then the available data are CSCO. If the statistician knows n and has access only to files of agency "H", then the available data are CSCNO. If Z and T are independent, then the data are *independent* CSC, and otherwise the data are *dependent* CSC.

Let us present the main probability formula used for analysis of independent CSC. Assuming that the monitoring time Z is a continuous variable, the joint mixed density (likelihood) of the pair (Z, Δ) is

$$f^{Z,\Delta}(z,\delta) = f^Z(z)[1 - S^T(z)]^\delta [S^T(z)]^{1-\delta}, \quad \delta \in \{0, 1\}. \tag{2.1.1}$$

Here f^Z is the density of monitoring time Z and S^T is the survival function of the lifetime of interest T. This formula explains all complications of the CSC. First of all, it shows that consistent estimation of the distribution of T over its support is possible if and only if the support of T is a subset of the support of Z, that is $X^T \subset X^Z$. In some applications the latter may not be the case, for instance, in the last example the doctor may not have access to individuals from specific age groups. It will be explained later what can be done in that case. Second, the likelihood depends on the survival function S^T, and this gives us a chance to estimate it. This sounds good, but note that estimation of S^T is based on knowing $f^{Z,\Delta}$ and f^Z. As we know from Chap. 1, optimal rate of estimation of the density f^Z is slower than n^{-1}. The same can be said about estimating $f^{Z,\Delta}$. Accordingly, the survival function S^T is estimable with the rate slower than n^{-1}. In statistics a setting implying slower rates, with respect to direct observations, is called *ill-posed*. We may conclude that the CSC makes the problem of distribution estimation more difficult, ill-posed, and comparable with density estimation for direct observations.

There is a special issue with CSCO and CSCNO subsamples that we need to be aware of. For CSC we can estimate f^Z using the available direct observations of Z. However, for CSCO, with $\Delta = 1$, and CSCNO, with $\Delta = 0$, the available monitoring times are distributed according to the conditional density

$$f^{Z|\Delta}(z|\delta) = \frac{f^Z(z) f^{\Delta|Z}(\delta|z)}{\mathbb{P}(\Delta = \delta)}. \tag{2.1.2}$$

As we know from Sect. 1.1.2, such a sampling (and the corresponding observation) is called biased with the biasing function $f^{\Delta|Z}(\delta|z)$. Because the biasing function is unknown and cannot be estimated based on the available data, the bias precludes us from consistent estimation of f^Z and hence consistent estimation of S^T. A possible remedy is an extra sample from Z, and this remedy is feasible because the extra sampling is unrelated to the lifetime of interest T. Then the main statistical issue is to explore the extra sample's size sufficient for matching the oracle who knows f^Z.

One more remark about CSCO and CSCNO is due. We can think about CSCO as a sample from $(\Delta Z, \Delta)$ and about CSCNO as a complementary sample from $(\Delta' Z, \Delta)$. Equivalently, we can describe the CSCO and CSCNO using the missing mechanism approach of Sect. 1.1.2. Namely, a CSCO subsample may be considered as a sample from (AZ, A) with $A := \Delta$ being the availability and $\mathbb{P}(\Delta = 1|Z = z)$ being the availability likelihood. Further, the missing is MNAR because the probability of missing depends on value of missing Z. The MNAR precludes us from consistent estimation, and this conclusion coincides with the above-presented explanation via the notion of biased data. The latter is not a surprise because the MNAR creates biased data. The CSCNO is described similarly only here the availability $A := \Delta'$. This remark about the underlying sampling models sheds a new light on the CSC.

The final remark about CSC is as follows. In survival analysis lifetimes of interest are typically bounded. For instance, in Example 5 the age of truckers is bounded by 55 years. Accordingly, in what follows without loss of generality we may assume that studied bounded lifetimes are supported on $[0, 1]$. This assumption will allow us to use the E-estimation algorithm defined in Chap. 1.

Now let us complement the above-presented introduction to CSC by a particular simulation. The left-top diagram in Fig. 2.1 presents by the circles a sample from T highlighted by the default R histogram. It also shows the underlying Bathtub density by the solid line (recall that it is the second corner function, see Fig. 1.8). The underlying cumulative distribution function F^T is shown by the dotted line. The density E-estimate is shown by the dashed line and the empirical cumulative distribution function by the dot-dashed line. It is fair to conclude that the sample is reasonable, the estimates are good, and the latter is supported by the indicated integrated squared errors (ISEs). The right-top diagram shows by the circles the CSC sample from $(Z, \Delta) = (Z, I(T \leq Z))$ where Z is Uniform$(0, 1)$. The sample of monitoring times is highlighted by the histogram and its perfect E-estimate (the wide long-dashed line). Note that no longer we can observe the lifetime of interest T. This diagram also shows the underlying cumulative distribution function, the density and their estimates by the same type of lines as in the left-top diagram. The estimates will be explained in the next section, and now let us only mention that the CSC estimates are dramatically worse than in the left-top diagram. The latter is highlighted by the ISEs. Recall that CSC distribution estimation is ill-posed, and we see the effect of that. The left-bottom diagram shows us by circles the CSCO subsample, that is observations with the status $\Delta = 1$. There are $N = \sum_{l=1}^{n} \Delta_l = 60$ CSCO observations. For the subsamples the density f^Z is known and hence consistent estimation is possible. The histogram highlights the available monitoring times that are skewed to the left. The quality of estimation is dramatically worse. The right-bottom diagram shows the CSCNO subsample, that is observations with the status $\Delta = 0$. Here the available monitoring times are skewed to the right, and despite the smaller number $N = \sum_{l=1}^{n} \Delta' = 40$ of observations, the estimates are better than for the CSCO.

As we will see shortly, it is an interesting, theoretically challenging and practically important topic of how to use CSCO or CSCNO subsamples for efficient estimation. Recall that the CSCO subsample can be described as a sample from $(\Delta Z, \Delta)$ while the CSCNO subsample can be described as a complementary sample from $(\Delta' Z, \Delta)$. Hence, with some obvious abuse of the notion, we may say that under the CSC there are three sampling models that are the CSCO, CSCNO and CSC. Further, it is of interest to understand how to use each of the subsamples and how to aggregate them for efficient nonparametric estimation. Accordingly, the following approach is used in the chapter. First the theory of efficient estimation for

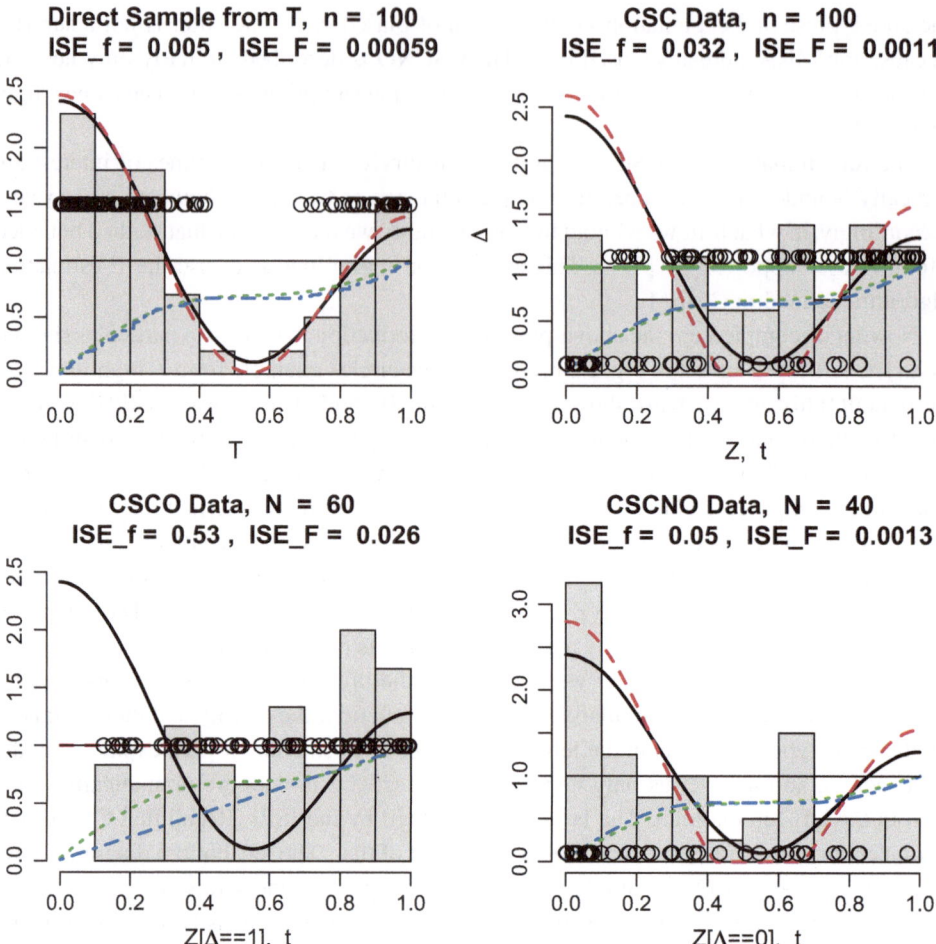

Fig. 2.1 Simulated sample from T and its CSC modification by the Uniform Z. Data are shown by the circles and highlighted by the histograms. In the right diagrams the circles are slightly shifted up for better visualization. The underlying density f^T is shown by the solid line, its estimate by the dashed line, the dotted and dot-dashed lines show F^T and its estimate. The wide long-dashed line in the right-top diagram is the estimate of f^Z, while this density is known for the CSCO and CSCNO subsamples to allow for consistent estimation. N is the size of the CSCO and CSCNO subsamples. The integrated squared errors of the density and cdf estimates are shown in the titles as ISE_f and ISE_F, respectively. [n = 100, corn = 2, cJ0 = 3, CJ1 = 0.8, cTH = 4]

CSCO and CSCNO is proposed, then the more difficult problem of efficient estimation (we may also say efficient aggregation of the two subsamples) for CSC is explored.

Now we are ready to discuss in turn classical problems of nonparametric estimation for CSC data.

2.2   Distribution Estimation

In this section, for a continuous lifetime of interest T, we simultaneously consider estimation of the cumulative distribution function (cdf) F^T (equivalently estimation of the survival function $S^T(t) = 1 - F^T(t)$) and the density $f^T(t) := dF^T(t)/dt = -dS^T(t)/dt$. Then estimation of the hazard rate is considered.

It is a teachable approach to consider estimation of cdf and density together because CSC is an important nonparametric setting for which cdf estimation "returns" to the family of nonparametric problems with slower than n^{-1} rate of the MISE convergence. Further, similarly to the discussed in Chap. 1 classical nonparametric estimation problems, rate of the MISE convergence depends on smoothness of the cumulative distribution function. Further, probability density is estimated with the rate corresponding to derivative of the cumulative distribution function, and vise versa. This is what makes the CSC so interesting with no other analog in the theory of nonparametric curve estimation. It will be also of interest to understand how to develop optimal estimators for CSCO and CSCNO subsamples and then, if possible, to aggregate them for optimal CSC estimation.

The developed theory will be complemented by real and practical examples. In particular, a discussion of how to deal with the case of dependent lifetime of interest T and monitoring time Z will be presented.

2.2.1   Independent Lifetime of Interest and Monitoring Time

We begin with the theory and then explore the E-estimator using simulations and real life example. Estimation of the cdf (survival function) and density for CSC, CSCO and CSCNO are considered simultaneously. This approach involves more complicated notations but makes the presentation more concise and insightful.

2.2.1.1 Asymptotic Theory
We begin with the oracle's approach when the oracle knows data, the nuisance density f^Z of the monitoring time Z, and smoothness of F^T.

Assumption 2.1 The lifetime of interest T and the monitoring time Z are independent continuous random variables, $\mathbb{P}(T \in [0, 1]) = 1$, density $f^Z(z)$ is continuous and positive on $[0, 1]$, and $\int_0^1 f^Z(z)dz = 1$.

The made assumption about independence of T and Z is standard, it yields validity of (2.1.1), and it will be discussed later in Sect. 2.2.1.3. Now we introduce the shrinking local Sobolev class of cumulative distribution functions supported on $[0, 1]$, that is cdfs satisfying $F(0) = 0$ and $F(1) = 1$. Recall global Sobolev function classes $\mathcal{S}_k(\alpha, Q)$ defined in (1.2.12), introduce the anchor cdf F_0 supported on $[0, 1]$, and define the sequence in n of shrinking local Sobolev classes of distributions,

$$\mathcal{S}_*(F_0, \alpha, Q, n) := \Big\{ F : F(t) = F_0(t) + g(t), t \in [0, 1]; \ F(0) = 0, \ F(1) = 1;$$

$$\frac{dF(t)}{dt} \geq 0, \ 0 < \zeta_1^* \leq \frac{dF_0(t)}{dt} \leq \zeta_2^* < \infty, \ t \in [0, 1];$$

$$g \in \mathcal{S}_1(\alpha, Q), \ \max_{t \in [0,1]} \left| \frac{dg(t)}{dt} \right| \leq \frac{\zeta_1^*}{\ln(\ln(n + 20))} \Big\}. \tag{2.2.1}$$

Note that $\ln(\ln(n + 20)) > 1$, and set for $k \in \{0, 1\}$

$$P_k(\alpha, Q) := \left[(Q(2\alpha + 1))^{\frac{2k+1}{2(\alpha-k)}} (2k + 1)^{-\frac{2\alpha+1}{2(\alpha-k)}} \frac{\alpha - k}{\pi(\alpha + 1 + k)} \right]^{\frac{2\alpha}{2\alpha+1}}. \tag{2.2.2}$$

Next we define coefficients of difficulty

$$d_O := \int_0^1 \frac{1 - S^T(t)}{f^Z(t)} dt, \quad d_{NO} := \int_0^1 \frac{S^T(t)}{f^Z(t)} dt,$$

$$d := \int_0^1 \frac{S^T(t)(1 - S^T(t))}{f^Z(t)} dt, \tag{2.2.3}$$

for the CSCO, CSCNO and CSC, respectively (note the added subscripts O and NO for the CSCO and CSCNO). As we see, we have the inequality $d < \min(d_O, d_{NO})$ as it should be for the CSC and its two subsamples.

In what follows notations $F^T(t) = 1 - S^T(t) =: (F^T(t))^{(0)}$ and $f^T(t) := dF^T(t)/dt =: (F^T(t))^{(1)}$ will help us to simultaneously consider estimation of the cdf and the density.

Theorem 2.1 *Suppose that Assumption 2.1 holds, a CSC sample of size n is available, and the anchor F_0 is such that $dF_0(t)/dt \in S_0(\alpha', Q'), \alpha' > \alpha, Q' < \infty$. Problems of estimating the cumulative distribution function $(F^T)^{(k)}, k = 0$ and the density $(F^T)^{(k)}, k = 1$ are considered for CSCO, CSCNO or CSC data. The oracle knows the parameter k, density f^Z, the shrinking local Sobolev class (2.2.1) with $\alpha \geq 1 + k$, and the underlying*

*data corresponding to CSCO, CSCNO, or CSC. Then the following lower bound for the
MISE holds for all oracle-estimators $\hat{F}^*_{(k)}$ of the estimand $(F^T)^{(k)}$,*

$$\inf_{\hat{F}^*_{(k)}} \sup_{F^T \in S_*(F_0, \alpha, Q, n)} \mathbb{E}\left\{ (n/d_*)^{2(\alpha-k)/(2\alpha+1)} \int_0^1 (\hat{F}^*_{(k)}(t) - (F^T(t))^{(k)})^2 dt \right\}$$

$$\geq P_k(\alpha, Q)(1 + o_n(1)). \tag{2.2.4}$$

*Here $k \in \{0, 1\}$, $P_k(\alpha, Q)$ is defined in (2.2.2), and d_* is defined in (2.2.3) for the underlying
type of data, that is CSCO, CSCNO, or CSC. Further, the lower bound is sharp and attainable
by a series oracle-estimator and by a blockwise-shrinkage series estimator for the local and
global Sobolev classes.*

Several comments are due. First, the theorem simultaneously consideres two estimands
and three types of available data (sampling models). Second, the result matches those in
Chap. 1 for the case of direct observations in terms of presenting the sharp lower bound and
its attainability by a series estimator. The latter will allow us to use the E-estimation algorithm
discussed in the next subsection. Third, while for direct observations the cdf can be estimated
with the parametric rate n^{-1} regardless of its smoothness, for CSC the rate slows down to the
classical nonparametric rate $n^{-2\alpha/(2\alpha+1)}$ defined by smoothness of the cdf. The latter is the
fascinating result because the CSC "returns" estimation of a nonparametric univariate cdf in
the realm of classical nonparametric univariate problems like regression. Accordingly, we
can say that the CSC distribution estimation problem is ill-posed, or equivalently that the CSC
data are ill-posed for distribution estimation with respect to direct data. Fourth, let us compare
the lower bound (2.2.4) for the cdf ($k = 0$) with a known sharp lower bound of Theorem
1.3 for the classical nonparametric regression $Y = m(X) + v(X)\xi$ discussed in Sect. 1.2.5.
For $m \in S_0(\alpha, Q)$ that lower bound is identical to (2.2.4) only the regression coefficient of
difficulty is $\int_0^1 [v^2(x)/f^X(x)]dx$. This allows us to relate the two models and understand
complexity of a CSC model via the simpler for visual interpretation regression model. For
instance, for the CSCO using $v^2(x) = F^T(x)$ and $f^X(x) = f^Z(x)$ yields equivalent models
for estimating $m(x)$, $x \in [0, 1]$ and $F^T(t)$, $t \in [0, 1]$. Fifth, let us comment on the rate of
density estimation. Because $f^T(t) = dF^T(t)/dt$, the density is $(\alpha - 1)$-fold differentiable.
Set $\beta := \alpha - 1$ for the number of derivatives of the density, and then for CSC we get the rate
$n^{-2\beta/(2\beta+3)}$ in place of $n^{-2\beta/(2\beta+1)}$ for direct observations. To appreciate the slower CSC
rate note that, according to Sect. 1.2.6, for direct observations the rate $n^{-2\beta/(2\beta+3)}$ is optimal
for estimation of a trivariate density with β derivatives in each variable, and then recall the
famous curse of dimensionality. This remark sheds additional light on the discovered "curse
of CSC" and makes it equivalent to the classical curse of dimensionality. Further, we may
say that the CSC makes the problem of density estimation ill-posed with respect to direct
data. Finally, formulas (2.2.3) for coefficients of difficulty point upon optimal monitoring

times Z that minimize them. It is directly verified by the Cauchy-Schwarz inequality that
the corresponding optimal monitoring densities are

$$f_{O*}^Z(t) := \frac{[1 - S^T(t)]^{1/2}}{\int_0^1 [1 - S^T(x)]^{1/2} dx} I(t \in [0, 1]), \tag{2.2.5}$$

$$f_{NO*}^Z(t) := \frac{[S^T(t)]^{1/2}}{\int_0^1 [S^T(x)]^{1/2} dx} I(t \in [0, 1]), \tag{2.2.6}$$

and

$$f_*^Z(t) := \frac{[S^T(t)(1 - S^T(t))]^{1/2}}{\int_0^1 [S^T(x)(1 - S^T(x))]^{1/2} dx} I(t \in [0, 1]). \tag{2.2.7}$$

2.2.1.2 E-Estimator for CDF and Density

Following the methodology of E-estimation presented in Chap. 1, it is sufficient to suggest
sample mean Fourier coefficient estimators for the cdf and density. We begin with the case
when f^Z is known. Note that f^Z is always known in controlled CSC experiments. Then
this assumption will be relaxed.

We begin with estimation of cdf F^T and consider the three sampling CSC models in turn.
The aim is to estimate Fourier coefficients $\theta_j = \int_0^1 F^T(t)\varphi_j(t)dt$ of the cdf F^T. For CSCO
we observe a sample of size n from $(\Delta Z, \Delta)$. According to (2.1.1) we have $f^{\Delta Z, \Delta}(z, 1) =
f^Z(z)F^T(z)$, and this allows us to write that

$$\theta_j = \mathbb{E}\left\{ \frac{\Delta \varphi_j(\Delta Z)}{f^Z(\Delta Z)} \right\}. \tag{2.2.8}$$

This formula yields the CSCO sample mean Fourier coefficient estimator

$$\hat{\theta}_{Oj}^* := n^{-1} \sum_{l=1}^n \frac{\Delta_l \varphi_j(\Delta_l Z_l)}{f^Z(\Delta_l Z_l)}. \tag{2.2.9}$$

As we will see shortly, this Fourier coefficient estimator is efficient. For CSCNO we observe
a sample of size n from $((1 - \Delta)Z, \Delta)$, and using a similar consideration we get the CSCNO
sample mean Fourier coefficient estimator

$$\hat{\theta}_{NOj}^* := I(j = 0) - n^{-1} \sum_{l=1}^n \frac{(1 - \Delta_l)\varphi_j((1 - \Delta_l)Z_l)}{f^Z((1 - \Delta_l)Z_l)}. \tag{2.2.10}$$

Again, as we will see shortly, this estimator is efficient for CSCNO.

Surprisingly, CSC data, which should be easier for suggesting an efficient Fourier coefficient estimator, is in reality a more challenging one. We are presenting two approaches
in turn. The first one is a straightforward aggregation of Fourier coefficient estimators for

CSCO and CSCNO subsamples. It is a relatively simple approach that sheds light on a classical linear aggregation. The second approach is a more sophisticated one.

The following lemma explains the first approach.

Lemma 2.1 *Let $\hat{\kappa}_O$ and $\hat{\kappa}_{NO}$ be two independent unbiased estimators of parameter κ with variances v_O^2 and v_{NO}^2, respectively. Then among all unbiased linear aggregated estimators $\lambda\hat{\kappa}_O + (1-\lambda)\hat{\kappa}_{NO}$, $\lambda \in [0, 1]$, the estimator with $\lambda^* := v_{NO}^2/(v_O^2 + v_{NO}^2)$ has the minimal variance $v_O^2 v_{NO}^2/(v_O^2 + v_{NO}^2)$.*

This lemma allows us to propose the corresponding linear aggregation of $\hat{\theta}_{Oj}^*$ and $\hat{\theta}_{NOj}^*$. The idea is appealing and may be recommended for small samples, however it does not yield efficient estimation. To utilize Lemma 2.1 for efficient estimation, a special sequence of bases is required, see the Notes.

The second approach for aggregation of CSCO and CSCNO subsamples is presented below in Theorem 2.2. Recall that for $j > k$ we have $\{0, 1, \dots, k\} \setminus \{j\} = \{0, 1, \dots, k\}$.

Theorem 2.2 *Let Assumption 2.1 hold and a CSC sample of size n be given. Introduce the pilot cdf estimator*

$$\check{F}^*_{-j}(t) := \sum_{i\in\{\{0,1,\dots,\lceil\ln(\ln(n+20))\rceil\}\setminus\{j\}\}} \hat{\theta}^*_{Oi}\varphi_i(t). \tag{2.2.11}$$

Then the Fourier coefficient estimator

$$\hat{\theta}^*_j := n^{-1}\sum_{l=1}^{n} \frac{(\Delta_l - \check{F}^*_{-j}(Z_l))\varphi_j(Z_l)}{f^Z(Z_l)} \tag{2.2.12}$$

of $\theta_j := \int_0^1 F^T(t)\varphi_j(t)dt$ is efficient, that is $\mathbb{E}\{(\hat{\theta}^*_j - \theta_j)^2\} = n^{-1}d(1 + o_n(1) + o_j(1))$ where d is defined in (2.2.3).

We observe an interesting procedure of aggregation. First, CSCO observations are used to evaluate the pilot cdf estimator (CSCNO may be used instead). Then both CSCO and CSCNO subsamples are aggregated by the estimator (2.2.12). The underlying idea of the pilot estimator $\check{F}^*_{-j}(t)$ is that it estimates the function

$$F_{-j}(t) := \sum_{i\in\{\{0,1,\dots,\lceil\ln(\ln(n+20))\rceil\}\setminus\{j\}\}} \theta_i\varphi_i(t). \tag{2.2.13}$$

This function approximates $F^T(t)$ as both n and j increase to infinity, and $\int_0^1 F_{-j}(x)\varphi_j(x)dx = \int_0^1 \check{F}^*_{-j}(t)\varphi_j(x)dx = 0$. Recall that the same idea of subtracting a pilot estimator was used in Remark 1.3 for efficient estimation of regression.

The Fourier coefficient estimator (2.2.12) is used for construction of the smooth cdf E-estimator for the CSC.

Now we are considering density estimation. Two feasible approaches can be used. The first one is to take derivative of the smooth cdf E-estimator. The second one, presented below, is to estimate the density via its Fourier coefficients. Accordingly, the aim is to propose efficient estimators of Fourier coefficients $\theta_j := \int_0^1 f^T(t)\varphi_j(t)dt$ for CSCO, CSCNO and CSC data. Complexity of the problem is that the likelihood (2.1.1) depends on the cdf while we are interested in estimating the density. Accordingly, we need to express Fourier coefficients of the density via functionals of the cdf.

Note that $\theta_0 = 1$ because the density is supported on $[0, 1]$. Using integration by parts, $F^T(0) = 0$ and $F^T(1) = 1$ we may write for $j \geq 1$,

$$\theta_j = \int_0^1 \varphi_j(t)dF^T(t)dt = \varphi_j(1)F^T(1) + (2^{1/2}\pi j)\int_0^1 \sin(\pi jt)F^T(t)dt. \quad (2.2.14)$$

This formula allows us to write for CSCO observations,

$$\theta_j = \varphi_j(1) + (2^{1/2}\pi j)\mathbb{E}\Big\{\Delta\frac{\sin(\pi j\Delta Z)}{f^Z(\Delta Z)}\Big\} = \varphi_j(1) + (\pi j)\mathbb{E}\{\Delta\frac{\psi_j(\Delta Z)}{f^Z(\Delta Z)}\}. \quad (2.2.15)$$

Here $\psi_j(x) := 2^{1/2}\sin(\pi jx)$, $j = 1, 2, \ldots$ are elements of the sine basis on $[0, 1]$.

In its turn, (2.2.15) yields the sample mean Fourier coefficient estimator

$$\tilde{\theta}_{Oj} := \varphi_j(1) + n^{-1}(\pi j)\sum_{l=1}^n \Delta_l\frac{\psi_j(\Delta_l Z_l)}{f^Z(\Delta_l Z_l)}. \quad (2.2.16)$$

Absolutely similarly, for CSCNO we get the following sample mean Fourier coefficient estimator ($\Delta' := 1 - \Delta$),

$$\tilde{\theta}_{NOj} := \varphi_j(0) - n^{-1}(\pi j)\sum_{l=1}^n \Delta'_l\frac{\psi_j(\Delta'_l Z_l)}{f^Z(\Delta'_l Z_l)}. \quad (2.2.17)$$

As we already know from the cdf estimation, efficient estimation for CSC is a bit more complicated. First, similarly to (2.2.11) we calculate the pilot series cdf estimator which uses the sine basis on $[0, 1]$,

$$\bar{F}^*_{-j}(t) := \sum_{i\in\{\{1,\ldots,\lceil\ln(\ln(n+20))\rceil\}\backslash\{j\}\}} \tilde{\eta}_{Oi}\psi_i(t), \quad (2.2.18)$$

where $\tilde{\eta}_{Oi} = (\pi i)^{-1}[-\varphi_i(1) + \tilde{\theta}_{Oi}]$. Then the proposed efficient CSC Fourier coefficient estimator is

$$\tilde{\theta}_j := \varphi_j(1) + n^{-1}(\pi j)\sum_{l=1}^n \frac{(\Delta_l - \bar{F}^*_{-j}(Z_l))\psi_j(Z_l)}{f^Z(Z_l)}. \quad (2.2.19)$$

Theorem 2.3 *Let Assumption 2.1 hold, f^Z is known, and a CSC sample of size n is given. Then for density estimation the Fourier coefficient estimators (2.2.16), (2.2.17) and (2.2.19) are efficient for the CSCO, CSCNO and CSC, respectively.*

Now let us consider the case of an unknown density f^Z of the monitoring time Z. The solution depends on the sampling model. For CSC we have n direct observations of Z that can be used for estimating f^Z. Set $\tilde{f}^Z(z) := \max(c_L/\ln(n + 20), \hat{f}^Z(z))$ with $\hat{f}^Z$ being the density E-estimator. Then $\tilde{f}^Z$ can be used in place of unknown f^Z. The situation is different for CSCO and CSCNO. Indeed, consider CSCO when only monitoring times with $\Delta = 1$ are observed and $f^{Z,\Delta}(z, 1) = f^Z(z)F^T(z)$. This yields that CSCO observations are biased and consistent estimation of F^T is possible either if f^Z is known or can be estimated using an extra sample from Z. For the latter to be possible, a cross-sectional study of the monitoring time Z must be conducted and an extra sample $Z_{E1}, \ldots, Z_{En_E}$ from Z be collected. The theory asserts that $n_E = O_n(1)n$ is sufficient for efficient estimation of cdf S^T and density f^T, see the Notes.

The proposed efficient Fourier coefficient estimators allow us to use the E-estimation algorithm of Sect. 1.2.3. Examples of density and cdf estimation are presented in the next subsection and Sect. 2.2.2, respectively.

2.2.1.3 Real and Simulated Examples of Density Estimation

We begin with example of aerobic treatment of municipal wastewater. The lifetime of interest T is the time when chemical pollutant appears at a sludge tank. Because it is impossible to observe the time directly, CSCNO and CSC experiments with uniform distribution of the monitoring time were conducted, see the Notes. The top diagram in Fig. 2.2 shows CSCNO data by the circles and the density E-estimate by the solid line. There are 27 observed monitoring times from $(\Delta'Z, \Delta)$. Let us look at the data and try to guess an underlying density. Recall that guessing an underlying density is often possible for direct observations. Unfortunately, visualization of an underlying density f^T is practically impossible for CSCNO data because $f^{Z|\Delta}(z|0) = f^Z(z)S^T(z)/\mathbb{P}(\Delta = 0)$. Indeed, even if f^Z is known, we first need to visualize the underlying S^T and then consider its derivative. That two steps are too complicated. The bottom diagram exhibits a larger CSC experiment for the same municipal plant, and we can expect the same underlying density f^T. The experiment is controlled with the uniform monitoring time Z. Visualization of the observations of Z indicates a uniformly distributed sample. The density E-estimate of f^Z (the dashed line) supports that conclusion. Now let us try to analyze the CSC data. Because $\Delta = I(T \leq Z)$ and all Δs with $Z > 0.65$ are equal to 1, the CSC data, together with the uniform monitoring time, tell us that an underlying density has a vanishing right tail. Next we note that the indicators for $Z < 0.3$ are zero and hence all corresponding underlying lifetimes of interest satisfy $T > Z$. This remark points upon a vanishing left tail. Unfortunately, there is nothing else that visualization of the CSC data can tell us. In particular, it is difficult to answer questions about

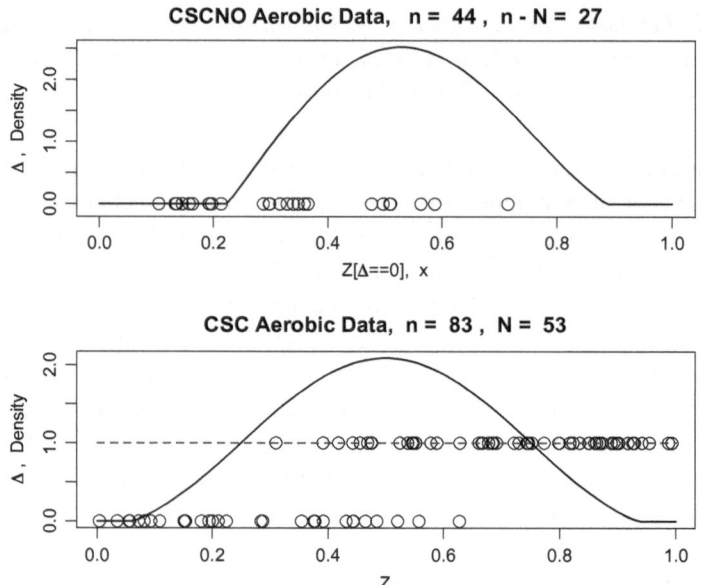

Fig. 2.2 Analysis of environmental CSCNO and CSC controlled experiments with the Uniform monitoring time. Monitoring times are rescaled onto [0, 1] and shown by the circles, $N := \sum_{l=1}^{n} \Delta_l$. The solid and dashed lines are the E-estimates of f^T and f^Z, respectively. {cL is the constant used in the lower bound $c_L / \ln(n + 20)$ for the estimator $\tilde{f}^Z$.} [cJ0 = 3, CJ1 = 0.8, cTH = 4, cL = 1]

symmetry or multimodality. This complexity is explained by formula (2.1.1) implying that to visualize an underlying density, one first need to visualize an underlying cdf and then visualize its derivative. The derivative evaluation step is too complicated for visual analysis. We may conclude that only statistical estimators may help us to gain understanding of CSC data. The E-estimate of f^T is shown by the solid line. We may conclude that the underlying distribution of T is unimodal and has light tails. The CSC density estimate also supports opinion of the unimodal CSCNO density E-estimate based on the smaller sample shown in the top diagram.

The presented analysis of the real CSC data is encouraging but the drawback is that we do not know the underlying distribution of the lifetime of interest T. The next figure is based on simulated CSC when the underlying distribution is known.

Figure 2.3 allows us to gain experience in dealing with the CSC and proposed E-estimators via numerical simulations. The underlying density is a known corner density and the monitoring time is the Uniform. Two simulations are conducted, and the software allows us to choose different underlying distributions and sample sizes, and the curves are explained in the caption. The particular simulations, shown in the figure, stress complexity of the CSC sampling and why the CSC distribution estimation problem is called ill-posed. The underlying corner densities are the Normal and the Bimodal. Overall, it is more difficult to estimate

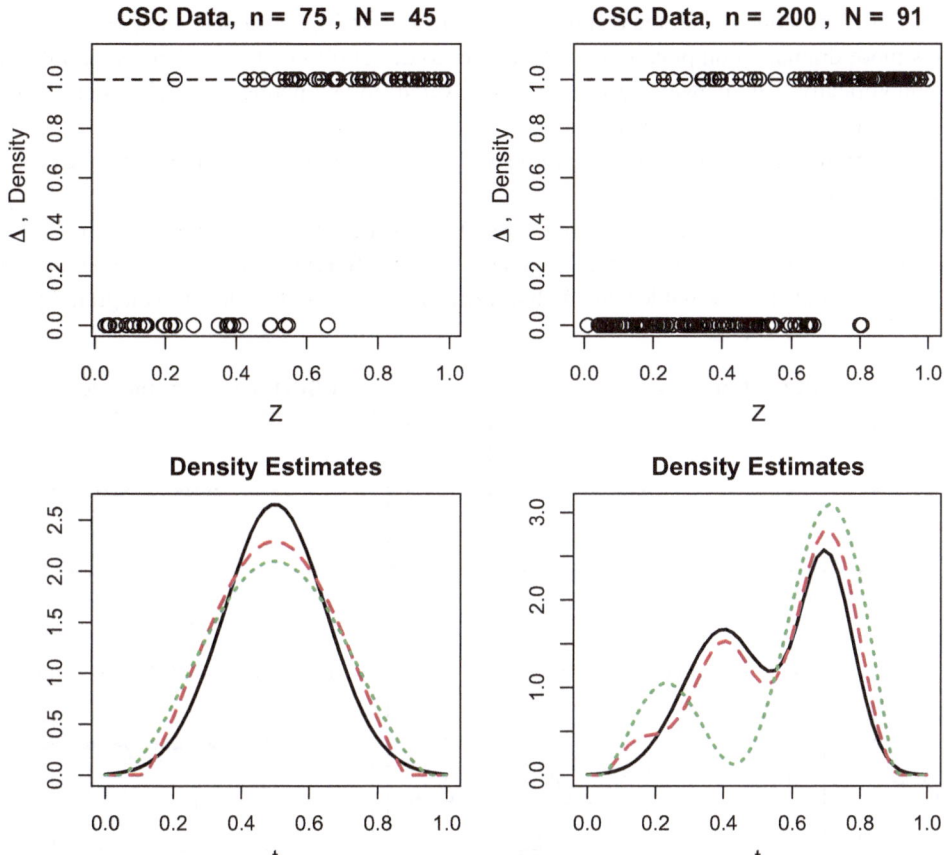

Fig. 2.3 Two simulated CSC examples. A top diagram shows CSC data by the circles, n is the sample size, $N := \sum_{l=1}^{n} \Delta_l$, the dashed line is the E-estimate of f^Z. A bottom diagram shows by the solid, dashed and dotted lines an underlying density of interest f^T, the density E-estimate based on an underlying sample from T, and the CSC density E-estimate, respectively. {This figure allows to choose two different sample sizes and two different underlying densities f^T using arguments n and *corn*, respectively.} [n = c(75, 200), corn = c(3, 4), cJ0 = 3, cJ1 = 0.8, cTH = 4, cL = 1]

the high-frequency Bimodal density than the low-frequency unimodal Normal density. Now let us repeat these simulations 5000 times, for each simulation and each estimate calculate its integrated squared error (ISE), calculate ratio of ISE of the CSC density E-estimate to ISE of the E-estimate based on direct observations of T (the oracle-estimate), and then calculate the sample mean of the 5000 ratios. The obtained results for the Normal density and sample sizes 100, 200, 300, 400 and 500 are 7.4, 12.3, 26.5, 43.1, 56.9 respectively. For the Bimodal density the corresponding sample mean ratios are 4.3, 7.3, 11.8, 14.4, 18.4. Let us examine the presented results. As we already know from the theory, small samples may present only onset of ill-posedness, and the results support this possibility. For larger samples CSC creates more dramatic complications with respect to direct observations. Namely, recall that asymptotically CSC density estimation is equivalent to estimating a trivariate density for direct data, and the results reflect this.

The interested reader can find more results of numerical studies in the references mentioned in the Notes.

Figure 2.4 is devoted to analysis of CSCO when the underlying distribution of f^Z is unknown. Its two columns allow us to visualize two different simulations. We begin with the left column. The available CSCO observations are shown in the left-top diagram by the

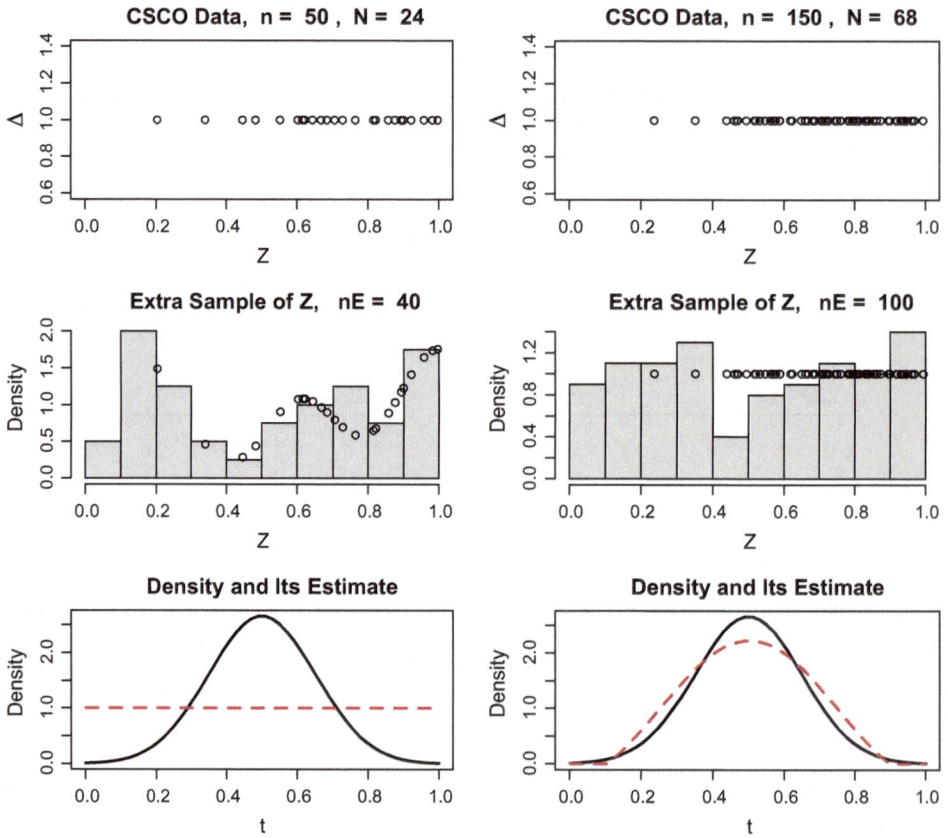

Fig. 2.4 Analysis of two simulated CSCO samples. Monitoring time Z is the Uniform. In a bottom diagram the underlying density f^T and its E-estimate are shown by the solid and dashed lines, respectively. {This figure allows to choose two different sample sizes, two different underlying densities f^T, and two sample sizes n_E for extra samples from Z using the arguments n, $corn$ and nE, respectively.} [n = c(50, 150), corn = c(3, 3), nE = c(40, 100), cJ0 = 3, CJ1 = 0.8, cTH = 4, cL = 1]

circles. The underlying sample size $n = 50$ is considered small even for direct observations, and for this particular simulation the E-estimator can use only $N := \sum_{l=1}^{n} \Delta_l = 24$ available observations of the monitoring time. By thinking about CSCO as the missing data, we are dealing with the very high rate of missingness. What we see in the diagram is that the observed monitoring times range from 0.2 to 1. As we already know from the theory, it is impossible to estimate an underlying distribution of the lifetime of interest T using solely CSCO data because the missing is MNAR and it creates biased observations. Indeed, here the underlying distribution is Normal with the mode at 0.5, and the CSCO subsample is clearly skewed to the left. Suppose that we can conduct an additional (extra) cross-sectional study of the monitoring time Z via its direct sampling. It is important to stress that this study does not involve observing T or even taking it into account. The extra sample of size n_E from Z is shown in the left-middle diagram via the default R histogram. The density E-estimate $\hat{f}^Z$ is shown by the circles at the monitoring times exhibited in the left-top diagram. Recall that only values of $f^Z(\Delta_l Z_l)$ with $\Delta_l = 1$ are used by the E-estimator of the underlying density of interest f^T. The E-estimate of f^Z is poor because we know that the underlying density is the Uniform, and at the same time it fairly well fits the extra sample. The main issue here is the small sample size $n_E = 40$ of the extra sample. The left-bottom diagram allows us to compare the underlying density of interest f^T (the solid line) with its estimate (the dashed line). The estimate is poor, and this particular outcome stresses complexity of the problem. A similar simulation, only with the increased $n = 150$ and $n_E = 100$, is shown in the right column of diagrams. The right-top diagram exhibits another simulated CSCO data with almost the same range of available monitoring times. In the right-middle diagram the extra sample from Z is again challenging, but here the density E-estimator does a good job in smoothing the histogram. In the right-bottom diagram, the density estimate of f^T (the dashed line) is reasonably good. Note how well it indicates the symmetric and unimodal shape of the underlying Normal density. Of course, it is worthwhile to stress that for a sample with $n = 150$ direct observations a typical density estimate is much better, but keeping in mind the ill-posed nature of the CSCO, the outcome is encouraging.

Let us complement the simulations in Fig. 2.4 by results from an intensive numerical study, based on 5000 repeated simulations of the figure, that compares the proposed density estimator with the oracle-estimator who knows f^Z. For brevity, let us present the range of sample mean ratios between the integrated squared errors of the estimator and the oracle-estimator when n_E increases from 25 to 400. For $n = 50$ the range is from 1.13 to 1.02, for $n = 100$ from 1.33 to 1.03, and for $n = 300$ from 2.14 to 1.29. As it could be expected, increase in n_E improves the estimator. An interesting outcome is that the estimator becomes more sensitive to estimation of f^Z as n increases. Overall, for relatively small n using n_E of order n is sufficient. Another important observation is that for all experiments the median ratio is close to 1. This tells us that there may be very poor estimates of f^Z that affect the mean ratio, like the one in the left–middle diagram of Fig. 2.4, but the median is stable over all studied n and n_E. More information about the study can be found in references mentioned in the Notes.

Overall, we can conclude that using an extra cross-sectional study of the monitoring time is a feasible remedy to overcome the destructive nature of the MNAR for CSCO and CSCNO. It is also important to stress that the recommended remedy of an extra sample from Z is unrelated to the lifetime of interest T.

2.2.2 Dependent Lifetime of Interest and Monitoring Time

This section begins with a motivating real-life example, then the theory and more examples are presented.

2.2.2.1 Motivating Example

We are interested in estimation of the distribution of age T at which a homosexual man is infected with gonorrhea. The presence or absence of the infectious disease is defined by an exam at age Z of the man. Accordingly, we know Z and the status $\Delta := I(T \leq Z)$ of the disease at age Z. The CSC data were collected in 1984–1985. The top and bottom diagrams in Fig. 2.5 show the CSC observations for men with and without college degree, for more about the data see the Notes. The solid lines show us the proposed in the previous section cdf E-estimates which are clearly not bona fide (not monotonically increasing). Is this due to poor performance of the estimator or the estimator sends us a message? To answer the question note that using (2.1.1) we get

$$F^T(z) = \frac{f^{Z,\Delta}(z, 1)}{f^Z(z)} = \mathbb{P}(\Delta = 1|Z = z) = \mathbb{E}\{\Delta|Z = z\}. \tag{2.2.20}$$

Accordingly, if (2.1.1) is valid, then the underlying cdf F^T may be thought as the regression of Δ on Z. Now let us return to the two diagrams in Fig. 2.5 and assess the data as the scattergrams and the curves as the regression functions. We definitely may conclude that the visualization supports the not monotonic shapes of the curves.

But why is that the case? Is it possible that some of the made assumptions about the CSC and implying (2.1.1) are violated? A reasonable conjecture is that T, the time of onset of the disease, and Z, the age at the time of exam, may be dependent.

If the conjecture is correct and T and Z are dependent, then in place of (2.1.1) we have the following formula for the mixed joint density (the likelihood)

$$f^{Z,\Delta}(z, \delta) = f^Z(z)[F^{T|Z}(z|z)]^\delta[1 - F^{T|Z}(z|z)]^{1-\delta}. \tag{2.2.21}$$

This formula implies that for dependent CSC (or we may say in a general setting) the cdf estimator, proposed in the previous section for independent CSC, estimates the conditional cdf $F^{T|Z}(z|z) := \mathbb{P}(T \leq z|Z = z)$. That conditional cdf is not necessarily monotone. This understanding, that in general we are estimating $F^{T|Z}$ and not F^T, sheds light on the estimates shown in Fig. 2.5. Further, let us add some useful information about the data.

The range in age among the men is about 30 years and we are dealing with men who were of college age from the 1950s through the 1980s. For men with college degree, we observe a sharp increase in likelihood to be infected for those who were of college age in the 1970s and 1980s. The outcome for those without college degree is more involved. We observe a significant deep in likelihood to be infected for the men who were of college age in the 1960s. Is this conclusion, based on analysis of the estimates, reasonable? We may compare the above-outlined trends with rates of reported cases of gonorrhea in the United States (per 100,000 population): 192 (the 1950s), 145 (the 1960s), 294 (the 1970s), 442 (the 1980s). At the same periods of time, the percentage of men with college degree was: 9% (the 1950s), 11% (the 1960s), 15% (the 1970s), 21% (the 1980s).

We may conclude that the made-above conjecture about dependence between T and Z has merits, and it sheds light on results presented in Fig. 2.5. Further, analysis of the univariate function $F^{T|Z}(z|z) = \mathbb{P}(T \leq z|Z = z)$ may be of a practical interest, and this is the characteristic that can be estimated from dependent CSC data.

Figure 2.6 helps us to get experience in "reading" the cdf E-estimates. Simulations are independent CSC, and accordingly the cdf E-estimator is consistent. Note that in the top diagram the estimate (the dashed line) is not monotone, this is a possibility, and here this is

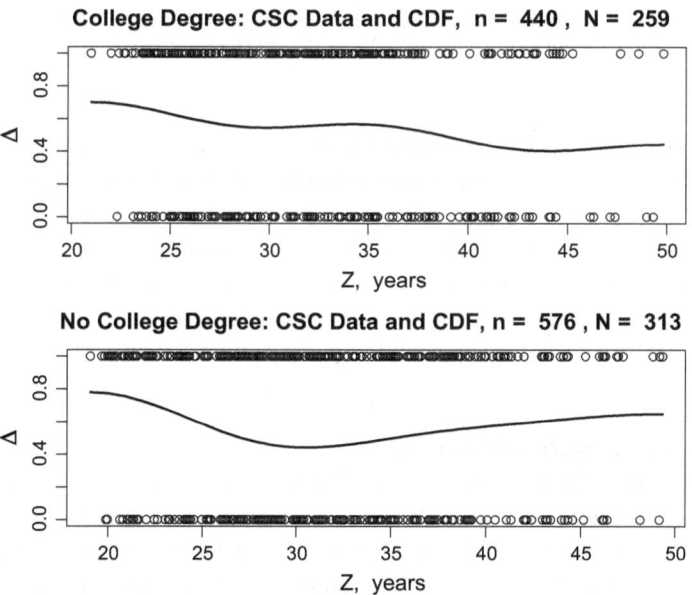

Fig. 2.5 CSC data and analysis of the age T when a homosexual man is infected with gonorrhea. The circles show CSC data with monitoring time Z being the age at the time of medical exam, $\Delta := I(T \leq Z)$, $N := \sum_{l=1}^{n} \Delta_l$. The solid line is the estimated cdf. [cJ0 $= 3$, CJ1 $= 0.8$, cTH $= 4$, cL $= 1$]

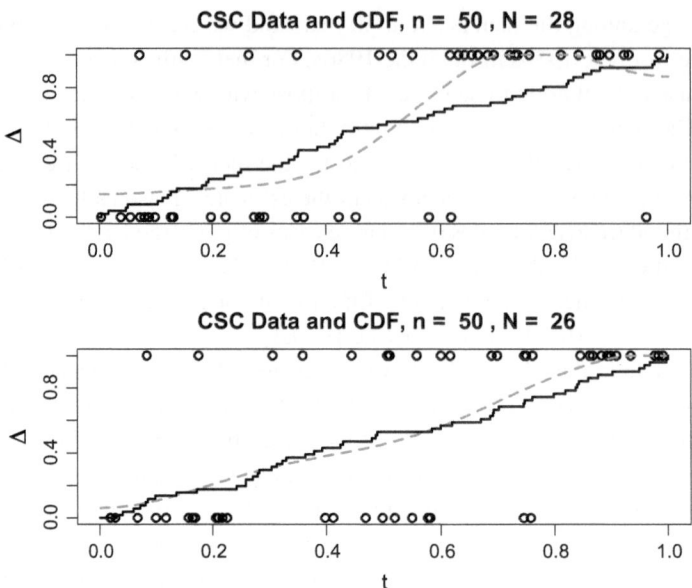

Fig. 2.6 Two simulated examples of independent CSC. The solid stepwise line is the empirical cdf based on the underlying sample from T, the dashed line is the cdf E-estimate based on the CSC data, $N := \sum_{l=1}^{n} \Delta_l$. [n = 50, corn = 1, cJ0 = 3, CJ1 = 0.8, cTH = 4, cL = 1]

due to the nature of the series approximation. Of course, we can implement the monotonic projection (the step 3 of the E-estimation algorithm) without visualization of the "underlying" series estimate. However, avoiding visualization of the series estimate precludes us from a possible analysis of the CSC for dependence. Accordingly, it is prudent to begin with visualization of the estimate. If the estimate is not bona fide, then we may check for a possible violation of assumptions. If the assumptions look reasonable, the projection may be used. The latter is a prudent approach for analysis of CSC data.

2.2.2.2 Theory for Dependent CSC

Independence of the lifetime of interest T and the monitoring time Z is a traditional assumption in the CSC literature. Sure enough, it is necessary for consistent estimation of distribution of the lifetime of interest. If the assumption about independence does not hold, then the setting is referred to as dependent CSC. Formula (2.2.21) implies that for dependent CSC the proposed cdf estimators estimate the univariate function $F^{T|Z}(z|z)$ in place of $F^T(z)$. Without additional information, $F^{T|Z}(z|z) := \mathbb{P}(T \leq Z|Z = z)$ is the only characteristic of the lifetime of interest T that can be estimated.

   A possible remedy is to find an auxiliary variable X such that T and Z are conditionally independent given X. Then the observed CSC sample is from the triplet (X, Z, Δ) where

$\Delta := I(T \leq Z)$. Suppose that T and Z are conditionally independent given X, and as before let us assume that each variable is supported on $[0, 1]$. Then the proposed solution is as follows. First of all, we note that the joint (mixed) density of the triplet (X, Z, Δ) is

$$f^{X,Z,\Delta}(x, z, \delta) = f^X(x) f^{Z|X}(z|x) [F^{T|X}(z|x)]^{\delta}$$

$$\times [1 - F^{T|X}(z|x)]^{1-\delta}, \quad (x, z) \in [0, 1]^2, \quad \delta \in \{0, 1\}. \tag{2.2.22}$$

Here $F^{T|X}(t|x) := \mathbb{P}(T \leq t | X = x)$, f^X is the density of X, and $f^{Z|X}$ is the conditional density of Z given X. Formula (2.2.22) allows us to appreciate complexity of the dependent CSC. The key difference between formula (2.2.22) for dependent CSC and formula (2.1.1) for independent CSC is that for dependent CSC we no longer have a direct access to the cdf of interest $F^T(t)$, instead the conditional cdf $F^{T|X}(t|x)$ is accessible.

Despite that obvious complication, let us explain how presented in the previous Sect. 2.2.1.2 methodology of E-estimation may be used for estimating $F^T(t)$. Using (1.1.4) we get the following formula,

$$F^T(t) = \mathbb{E}\{F^{T|X}(t|X)\}. \tag{2.2.23}$$

This formula and (2.2.22) allow us write down Fourier coefficients of F^T as the expectation,

$$\theta_j := \int_0^1 F^T(z) \varphi_j(z) dz$$

$$= \int_0^1 \mathbb{E}\{F^{T|X}(z|X)\} \varphi_j(z) dz = \int_0^1 \mathbb{E}\left\{ \frac{f^{X,Z,\Delta}(X, z, 1)}{f^X(X) f^{Z|X}(z|X)} \right\} \varphi_j(z) dz$$

$$= \int_{[0,1]^2} \frac{f^{X,Z,\Delta}(x, z, 1) \varphi_j(z)}{f^{Z|X}(z|x)} dx dz = \mathbb{E}\left\{ \frac{\Delta \varphi_j(Z)}{f^{Z|X}(Z|X)} \right\}. \tag{2.2.24}$$

This equation yields the following sample mean Fourier coefficient estimator of θ_j for dependent CSCO,

$$\hat{\theta}_{oj} := n^{-1} \sum_{l=1}^n \frac{\Delta_l \varphi_j(\Delta_l Z_l)}{f^{Z|X}(\Delta_l Z_l | X_l)}. \tag{2.2.25}$$

Writing down formulas for dependent CSCNO and dependent CSC is left as an exercise (and also see the Notes). Further, the same technique allows us to estimate Fourier coefficients of the density f^T.

The sample mean Fourier coefficient estimators allow us to use the E-estimation algorithm. If $f^{Z|X}$ is unknown, then its estimate should be plugged in. For CSC the sample from (Z, X) is available, for CSCO and CSCNO an extra sample from (Z, X) should be provided. A corresponding example will be presented shortly.

Assumption 2.2 The lifetime of interest Z and the auxiliary variable X are continuous random variables and each is supported on $[0, 1]$. The density $f^X(x)$ of X is continuous and positive on $[0, 1]$. The lifetime of interest T and the monitoring time Z are conditionally independent given X. A known conditional density $f^{Z|X}(z|x)$ is continuous and positive on $[0, 1]^2$.

Given this assumption and considering a shrinking Sobolev class of underlying $F^{T|X}$, the cdf F^T can be estimated with the rate known for the independent CSC. This is a remarkable theoretical result. Moreover, even constants are known, see the Notes.

2.2.2.3 Numerical Example

The example is based on a simulation motivated by the following real-life example. Undergraduate students with major in actuarial science take a general college probability class (together with students from other majors) where their progress is evaluated by a score X ranging from 0 to 100. After taking the probability class, actuarial students are advised to prepare for the society actuarial exam by solving an additional number Z of recommended problems. We would like to evaluate distribution of the minimal number T of additional problems required for passing the society exam. In other words, T is the necessary and sufficient number of additional problems to be solved for passing the society exam. That minimal required number of problems T cannot be directly observed, and a CSC study may present an opportunity to shed light on the distribution of T. Let us explain what information may be collected. Actuarial exams are expensive, and grants awarded to the college actuarial program are used to reimburse students who passed the society exam. In other words, the program reimburses students with the status $\Delta = I(T \le Z) = 1$. As a result, the program knows all students who passed the society exam, and all these students are willing to share information about the pair (X, Z). Accordingly, we get CSCO observations $(\Delta_1 X_1, \Delta_1 Z_1, \Delta_1), \ldots, (\Delta_n X_n, \Delta_n Z_n, \Delta_n)$ for all n actuarial students who took the college probability class. The scattergram, of rescaled onto $[0, 1]^2$ pairs (X_l, Z_l) of observations for students who passed the society exam (the CSCO data), is shown in the left-top diagram of Fig. 2.7.

It is unrealistic to assume that T and Z are independent. Let us assume that T and Z are conditionally independent given score X in the probability class. According to the above-presented theory, to use the proposed estimation procedure for dependent CSCO we need to estimate the conditional density $f^{Z|X}$. The latter can be done via sampling the pair (X, Z), that is via asking actuarial students about their score in the probability class and the number of solved additional problems. The sampling of (X, Z) is unrelated to passing/taking the society exam, and it is performed via an anonymous census of actuarial students. The obtained "extra" sample $\{(X_{El}, Z_{El}), l = 1, \ldots, n_E\}$ is exhibited in the right-top diagram in Fig. 2.7. Let us stress one more time that collecting information about (X, Z) does not involve information about how a student performed on the society exam. The left-bottom

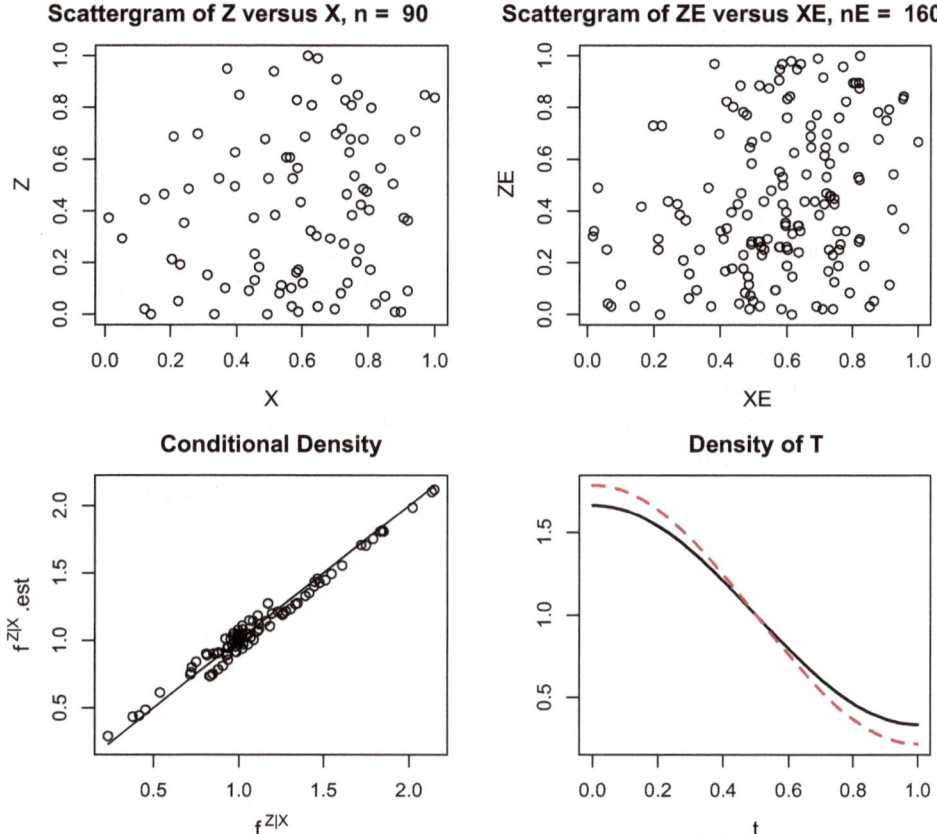

Fig. 2.7 Simulated dependent CSCO data motivated by an actuarial society exam. The left-top diagram shows simulated CSCO data, the right-top diagram shows the extra sample from (X, Z). The left-bottom diagram shows the scattergram of estimates $\hat{f}^{Z|X}(Z_l|X_l)$ versus underlying $f^{Z|X}(Z_l|X_l)$ for observations with $\Delta_l = 1$. The solid and dashed lines in the right-bottom diagram are the underlying density f^T and its E-estimate. [n = 90, nE = 160, cJ0 = 3, CJ1 = 0.8, cTH = 4, cL = 1]

diagram allows us to appreciate accuracy in estimating the conditional density $f^{Z|X}$. This diagram can be used for understanding the effect of n_E on the estimation. Finally, the right-bottom diagram shows the underlying density of interest f^T by the solid line and its E-estimate by the dashed line. The result is impressive keeping in mind complexity of the considered problem.

2.3   Univariate Regression

Consider a classical problem of estimating the regression function

$$m(x) := \mathbb{E}\{T|X = x\}, \quad x \in [0, 1], \tag{2.3.1}$$

where T is the continuous lifetime of interest (time to an event of interest) and X is the continuous random predictor supported on $[0, 1]$. As we already know from Chap. 1, for the case of a direct sample of size n from pair (X, T) and an α-fold differentiable regression function, the univariate regression function m can be estimated with the optimal rate $n^{-2\alpha/(2\alpha+1)}$ of the MISE (mean integrated squared error) convergence.

For direct observations of (X, T) there exists the *principle of equivalence* between nonparametric estimation of α-fold differentiable probability density f^T and nonparametric estimation of α-fold differentiable regression m. The principle states that the same asymptotic results must hold for these two problems, and we observed the equivalence in the asymptotic results presented in Chap. 1.

Now consider the *CSC regression* when we observe a sample of size n from the triplet $(X, Z, \Delta) := (X, Z, I(T \leq Z))$ and want to estimate the regression (2.3.1). If the equivalence principle holds for the CSC, then according to the previous section the optimal rate of regression estimation must slow down to $n^{-2\alpha/(2\alpha+3)}$. Fortunately and surprisingly, the latter is not the case for CSC regression. As we will see shortly, the optimal rate remains the same $n^{-2\alpha/(2\alpha+1)}$ as for direct sampling from (X, T). The latter is the main message of this section. Accordingly, whenever feasible, analysis of CSC should be done via regression.

Let us stress that this promising theoretical result does not diminish complexity of CSC regression. To shed light on the complexity, Fig. 2.8 presents real-life example of CSC regression data (the scattergram) that will be explained and analyzed shortly. Note that we do not observe a sample from the pair of interest (X, T) where X is the predictor and T is the response. Instead, each observation (X_l, Z_l, Δ_l), $l = 1, 2, \ldots, n$ only tells us about the predictor X_l and the status Δ_l of T_l with respect to the known monitoring time Z_l. Accordingly, it is a difficult task to visualize an underlying regression function in the CSC scattergram, while the latter is feasible for a standard regression. As we will see shortly, special methods and a corresponding software are needed for analysis of CSC regression.

Now let us explain the data. Anaerobic digestion of organic municipal solid waste is considered as a key element in sustainable municipal waste management due to its benefits for energy, environment, and economy. This process dramatically reduces emission of greenhouse gases, generates renewable natural gas, and produces fertilizers and soil amendments. The lifetime of interest T is the minimal time required for anaerobic digestion of municipal sludges with different thickness X which is created by either the gravity thickening, or gravity belt thickening, or a centrifuge. Anaerobic digestion happens in the absence of oxygen in a sealed, oxygen-free tank called an anaerobic digester, and the word anaerobic means "in the absence of oxygen". Because an anaerobic digester is sealed, the minimal

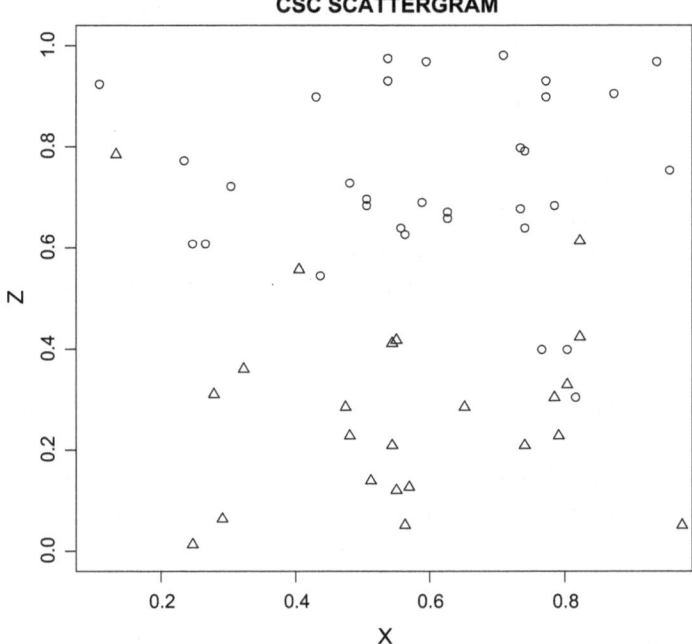

Fig. 2.8 Environmental example of CSC regression with $n = 58$ and $N := \sum_{l=1}^{n} \Delta_l = 32$. The circles and the triangles show pairs (X, Z) corresponding to status 1 and 0, respectively. In other words, CSCO and CSCNO subsamples are shown by the circles and the triangles, respectively. Observations are rescaled onto the unit square

time T to desired digestion cannot be directly observed. Instead, an anaerobic process may be terminated at a monitoring time Z, and then a laboratory analysis of the treated sludge will show its status $\Delta = I(T \leq Z)$. Results of $n = 58$ experiments are shown in Fig. 2.8, and regression analysis of the CSC data will be presented shortly. But for now please try to analyze the CSC scattergram and visualize the underlying regression $\mathbb{E}\{T|X = x\}$.

Apart of the regression, it will be useful and interesting to consider in this section several related problems. Introduce the *conditional survival function* $S^{T|X}(t|x) := \mathbb{P}(T > t|X = x)$. Using the integration by parts, the regression (2.3.1) can be written as

$$m(x) = \int_0^\infty S^{T|X}(t|x)dt, \quad x \in [0, 1]. \tag{2.3.2}$$

Accordingly, regression is the linear functional of the conditional survival. In particular, in survival analysis the integral $\int_0^\infty S^T(t)dt$ is called the *mean survival time*.

Censoring always causes issues with estimation of right tail of the distribution, and to remedy the issue it has been proposed to consider restricted linear functionals, like the *restricted mean survival time* (RMST) $\int_0^r S^T(t)dt$ and the *conditional restricted mean survival time* (CRMST)

$$\mu_r(x) := \int_0^r S^{T|X}(t|x)dt, \quad x \in [0, 1], \ 0 < r < \infty. \tag{2.3.3}$$

To simplify the terminology, in what follows we are referring to the μ_r and to the constant r as the *restricted regression* and the *restriction*, respectively. Estimation of more general restricted linear functionals will be explored later in Sect. 2.5.

If T and Z are conditionally independent given X, and this is a standard assumption, then the following formula holds (recall that $\Delta' := I(T > Z) = 1 - \Delta$),

$$\mathbb{P}(\Delta' = 1|X = x, Z = t) = S^{T|X}(t|x). \tag{2.3.4}$$

This conditional probability, and equivalently the conditional survival function, is of a special interest in all the above-presented practical examples. For instance, in the cancer example this is the conditional likelihood to be free of cancer at time t after the surgery. The probability (2.3.4) is also of the central interest for the econometric binary choice model. Accordingly, we will discuss nonparametric estimation of $S^{T|X}$.

2.3.1 Theory

The aim is to explain what can and cannot be done, based on CSC, CSCO and CSCNO samples, for estimating regression (2.3.2) and restricted regression (2.3.3). Recall that we observe a sample from (X, Z, Δ) for CSC, a sample from $(\Delta X, \Delta Z, \Delta)$ for CSCO, and a sample from $(\Delta'X, \Delta'Z, \Delta)$ for CSCNO. Here $\Delta := I(T \le Z)$ and $\Delta' := 1 - \Delta$. The approach is to consider the oracle who knows more than the statistician, and then develop a sharp lower minimax bound for the mean integrated squared error (MISE) of oracle-estimators. The idea is that the statistician cannot solve a problem better than the oracle, but may try to match performance of the oracle.

We begin with assumptions. The first assumption relaxes independence between T and Z by the conditional independence given the predictor X.

Assumption 2.3 The predictor X is a continuous random variable supported on $[0, 1]$. The monitoring time Z and the lifetime of interest T are nonnegative continuous random variables. Given predictor X, the lifetime of interest T is conditionally independent of the monitoring time Z, that is $S^{T|X,Z}(t|x, z) = S^{T|X}(t|x)$. The pair (X, Z) may be dependent

The next two assumptions allow us to develop oracle's lower bounds. To explain the assumptions, we begin with several preliminary remarks. First, note that the regression and the restricted regression are linear functionals of an underlying conditional survival function, namely

$$m(x) = \int_0^\infty S^{T|X}(t|x)dt, \quad \mu_r(x) = \int_0^r S^{T|X}(t|x)dt, \quad x \in [0, 1]. \qquad (2.3.5)$$

Second, for the oracle a minimax lower bound for estimating μ_r is always smaller than for estimating regression m because the oracle may set $S^{T|X}(t|x)$ to be known for $t > r$. Therefore we begin with a lower bound for restricted regression. Third, it is natural to obtain a lower bound via appropriate perturbations of $S^{T|X}(t|x)$ for $(x, t) \in \mathcal{R}_r := [0, 1] \times [0, r]$. Accordingly, a minimax lower bound is developed for a special class of conditional survival functions. Further, as we will see shortly, a sharp minimax constant is a functional of $S^{T|X}$. This is why a local minimax approach is used when all considered conditional survival functions converge in L_∞-norm to an anchor $S_0^{T|X}$ as $n \to \infty$. Let us stress that the anchor is not an underlying conditional survival function and its primary role is to let the oracle know that all underlying conditional survival functions are near the anchor known to the oracle. The latter also implies that all underlying restricted regressions μ_r are near the restricted regression anchor $\mu_{r0}(x) := \int_0^r S_0^{T|X}(t|x)dt$, $x \in [0, 1]$. The regression anchor m_0 is defined similarly as $m_0(x) := \int_0^\infty S_0^{T|X}(t|x)dt$, $x \in [0, 1]$.

The above-made remarks explain the following two assumptions. Recall that $\varphi_0(x) = 1$, $\varphi_j(x) = 2^{1/2} \cos(\pi j x)$, $j = 1, 2, \ldots$ are elements of the cosine basis on $[0, 1]$, and $\eta_{rj}(x) := r^{-1/2} \varphi_j(x/r)$ are elements of the cosine basis on $[0, r]$.

Assumption 2.4 The anchor $S_0^{T|X}(t|x)$, $(x, t) \in [0, 1] \times [0, \infty)$ is known to the oracle. The anchor is continuous in x and differentiable in t on $\mathcal{R}_r$, and

$$\min_{(x,t)\in\mathcal{R}_r} S_0^{T|X}(t|x) \geq \zeta_1(r) > 0, \quad \max_{(x,t)\in\mathcal{R}_r} \frac{\partial S_0^{T|X}(t|x)}{\partial t} \leq -\zeta_2(r) < 0. \qquad (2.3.6)$$

The next assumption introduces a shrinking (toward the anchor) local Sobolev class of underlying conditional survival functions.

Assumption 2.5 An underlying conditional survival function $S^{T|X}$ belongs to a sequence in n of classes

$$\mathcal{F}_n(S_0^{T|X}, \alpha, Q, r) := \{S^{T|X} : \int_0^r S^{T|X}(t|x)dt \in \mathcal{M}_n(\mu_{r0}, \alpha, Q, r)\}, \qquad (2.3.7)$$

where

$$\mathcal{M}_n(\mu_{r0}, \alpha, Q, r) := \Big\{\mu_r : \mu_r(x) = \mu_{r0}(x) + g(x),$$

$$g \in S_0(\alpha, Q), |g(x)| \leq 1/\ln(\ln(n + 20)), x \in [0, 1]\Big\}. \qquad (2.3.8)$$

In (2.3.8) the restricted regression anchor $\mu_{r0}(x) := \int_0^r S_0^{T|X}(t|x)dt$, $\alpha \geq 1$ and

$$\mathcal{S}_0(\alpha, Q)$$

$$:= \left\{ g : g(x) = \sum_{j=0}^{\infty} \theta_j \varphi_j(x), \; \sum_{j=0}^{\infty}[1 + (\pi j)^{2\alpha}]\theta_j^2 \leq Q < \infty, \; x \in [0, 1] \right\} \quad (2.3.9)$$

is the global Sobolev class (ellipsoid).

Let us comment on the relationship between the conditional survival function $S^{T|X}$ and the restricted regression μ_r that sheds extra light on Assumption 2.5. For the conditional survival we can write using the Fourier theorem,

$$S^{T|X}(t|x) = \sum_{j,i=0}^{\infty} \kappa_{ji} \varphi_j(x) \eta_{ri}(t)$$

$$= r^{-1/2} \sum_{j=0}^{\infty} \kappa_{j0} \varphi_j(x) + \sum_{j=0}^{\infty} \sum_{i=1}^{\infty} \kappa_{ji} \varphi_j(x) \eta_{ri}(t), \quad (x, t) \in \mathcal{R}_r. \quad (2.3.10)$$

Here

$$\kappa_{ji} := \int_{\mathcal{R}_r} S^{T|X}(t|x)\varphi_j(x)\eta_{ri}(t)dtdx \quad (2.3.11)$$

are Fourier coefficients of the conditional survival. Using $\int_0^r \eta_{ri}(t)dt = 0$ for $i \geq 1$, we conclude that

$$\mu_r(x) = r^{1/2} \sum_{j=0}^{\infty} \kappa_{j0} \varphi_j(x), \quad x \in [0, 1]. \quad (2.3.12)$$

This and (2.3.10) imply that the restricted regression is proportional to the univariate Fourier component of $S^{T|X}(t|x)$ in x.

Remark 2.1 In what follows a class $\mathcal{F}_n(S_0^{T|X}, \alpha, Q, \infty)$ is formally defined by replacing in (2.3.7) the restricted regression μ_r and the restricted anchor regression μ_{r0} by the regression m and the anchor regression m_0, respectively. This class will be used for analysis of estimators proposed for unbounded T.

Our final assumption is devoted to the joint density $f^{X,Z}(x, z)$ which is known to the oracle and to the statistician for the case of a controlled CSC study.

Assumption 2.6 The joint density $f^{X,Z}$ is known, continuous and positive on $\mathcal{R}_r$.

Now we are in a position to formulate lower bounds for oracle-estimators using CSC observations. Recall that we may write $\mathbb{E}_{S^{T|X}}\{\cdot\}$ to stress that the expectation is calculated using the given conditional survival function $S^{T|X}$.

Theorem 2.4 (Lower bounds for CSC) *(i) Suppose that Assumptions 2.3–2.6 hold and a CSC sample of size n from (X, Z, Δ) is given. Then*

$$\inf_{\tilde{\mu}_r^*} \sup_{S^{T|X} \in \mathcal{F}_n(S_0^{T|X}, \alpha, Q, r)} \left\{ [n/d(S^{T|X}, f^{X,Z}, r)]^{2\alpha/(2\alpha+1)} \right.$$

$$\left. \times \mathbb{E}_{S^{T|X}} \left\{ \int_0^1 (\tilde{\mu}_r^*(x) - \mu_r(x))^2 dx \right\} \right\} \geq P(\alpha, Q)(1 + o_n(1)). \tag{2.3.13}$$

Here the infimum is taken over all possible oracle-estimators that know data, the function class $\mathcal{F}_n(S_0^{T|X}, \alpha, Q, r)$ and the joint design density $f^{X,Z}(x, z)$,

$$P(\alpha, Q) := Q^{1/(2\alpha+1)}(2\alpha + 1)^{1/(2\alpha+1)}[\alpha/(\pi(\alpha + 1))]^{2\alpha/(2\alpha+1)}, \tag{2.3.14}$$

and

$$d(S^{T|X}, f^{X,Z}, r) := \int_{\mathcal{R}_r} \frac{(1 - S^{T|X}(t|x))S^{T|X}(t|x)}{f^{X,Z}(x, t)} dt dx \tag{2.3.15}$$

is the coefficient of difficulty for CSC restricted regression.
(ii) Let us additionally assume that $d(S^{T|X}, f^{X,Z}, \infty) < \infty$ and Assumption 2.4 holds for any finite r. Then

$$\inf_{\tilde{m}^*} \sup_{S^{T|X} \in \mathcal{F}_n(S_0^{T|X}, \alpha, Q, \infty)} \left\{ [n/d(S^{T|X}, f^{X,Z}, \infty)]^{2\alpha/(2\alpha+1)} \right.$$

$$\left. \times \mathbb{E}_{S^{T|X}} \left\{ \int_0^1 (\tilde{m}^*(x) - m(x))^2 dx \right\} \right\} \geq P(\alpha, Q)(1 + o_n(1)). \tag{2.3.16}$$

Let us make several comments about the result. First, the lower bounds are sharp and attained by oracle-estimators and adaptive blockwise-shrinkage estimators. Second, similarly to the classical case of a direct sample from (X, T) and the regression model $T = m(X) + \sigma(X)\xi$ discussed in Sect. 1.2.5, rate of the MISE convergence is the classical $n^{-2\alpha/(2\alpha+1)}$. Accordingly, CSC regression avoids the curse of CSC distribution estimation when the rate is slower than for direct observations. This also shows that the familiar equivalence between density and regression estimation, known for direct observations, breaks down for CSC. Third, complexity of a CSC regression model is captured by its coefficient of difficulty (2.3.15). Fourth, the coefficient of difficulty sheds light on the role of restriction r which allows us to avoid improper integrals.

The oracle's lower bound warns us that even the oracle may not be able to propose a consistent regression estimator for unbounded lifetimes T or if the support of Z is a subset of the support of T. Fortunately, in survival analysis an unbounded lifetime is a rare phenomenon, and in a majority of statistical applications a lifetime of interest is bounded by a known value.

Now let us consider CSCO and CSCNO regressions.

Theorem 2.5 (Lower bounds for CSCO and CSCNO) *Suppose that Assumptions 2.3–2.6 hold. Set*

$$d_*(S^{T|X}, f^{X,Z}, r) := \int_{\mathcal{R}_r} \frac{S^{T|X}(t|x)}{f^{X,Z}(x,t)} dt dx. \qquad (2.3.17)$$

Then the assertion of Theorem 2.3.1 holds with $d(S^{T|X}, f^{X,Z}, r)$ being replaced by $d_*(1 - S^{T|X}, f^{X,Z}, r)$ and $d_*(S^{T|X}, f^{X,Z}, r)$ for CSCO and CSCNO, respectively.

The main theoretical conclusion of the theorem is that neither CSCO nor CSCNO slows down rate of the MISE convergence but makes the sharp constant larger than for the CSC. In other words, using a subsample (or we may say missing data) affects only constant of the MISE convergence. The ratio $(d_*/d)^{2\alpha/(2\alpha+1)}$ defines the effect of missing on the accuracy of estimation under the MISE criterion. Here and in what follows we may use notations d and d_* for the functionals (2.3.15) and (2.3.17) whenever no confusion occurs.

Remark 2.2 In the following subsection upper bounds for minimax MISEs will be presented. Under a minimax approach, like the one in (2.3.13) or (2.3.16), the supremum is taken over a class $\mathcal{F}$ of underlying distributions $S^{T|X}$ and the infimum over a class of estimators. Accordingly, in a lower bound it is desirable to consider a smaller $\mathcal{F}$ and in the upper bound a larger $\mathcal{F}$. In the above-presented lower bounds we are considering sequences in n of shrinking, toward a specific anchor distribution, classes $\mathcal{F}_n$ of distributions. To establish sharpness of those lower bounds, we will present oracle-estimators and estimators that attain the lower bounds for $S^{T|X} \in \mathcal{F}_n$. As we know from Chap. 1, it is also a tradition in nonparametric curve estimation to analyze MISE over a class of estimands (in our setting over a class of regressions or restricted regressions). Traditional classes of estimands are global Sobolev classes $\mathcal{S}_0(\alpha, Q)$ defined in (2.3.9). Accordingly, in upper bounds we may simultaneously consider supremums over $S^{T|X} \in \mathcal{F}_n$ and over an estimand (m or μ_r) from $\mathcal{S}_0(\alpha, Q)$. As an example, we may write $\sup_{S^{T|X} \in \mathcal{F}_n, \mu_r \in \mathcal{S}_0(\alpha, Q)}$.

2.3.2   Estimators

We begin with the case of a controlled study when the joint design density $f^{X,Z}$ is known. Then the case of an observational study, when the design density is unknown, will be considered. We begin with Sect. 2.3.2.1 devoted to the heuristic. Here the oracle explains how the regression can be estimated. The cases of CSCO, CSCNO and CSC are considered in turn.

2.3.2.1 Heuristic of Oracle-Estimators

The main aim of this subsection is to recall and explain in more details the used methodology of efficient nonparametric series estimation. The oracle, who knows everything about the regression, will help us to understand the methodology via analysis of several series oracle-estimators.

We begin with a simple technical result. Recall that ζ denotes generic positive constants.

Lemma 2.2 *Suppose that function g belongs to the global Sobolev class $S_0(\alpha, Q)$ defined in (2.3.9). Suppose that Fourier coefficients $\kappa_j := \int_0^1 g(x)\varphi_j(x)dx$ of g can be estimated by $\check{\kappa}_j$ satisfying*

$$\mathbb{E}\{(\check{\kappa}_j - \kappa_j)^2\} \le \zeta n^{-1}. \tag{2.3.18}$$

Set $J_n' := n^{1/(2\alpha+1)}$, and introduce the nonparametric projection oracle-estimator

$$\check{g}^*(x) = \sum_{j=0}^{J_n'} \check{\kappa}_j \varphi_j(x). \tag{2.3.19}$$

Then

$$\sup_{g \in S_0(\alpha, Q)} \mathbb{E}\{\int_0^1 (\check{g}^*(x) - g(x))^2 dx\} \le \zeta n^{-2\alpha/(2\alpha+1)}. \tag{2.3.20}$$

This assertion and its proof are simple and insightful. Note that the oracle-estimator is rate-optimal and depends only on the nuisance parameter α. Keeping in mind that in statistical practice it is often assumed that $\alpha = 1$ or $\alpha = 2$, we get a simple estimator. The proof of Lemma 2.2 is based on the Parseval identity. Write,

$$\sup_{g \in S_0(\alpha, Q)} \mathbb{E}\{\int_0^1 (\check{g}^*(x) - g(x))^2 dx\} = \sup_{g \in S_0(\alpha, Q)} [\sum_{j=0}^{J_n'} \mathbb{E}\{(\check{\kappa}_j - \kappa_j)^2\} + \sum_{j > J_n'} \kappa_j^2]$$

$$\le \zeta[n^{-1}J_n' + (J_n')^{-2\alpha}] \le \zeta n^{-2\alpha/(2\alpha+1)}.$$

Lemma 2.2 is proved.

After this warming up, let us consider a more sophisticated assertion which reminds us the result presented in Chap. 1 for density estimation based on direct observations.

Lemma 2.3 *(i) Suppose that a function of interest g belongs to the global Sobolev class $S_0(\alpha, Q)$ defined in (2.3.9). Suppose that Fourier coefficients $\kappa_j := \int_0^1 g(x)\varphi_j(x)dx$ of g can be estimated by $\tilde{\kappa}_j$ satisfying*

$$\mathbb{E}\{\tilde{\kappa}_j\} = \kappa_j, \ \mathbb{V}\{\tilde{\kappa}_j\} \le dn^{-1}(1 + o_n(1) + o_j(1)), \ 0 < d < \infty. \tag{2.3.21}$$

Introduce the nonparametric oracle-estimator

$$\tilde{g}(x) := \sum_{j=0}^{J_n} \tilde{\kappa}_j I(\tilde{\kappa}_j^2 > c_{TH} \mathbb{V}\{\tilde{\kappa}_j\}) \varphi_j(x) + \sum_{j=J_n+1}^{J_n^*} (1 - (j/J_n^*)^{\alpha}) \tilde{\kappa}_j \varphi_j(x), \qquad (2.3.22)$$

where $J_n = \lceil c_{J0} + c_{J1} \ln(n) \rceil$, c_{TH}, c_{J0} and c_{J1} are nonnegative constants, and $J_n^* := J_n + \lceil [(n/d) Q \pi^{-2\alpha} (\alpha + 1)(2\alpha + 1)/\alpha]^{1/(2\alpha+1)} \rceil$. Then the following upper bound is valid for MISE of this oracle-estimator,

$$\sup_{g \in S_0(\alpha, Q)} [n/d]^{2\alpha/(2\alpha+1)} \mathbb{E}\{ \int_0^1 (\tilde{g}(x) - g(x))^2 dx \} \le P(\alpha, Q)(1 + o_n(1)). \qquad (2.3.23)$$

Here $P(\alpha, Q)$ is defined in (2.3.14).
(ii) Let the function of interest be $g = g_0 + g_*$ where $g_0 \in S_0(\alpha', Q')$, $\alpha' > \alpha$, $Q' < \infty$ is the anchor function and $g_* \in S_0(\alpha, Q)$. Suppose that (2.3.21) holds. Then MISE of the oracle-estimator (2.3.22), that does not use the anchor g_0, satisfies the following upper bound,

$$\sup_{g_* \in S_0(\alpha, Q)} [n/d]^{2\alpha/(2\alpha+1)} \mathbb{E}\{ \int_0^1 (\tilde{g}(x) - g(x))^2 dx \} \le P(\alpha, Q)(1 + o_n(1)). \qquad (2.3.24)$$

Note that the second part (ii) of the lemma does not follow from the first one because in part (ii) the estimand g no longer belongs to $S_0(\alpha, Q)$.

The main conclusion of Lemma 2.3 is that to construct an efficient nonparametric oracle-estimator it is sufficient to propose a Fourier coefficient estimator satisfying (2.3.21) with d being an appropriate coefficient of difficulty. Let us show how this can be done for CSCNO and CSCO samples. We are considering estimation of Fourier coefficients for restricted regression μ_r, and then explore the case of regression.

Suppose that Assumption 2.3 holds and assume that the coefficient of difficulty $d_* := d_*(S^{T|X}, f^{X,Z}, r)$, defined in (2.3.17), is finite. Recall our notation $\mu_r(x) := \int_0^r S^{T|X}(t|x) dt$ for restricted regression. Using the Fourier theorem we get

$$\mu_r(x) = \int_0^r S^{T|X}(t|x) dt = \sum_{j=0}^{\infty} \theta_j \varphi_j(x), \quad x \in [0, 1]. \qquad (2.3.25)$$

Here θ_j are Fourier coefficients of μ_r,

$$\theta_j := \int_0^1 \mu_r(x) \varphi_j(x) dx = \int_0^1 \left[\int_0^r S^{T|X}(t|x) dt \right] \varphi_j(x) dx. \qquad (2.3.26)$$

Using Assumption 2.3 we can write

$$f^{X,Z,\Delta}(x, t, \delta) = f^{X,Z}(x, t)[1 - S^{T|X}(t|x)]^\delta[S^{T|X}(t|x)]^{1-\delta}. \tag{2.3.27}$$

Note that $f^{X,Z,\Delta}(x, t, 0) = 0$ if $f^{X,Z}(x, t) = 0$ or $S^{T|X}(t|x) = 0$. Also recall notation $\Delta' := 1 - \Delta = I(T > Z)$. This allows us to continue (2.3.26),

$$\theta_j = \int_0^1 \int_0^r \frac{I(f^{X,Z}(x, t) > 0) f^{X,Z,\Delta}(x, t, 0)\varphi_j(x)}{f^{X,Z}(x, t)} dt dx$$

$$= \mathbb{E}\left\{ \frac{\Delta' I(\Delta' Z \le r)\varphi_j(\Delta' X)}{f^{X,Z}(\Delta' X, \Delta' Z)} \right\}. \tag{2.3.28}$$

The joint density $f^{X,Z}$ is known to the oracle. Accordingly, for CSCNO formula (2.3.28) yields the sample mean Fourier estimator

$$\check{\theta}_j := n^{-1} \sum_{l=1}^n \frac{\Delta'_l I(\Delta'_l Z_l \le r)\varphi_j(\Delta'_l X_l)}{f^{X,Z}(\Delta'_l X_l, \Delta'_l Z_l)}. \tag{2.3.29}$$

Further, by a direct calculation we get

$$\mathbb{E}\{\check{\theta}_j\} = \theta_j, \quad \mathbb{V}\{\check{\theta}_j\} = d_*(S^{T|X}, f^{X,Z}, r)n^{-1}(1 + o_j(1)). \tag{2.3.30}$$

Now we can invoke Lemma 2.2 and conclude that the oracle-estimator (2.3.19), using the Fourier coefficient estimator (2.3.29), is rate optimal. Further, Lemma 2.3 implies that the oracle-estimator (2.3.22), using the Fourier coefficient estimator (2.3.29), is efficient according to Theorem 2.5.

Next we are considering the CSCO when the sample is from $(\Delta X, \Delta Z, \Delta)$. Using (2.3.27) with $\delta = 1$ we get,

$$S^{T|X}(t|x) = 1 - \frac{f^{X,Z,\Delta}(x, t, 1)}{f^{X,Z}(x, t)}. \tag{2.3.31}$$

This and $\int_0^1 \varphi_j(x)dx = I(j = 0)$ allow us to continue (2.3.26) and get

$$\theta_j = \int_0^1 [\int_0^r (1 - \frac{f^{X,Z,\Delta}(x, t, 1)}{f^{X,Z}(x, t)})dt]\varphi_j(x)dx$$

$$= rI(j = 0) - \mathbb{E}\{\frac{\Delta I(\Delta Z \le r)\varphi_j(\Delta Z)}{f^{X,Z}(\Delta X, \Delta Z)}\}. \tag{2.3.32}$$

This yields the following sample mean Fourier coefficient estimator for CSCO,

$$\check{\theta}'_j := rI(j = 0) - n^{-1} \sum_{l=1}^n \frac{\Delta_l I(\Delta_l Z_l \le r)\varphi_j(\Delta_l Z_l)}{f^{X,Z}(\Delta_l X_l, \Delta_l Z_l)}, \tag{2.3.33}$$

with the following properties

$$\mathbb{E}\{\check{\theta}'_j\} = \theta_j, \quad \mathbb{V}\{\check{\theta}'_j\} = d_*(1 - S^{T|X}, f^{X,Z}, r)(1 + o_j(1)). \tag{2.3.34}$$

Note the difference in the variances for the CSCNO and the CSCO.

Now we are considering a CSC sample of size n from (X, Z, Δ). The aim is to propose an efficient Fourier coefficient estimator. As we will see shortly, this is a more complicated problem than the above-considered CSCNO and CSCO. It is possible to directly aggregate the above-proposed Fourier coefficient estimators for CSCO and CSCNO subsamples, however efficient aggregation is not simple and requires using special bases, see the Notes. Instead, in what follows a different type of aggregation is used to construct an efficient Fourier coefficient estimator. The used procedure of aggregation is motivated by the efficient regression methodology proposed in Remark 1.3 of Chap. 1.

The underlying idea of the efficient CSC estimation is to modify the above-proposed CSCNO Fourier coefficient estimator using an estimator of the conditional survival function $S^{T|X}$. Namely, consider Fourier coefficients of the conditional survival function on $\mathcal{R}_r :=$ $[0, 1] \times [0, r]$,

$$\beta_{ki} := \int_{\mathcal{R}_r} S^{T|X}(t|x)\varphi_k(x)\eta_{ri}(t)dtdx. \tag{2.3.35}$$

These Fourier coefficients allow us to introduce a special Fourier approximation of $S^{T|X}$ with a skipped subset of jth Fourier coefficients $\{\beta_{ji}, i = 0, 1, \ldots\}$,

$$S(j, n, t, x) := \sum_{k \in \{\{0,1,\ldots,J_n\}\setminus\{j\}\}} \sum_{i=0}^{J_n} \beta_{ki} \, \varphi_k(x)\eta_{ri}(t), \quad (t, x) \in \mathcal{R}_r. \tag{2.3.36}$$

Note that $\int_0^1 S(j, n, t, x)\varphi_j(x)dx = 0$ and $S(j, n, t, x)$ converges to $S^{T|X}(t|x)$ as j and n increase. The oracle suggests a new unbiased Fourier coefficient estimator

$$\tilde{\theta}^*_j := n^{-1} \sum_{l=1}^n \frac{[\Delta'_l - S(j, n, Z_l, X_l)]\varphi_j(X_l)}{f^{X,Z}(X_l, Z_l)}. \tag{2.3.37}$$

Note how the oracle proposes to aggregate the CSCO and CSCNO subsamples. The oracle's rationale is that (recall the d defined in (2.3.15))

$$\mathbb{V}\{\tilde{\theta}^*_j\} = d(S^{T|X}, f^{X,Z}, r)n^{-1}(1 + o_n(1) + o_j(1)). \tag{2.3.38}$$

Accordingly, the proposed Fourier coefficient oracle-estimator $\tilde{\theta}^*_j$ is efficient and can be used for sharp minimax estimation of the restricted regression μ_r. Of course, the conditional survival function $S^{T|X}$ is unknown, but it may be estimated with sufficient accuracy to match the oracle as we will see shortly in Sect. 2.5.

Finally, if $m_r \in S_0(\alpha, Q)$, then how can a series estimator adapt to unknown nuisance parameters (α, Q) and d? Let us briefly recall heuristic of the blockwise shrinkage that

performs the desired sharp adaptation to the nuisance parameters. The adaptive estimator
will be presented in the next subsection.

We begin with the following classical result in point estimation. If $\bar{\theta}_j$ is unbiased estimator
of parameter θ_j, then it may be beneficial to look at the shrinking estimator $\lambda_j \bar{\theta}_j, \lambda_j \in [0, 1]$
which minimizes the mean squared error $\mathbb{E}\{(\lambda_j \bar{\theta}_j - \theta_j)^2\}$. The oracle's solution, known as
the Wiener filter, is

$$\lambda_j^* = \frac{\theta_j^2}{\theta_j^2 + \sigma_j^2 n^{-1}}, \quad \sigma_j^2 := n\mathbb{E}\{(\bar{\theta}_j - \theta_j)^2\}. \tag{2.3.39}$$

It may be tempting to plug in appropriate estimates of θ_j^2 and σ_j^2 and replace λ_j^* by the
corresponding estimate. Unfortunately, this idea is not feasible because θ_j^2 is estimable with
the parametric rate n^{-1}. On the other hand, we can see from the high-frequency component
of the oracle-estimator (2.3.22) that the shrinking weights $1 - (j/J_n^*)^\alpha$ are close to each
other for adjacent indexes j. This leads us to the idea of using a single optimal shrinking
weight for a block of adjacent indexes j. Namely, let $B := \{i + 1, \ldots, i + L\}$ be a block of
length L of positive integers. Then the Λ^*, which minimizes $\mathbb{E}\{\sum_{j \in B}(\Lambda \bar{\theta}_j - \theta_j)^2\}$, is

$$\Lambda^* = \frac{L^{-1} \sum_{j \in B} \theta_j^2}{L^{-1} \sum_{j \in B} \theta_j^2 + [L^{-1} \sum_{j \in B} \sigma_j^2] n^{-1}} =: \frac{\Theta}{\Theta + \sigma^2 n^{-1}}. \tag{2.3.40}$$

In (2.3.40)

$$\Theta := L^{-1} \sum_{j \in B} \theta_j^2 \tag{2.3.41}$$

is the classical Sobolev functional which is the focal point of the blockwise nonparametric
adaptation. The theory of estimating Sobolev functionals is well developed, and while θ_j^2
may be estimated with the classical parametric rate n^{-1}, the Sobolev functional is estimable
with the same rate but the constant decreases as L (the length of block) increases. This
is what creates the opportunity for estimating Λ^* with sufficient accuracy for matching
oracle-estimators.

2.3.2.2 Efficient Adaptive Estimation of Restricted Regression for CSCNO

In this subsection the lifetime of interest may be bounded or unbounded, and these two
cases are considered simultaneously. Further, recall that for a bounded lifetime a restricted
regression, with the restriction r equal to or larger than that bound, is the underlying regression. The available sample of size n is from $(\Delta'X, \Delta'Z, \Delta)$, and we estimate the restricted
regression $\mu_r(x) = \int_0^r S^{T|X}(t|x)dt$. Because the restriction is finite, we can simultaneously
consider bounded and unbounded lifetimes T given Assumption 2.6. Indeed, that assumption is not tied to the support of T and only requires that the known (recall that in this
section we study the controlled sampling) design density $f^{X,Z}$ is continuous and positive
on $\mathcal{R}_r := [0, 1] \times [0, r]$.

Recall the Fourier coefficient estimator for CSCNO observations is

$$\check{\theta}_j := n^{-1} \sum_{l=1}^{n} \frac{\Delta' I(\Delta'_l Z_l \leq r)\varphi_j(\Delta'_l X_l)}{f^{X,Z}(\Delta'_l X_l, \Delta'_l Z_l)}. \tag{2.3.42}$$

As we will see shortly, it can be used to construct efficient restricted regression estimator. Our next step, according to the heuristic of Sect. 2.3.2.1, is to define a blockwise adaptive estimator. For $j > J_n$ introduce consecutive and non-overlapping blocks B_k, $k = 1, 2, \ldots$ of length $L_k := \lceil (1 + 1/\ln(\ln(n + 20)))^k \rceil$, that is $B_1 := \{J_n + 1, \ldots, J_n + L_1\}$, $B_2 := \{J_n + L_1 + 1, \ldots, J_n + L_1 + L_2\}$, etc. Then for each block we calculate two statistics. The first one is the U-statistic

$$\tilde{\Theta}_k := \frac{2}{L_k n(n-1)} \sum_{1 \leq l_1 < l_2 \leq n} \sum_{j \in B_k} \prod_{i=1}^{2} \frac{\Delta'_{l_i} I(\Delta'_{l_i} Z_{l_i} \leq r)\varphi_j(\Delta'_{l_i} X_{l_i})}{f^{X,Z}(\Delta'_{l_i} X_{l_i}, \Delta'_{l_i} Z_{l_i})}. \tag{2.3.43}$$

The second statistic is based on the Fourier coefficient estimator $\check{\theta}_j$ defined in (2.3.42),

$$\tilde{\Theta}'_k := L_k^{-1} \sum_{j \in B_k} \check{\theta}_j^2. \tag{2.3.44}$$

Let k_n be the smallest integer such that $\sum_{k=1}^{k_n} L_k > n^{1/(2\alpha_0+1)} \ln(\ln(n + 20))$ where α_0 is the smallest assumed value of parameter α. Recall that Assumption 2.5 sets $\alpha_0 = 1$, but other values also may be specified. For instance, $\alpha_0 = 2$ implies that the restricted regression is twice differentiable, and this is another traditional choice.

The proposed efficient (sharp-minimax) adaptive estimator is

$$\tilde{\mu}_r(x) := \tilde{\mu}_r(x, f^{X,Z}) := \sum_{j=0}^{J_n} \check{\theta}_j I(\check{\theta}_j^2 > c_{TH}\check{\sigma}_{jn}^2)\varphi_j(x)$$

$$+ \sum_{k=1}^{k_n} [\tilde{\Theta}_k/\tilde{\Theta}'_k] I(\tilde{\Theta}_k > 1/[n\ln(k+3)]) \sum_{j \in B_k} \check{\theta}_j \varphi_j(x). \tag{2.3.45}$$

Here $\check{\sigma}_{jn}^2$ is the sample mean estimator of the variance $\mathbb{V}\{\check{\theta}_j\}$, and recall that the variance is of order n^{-1}.

Theorem 2.6 *Let Assumptions 2.3 and 2.6 hold, and the anchor μ_{r0} belongs to a Sobolev class $S(\alpha', Q')$ with $\alpha' > \alpha$ and $Q' < \infty$. Consider a CSCNO sample of size n from $(\Delta'X, \Delta'Z, \Delta)$. Then the following upper bound holds for MISE of the adaptive estimator (2.3.45),*

$$\sup_{\{S^{T|X} \in \mathcal{F}_n(S_0^{T|X}, \alpha, Q, r), \, \mu_r \in \mathcal{S}_0(\alpha, Q)\}} \left\{ [n/d_*(S^{T|X}, f^{X,Z}, r)]^{2\alpha/(2\alpha+1)} \right.$$

$$\times \mathbb{E}\left\{ \int_0^1 (\tilde{\mu}_r(x) - \mu_r(x))^2 dx \right\} \bigg\} \le P(\alpha, Q)(1 + o_n(1)). \tag{2.3.46}$$

We can conclude that the lower bound of Theorem 2.5 is sharp. Further, the lower bound is attainable for μ_r from the global Sobolev class $\mathcal{S}_0(\alpha, Q)$. Accordingly, we get the same result as for the case of regressions based on direct observations from (X, T) where efficient adaptive estimators are proposed for global Sobolev classes, see Sect. 1.2.5.

The case of CSCO is considered similarly, and it is left as an exercise.

2.3.2.3 Efficient Estimation of Restricted Regression for CSC

In this subsection the lifetime of interest may be bounded or unbounded, and these two cases are considered simultaneously. Further, recall that if T is bounded and we set the restriction r to be equal to or larger than the upper bound for the support of T, then the restricted regression coincides with the regression.

We are considering a CSC sample of size n from (X, Z, Δ) and propose an efficient estimator of μ_r. Set $\mathcal{J}(j, n) := \{0, 1, \ldots, J_n\} \setminus \{j\}$. Following the heuristic of Sect. 2.3.2.1, introduce an estimator of $S^{T|X}$ with subtracted projection on φ_j,

$$\tilde{S}(j, n, z, x)$$

$$:= n^{-1} \sum_{l=1}^n \sum_{k \in \mathcal{J}(j,n)} \sum_{i=0}^{J_n} \frac{\Delta'_l I(Z_l \le r) \eta_{ri}(Z_l) \varphi_k(X_l) \eta_{ri}(z) \varphi_k(x)}{f^{X,Z}(X_l, Z_l)}, \tag{2.3.47}$$

the Fourier coefficient estimator

$$\hat{\theta}_j := n^{-1} \sum_{l=1}^n \frac{(\Delta'_l - \tilde{S}(j, n, Z_l, X_l)) I(Z_l \le r) \varphi_j(X_l)}{f^{X,Z}(X_l, Z_l)}, \tag{2.3.48}$$

and the sample mean estimator $\hat{\sigma}_{jn}^2$ of the variance $\mathbb{V}\{\hat{\theta}_j\}$. Recall that $\mathbb{E}\{\hat{\sigma}_{jn}^2\}$ is of order n^{-1}.

For adaptation to unknown smoothness of μ_r, we again use blocks B_k of length L_k, introduced below line (2.3.42), and the sequence k_n defined below line (2.3.44). For each block we calculate two statistics. The first one is the U-statistic

$$\hat{\Theta}_k := \frac{2}{L_k n(n-1)}$$

$$\times \sum_{1 \le l_1 < l_2 \le n} \sum_{j \in B_k} \prod_{i=1}^2 \frac{(\Delta'_{l_i} - \tilde{S}(j, n, Z_{l_i}, X_{l_i})) I(Z_{l_i} \le r) \varphi_j(X_{l_i})}{f^{X,Z}(X_{l_i}, Z_{l_i})}. \tag{2.3.49}$$

The second statistic is based on Fourier estimates $\hat{\theta}_j$ defined in (2.3.48),

$$\hat{\Theta}'_k := L_k^{-1} \sum_{j \in B_k} \hat{\theta}_j^2. \tag{2.3.50}$$

The proposed estimator is

$$\hat{\mu}_r(x) := \hat{\mu}_r(x, f^{X,Z}) := \sum_{j=0}^{J_n} \hat{\theta}_j I(\hat{\theta}_j^2 > c_{TH} \hat{\sigma}_{jn}^2) \varphi_j(x)$$

$$+ \sum_{k=1}^{k_n} [\hat{\Theta}_k / \hat{\Theta}'_k] I(\hat{\Theta}_k > 1/[n \ln(k+3)]) \sum_{j \in B_k} \hat{\theta}_j \varphi_j(x). \tag{2.3.51}$$

Theorem 2.7 *Let Assumptions of Theorem 2.4 hold. Consider a CSC sample of size n from (X, Z, Δ). Then the following upper bound holds for the MISE of estimator (2.3.51),*

$$\sup_{\{S^{T|X} \in \mathcal{F}_n(S_0^{T|X}, \alpha, Q, r), \ \mu_r \in S_0(\alpha, Q)\}} \left\{ [n/d(S^{T|X}, f^{X,Z}, r)]^{2\alpha/(2\alpha+1)} \right.$$

$$\times \mathbb{E}\left\{ \int_0^1 (\hat{\mu}_r(x) - \mu_r(x))^2 dx \right\} \le P(\alpha, Q)(1 + o_n(1)). \tag{2.3.52}$$

Theorem 2.7, together with the lower bound of Theorem 2.4, allow us to conclude that not only the optimal rate $n^{-2\alpha/(2\alpha+1)}$ is preserved, but also the sharp constant is attainable. Further, the proposed estimator attains the same sharp constant over the global Sobolev classes. In short, for the CSC regression we have the same bouquet of results as for the case of direct observations.

2.3.2.4 Estimation of Regression for Unbounded T

Recall the coefficients of difficulty defined in (2.3.15) and (2.3.17), and it will be convenient to use notations $D := d(S^{T|X}, f^{X,Z}, \infty)$ and $D_* := d_*(S^{T|X}, f^{X,Z}, \infty)$.

Assumption 2.7 Conditional survival function $S^{T|X}$ and joint density $f^{X,Z}$ satisfy

$$D_* := \int_0^1 \int_0^\infty \frac{S^{T|X}(t|x)}{f^{X,Z}(x,t)} dt dx \le \zeta < \infty. \tag{2.3.53}$$

Assumption 2.7 yields that D is also bounded. At the same time, if D is bounded then D_* may be unbounded if $1/f^{X,Z}(x,t)$ is not integrable for small t while $(1 - S^{T|X}(t|x))/f^{X,Z}(x,t)$ is.

Set $J_n^* := J_n + \lceil [(n/D_*)Q\pi^{-2\alpha}(\alpha + 1)(2\alpha + 1)/\alpha]^{1/(2\alpha+1)} \rceil$,

$$\bar{\theta}_j := n^{-1} \sum_{l=1}^{n} \frac{\Delta_l' \varphi_j(\Delta_l' X_l)}{f^{X,Z}(\Delta_l' X_l, \Delta_l' Z_l)}, \tag{2.3.54}$$

and introduce the regression oracle-estimator motivated by Lemma 2.3,

$$\bar{m}^*(x) := \sum_{j=0}^{J_n} \bar{\theta}_j I(\bar{\theta}_j^2 > c_{TH} \mathbb{V}\{\bar{\theta}_j\})\varphi_j(x) + \sum_{j=J_n+1}^{J_n^*} (1 - (j/J_n^*)^\alpha)\bar{\theta}_j \varphi_j(x). \tag{2.3.55}$$

Theorem 2.8 *Suppose that Assumptions 2.3 and 2.7 hold, and the anchor $m_0 \in S_0(\alpha', Q')$ where $\alpha' > \alpha$ and $Q' < \infty$. Then MISE of oracle-estimator (2.3.55) satisfies the inequality*

$$\sup_{S^{T|X} \in \mathcal{F}_n(S_0^{T|X}, \alpha, Q, \infty), \, m \in S_0(\alpha, Q)} \left\{ [n/D_*]^{2\alpha/(2\alpha+1)} \right.$$

$$\left. \times \mathbb{E}\left\{ \int_0^1 (\bar{m}^*(x) - m(x))^2 dx \right\} \right\} \le P(\alpha, Q)(1 + o_n(1)). \tag{2.3.56}$$

Corollary 2.1 *The oracle-estimator (2.3.55) is asymptotically efficient for CSCNO. Its MISE is within factor $(D_*/D)^{\frac{2\alpha}{2\alpha+1}}$ from the lower bound (2.3.16) for CSC. Accordingly, the rougher the regression the smaller the factor.*

The blockwise-shrinkage methodology of Sect. 2.3.2.2 allows us to propose an efficient estimator for CSCNO that matches MISE of the oracle-estimator. The efficient estimator for CSCO is constructed similarly. At the same time, efficient estimation for CSC is an interesting and open problem to explore.

2.3.2.5 Estimation for an Observational Study

So far we have considered the case of a known design density $f^{X,Z}$. In general this bivariate density may be unknown, and this is the setting that we are considering. It will be shown that under a mild assumption a plug in methodology is feasible. For CSC we can use the available sample of size n from (X, Z) to estimate $f^{X,Z}$. As we already know, for CSCNO and CSCO the joint density $f^{X,Z}$ cannot be consistently estimated because the data is biased. Accordingly, for CSCNO and CSCO it is assumed that an extra sample of size n from (X, Z) is available.

Assumption 2.8 An underlying joint probability density $f^{X,Z}$ has a continuous mixed derivative $\partial^2 f^{X,Z}(x,z)/\partial x \partial z$ on $\mathcal{R}_r := [0,1] \times [0,r]$ and

$$\int_{\mathcal{R}_r} \left[[\partial f^{X,Z}(x,z)/\partial x]^2 + [\partial f^{X,Z}(x,z)/\partial z]^2 + [\partial^2 f^{X,Z}(x,z)/\partial x \partial z]^2 \right] dx dz$$

$$\leq \zeta < \infty. \tag{2.3.57}$$

An underlying conditional survival function $S^{T|X}(t|x)$ has a continuous mixed derivative $\partial^2 S^{T|X}(t|x)/\partial t \partial x$ on $\mathcal{R}_r$ and

$$\int_{\mathcal{R}_r} \left[[\partial S^{T|X}(t|x)/\partial x]^2 + [\partial S^{T|X}(t|x)/\partial t]^2 + [\partial^2 S^{T|X}(t|x)/\partial t \partial x]^2 \right] dx dt$$

$$\leq \zeta < \infty. \tag{2.3.58}$$

The left sides of (2.3.57) and (2.3.58) are classical Sobolev functionals. It is important to stress that these Sobolev functionals do not involve second-order derivatives with respect to either of the arguments.

Given Assumption 2.8, for CSC model introduce the bivariate density estimator

$$\hat{f}^{X,Z}(x,z) := \max \left(c_L / \ln(\ln(n+20)), \right.$$

$$\left. n^{-1} \sum_{i,s=0}^{\lceil 1+n^{1/4} \rceil} \sum_{l=1}^{n} \varphi_i(X_l) \eta_{rs}(Z_l) \varphi_i(x) \eta_{rs}(z) \right). \tag{2.3.59}$$

For CSCNO and CSCO the estimator (2.3.59) is based on the extra sample from (X, Z). Then the estimator of the joint design density $f^{X,Z}$ can be used in place of an underlying $f^{X,Z}$. In other words, we use the plug-in methodology. Note that the estimator is separated from zero by $c_L / \ln(\ln(n+20))$ and hence may be used in a denominator.

Theorem 2.9 *Let Assumption 2.8 and the assumption of Theorem 2.4 hold, only now the joint density $f^{X,Z}$ is unknown and $\alpha \geq \alpha_0 = 2$. Introduce the plug-in estimators $\tilde{\mu}_r(x, \hat{f}^{X,Z})$ and $\hat{\mu}_r(x, \hat{f}^{X,Z})$ defined in (2.3.45) and (2.3.51), respectively. Then the assertions of Theorems 2.6 and 2.7 hold for the plug-in estimators.*

We may conclude that the data-driven estimation is possible and the lower bounds of Sect. 2.3 are sharp. Further, CSC does not slow down rate of the regression estimation with respect to direct data observations.

Now we are in a position to complement the theory by examples.

2.3.2.6 Analysis of Simulated and Real Examples

Here we begin with visual analysis of two simulated CSC datasets when underlying regression functions are known. In the former simulation X and Z are independent, and in the latter they are dependent. For these simulations the above-presented estimates are shown and discussed. Then the more complicated simulation, with dependent predictor and monitoring time, is repeated 10 times and residuals of the data-driven CSC estimator are shown. This experiment sheds light on the bias and variance of the regression estimator. Then results of an intensive numerical study are presented via histograms of ratios between integrated squared errors of studied estimators. Finally, regression analysis of real data is presented.

Two particular simulations are shown in the two columns of Fig. 2.9. The top diagrams show by crosses the underlying direct observations of (X, T). The bottom diagrams show corresponding CSC samples from (X, Z, Δ) with triangles and circles indicating observations with $\Delta = 0$ and $\Delta = 1$, respectively. Full description of the underlying experiments and the diagrams can be found in the caption.

Let us comment on the scattergrams and the estimates shown in Fig. 2.9. Similarly to other figures, the particular simulations are shown to explain some interesting specifics. The two direct data scattergrams are complicated due to the strong heteroscedasticity and the fact that all lifetimes are from $[0, 1]$. This creates very special scattergrams that the reader may get used to via repeated simulations. Still, it is possible to visualize the underlying regression as a curve that goes through the "middle" of data. The reader may try to make a guess about the regression, and then compare the guess with the solid line (the underlying regression) and the dashed line (the data-driven nonparametric E-estimate). Note how the E-estimates fit the data. The bottom diagrams show CSC modifications of the direct data. Here the triangles show CSCNO observations and the circles show CSCO observations. Overall, if not due to the status Δ of a CSC observation, it is impossible to visualize the underlying regression. Indeed, in the both bottom diagrams the observations are spread over the unit square and the best bet may be a horizontal line. Now let us look at the underlying regression (the solid line) and the data paying attention to the triangles and circles. Overall, with the help of the solid line, it is possible to appreciate the special structure of CSC scattergrams and even get a "feeling" of the underlying regression. With some training in visualization of different CSC scattergrams, it is possible to get a general feeling of the shape of an underlying regression, but overall this is a complicated task. Only a special software can estimate the CSC regression because here one needs first to figure out an underlying conditional survival and then evaluate its integral.

The two CSC estimates in the bottom diagrams, shown by the dashed and dotted lines, are the estimates for controlled (the design density $f^{X,Z}$ is known) and observational (the design density $f^{X,Z}$ is unknown) CSC models, respectively. The chosen simulations show us several possible outcomes. The left-bottom diagram exhibits a very good outcome for the controlled regression with known $f^{X,Z}$. To appreciate this estimate, compare its ISEcontr with the ISE for direct data. The case of unknown design density is still reasonable keeping in mind the small sample size. The outcome is more nuanced for the data in the right-bottom

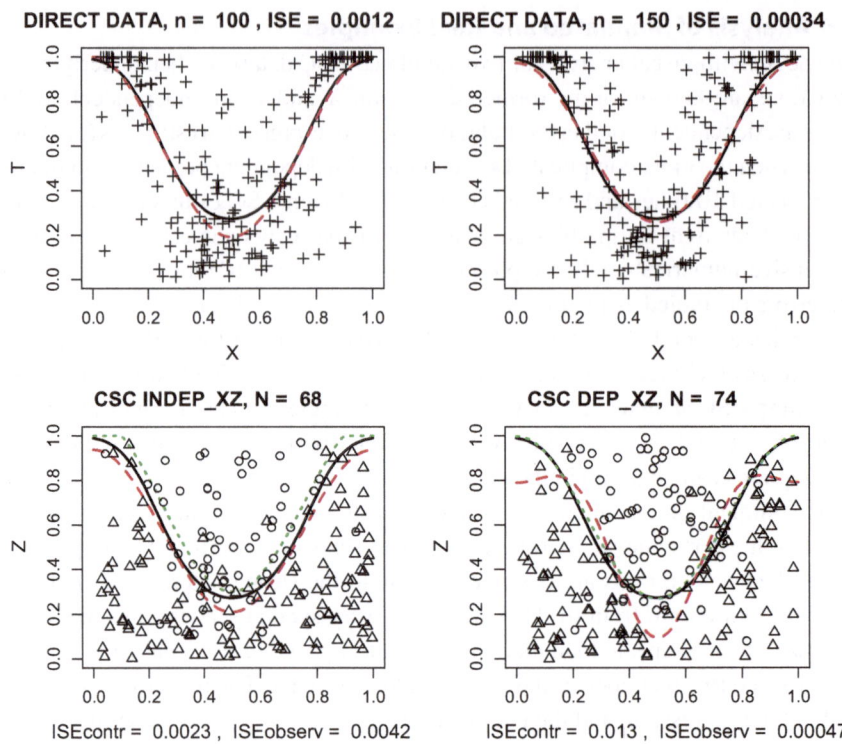

Fig. 2.9 Two simulated CSC regressions with independent and dependent variables X and Z. Underlying scattergrams from (X, T) and the corresponding CSC scattergrams from (X, Z, Δ) are shown in the top and bottom diagrams, respectively. Direct observations are shown by the crosses. CSC observations are shown by the triangles and the circles for $\Delta = 0$ and $\Delta = 1$, respectively. The left column presents the experiment with independent X and Z. Here the predictor X is distributed according to the density $f^X(x) = (4/5)(1 + 0.5x)I(x \in [0, 1])$, and the conditional distribution of the response T given the predictor is Beta$(1, f_3(x))$. As a result, $m(x) = 1/[1 + f_3(x)]$ for $x \in [0, 1]$. Function f_3 is the third (Normal) corner function. The underlying monitoring lifetime Z is independent of the predictor X and $f^Z(z) = (1 + 0.5\cos(\pi z))I(z \in [0, 1])$. The right column presents the experiment with the same conditional distribution of T given X but with dependent X and Z. Here the design density is $f^{X,Z}(x, z) = [1 + (1/4)\cos(\pi x) + (1/2)\cos(\pi x)\cos(\pi z)]I((x, z) \in [0, 1]^2)$. In all diagrams the solid line is the underlying regression. In the top diagrams the dashed lines are the regression E-estimates of Chap. 1. In the bottom diagrams the dashed lines are CSC E-estimates for controlled regression with known $f^{X,Z}$, and the dotted lines are data-driven E-estimates for observational regression when $f^{X,Z}$ is unknown. The estimates use the information $0 \le m(x) \le 1$ for $x \in [0, 1]$. The titles show the type of regression, sample size n, and $N := \sum_{l=1}^n \Delta_l$. The empirical integrated squared errors (ISE) of the estimates are denoted as ISE, ISEcontr, and ISEobserv for the estimates based on direct data, controlled ($f^{X,Z}$ is known) CSC data and observational CSC data, respectively. {This figure allows to choose for the left column the shapes of f^X and f^Z by arguments *desdenX* and *desdenZ*. The regression $m(x)$ is controlled by the argument *corn*, sample sizes by argument *n*.} [n = c(100, 150), corn = 3, desdenX ="1 + 0.5 * x", desdenZ ="1 + 0.5 * cos(pi * x)", cJ0 = 3, CJ1 = 0.8, cTH = 4, cL = 1]

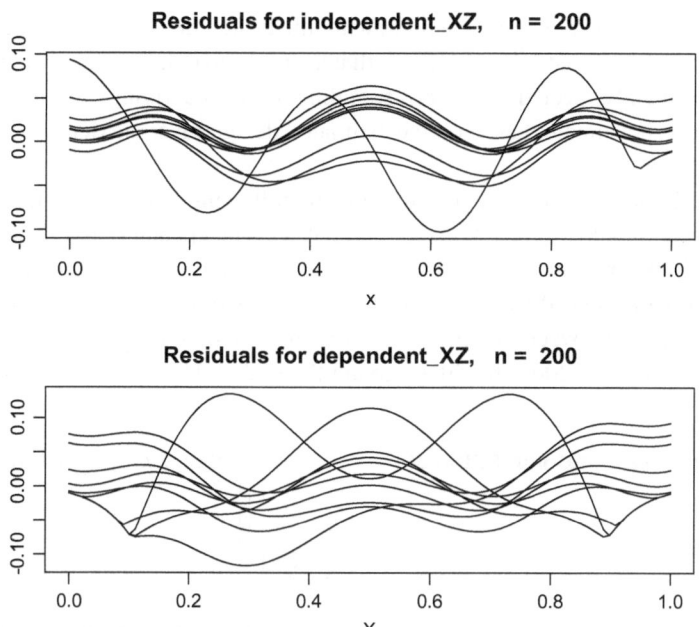

Fig. 2.10 Residuals $m(x) - \hat{m}(x), x \in [0, 1]$. The underlying experiments are the same as in Fig. 2.9. {When this figure is used, first a diagram, similar to Fig. 2.9, is shown to verify the underlying simulation. Then, at the R prompt *Browser*[1] >, enter c. This will show the residuals. Argument *nsim* controls the number of shown residuals.} [n = c(200, 200), nsim = 10, corn = 3, desdenX $= "1 + 0.5 * x"$, desdenZ $= "1 + 0.5 * cos(pi * x)"$, cJ0 = 3, CJ1 = 0.8, cTH = 4, cL = 1]

diagram where X and Z are dependent. Here the estimate for observational regression dramatically outperforms the estimate for controlled regression. Further, its performance is comparable with the estimate based on direct observations. Is this an abnormal simulation? A surprising answer will be presented shortly.

Figure 2.10 sheds additional light on performance of the proposed data-driven estimator for observational CSC data when the design density $f^{X,Z}$ is unknown. The underlying CSC models are the same as in Fig. 2.9, only now the sample size $n = 200$ is used. Ten simulations are performed, and then the residuals are shown. The residuals help us to visualize the bias and variance of the estimates. This figure is useful for choosing parameters of the data-driven estimator, and it also sheds a special light on the used E-estimator.

Now we return to the earlier formulated question of why a data-driven regression estimator may perform better than an estimator that uses an underlying design density $f^{X,Z}$. Further, we would like to shed light on relative performance, measured by the ISE (integrated squared error), of the three above-discussed regression estimators. Namely, we would like to compare the estimators based on direct data, controlled CSC data when $f^{X,Z}$ is known, and observational CSC data when $f^{X,Z}$ is unknown and should be estimated. We already observed in Fig. 2.9 the outcome when knowing the design density was not helpful. A reasonable explanation of that phenomenon is that a simulated data, especially a bivariate one, may be far from an "expected" data corresponding to the underlying design density. In that case a data-driven estimate, based on the estimated design bivariate density, may be better and yield a smaller ISE.

Let us check the above-made conjecture via the following intensive simulations. We use the two underlying CSC experiments of Fig. 2.9 with independent and dependent X and Z, repeat each experiment 1000 times, for each simulation calculate the above-explained ISEDirect, ISEContr and ISEObserv for the 3 corresponding estimates and then visualize histograms of ratios between the three ISEs. Figure 2.11 shows us the histograms of ratios for the sample size $n = 200$. The titles explain the underlying ratios. Similarly to Fig. 2.9,

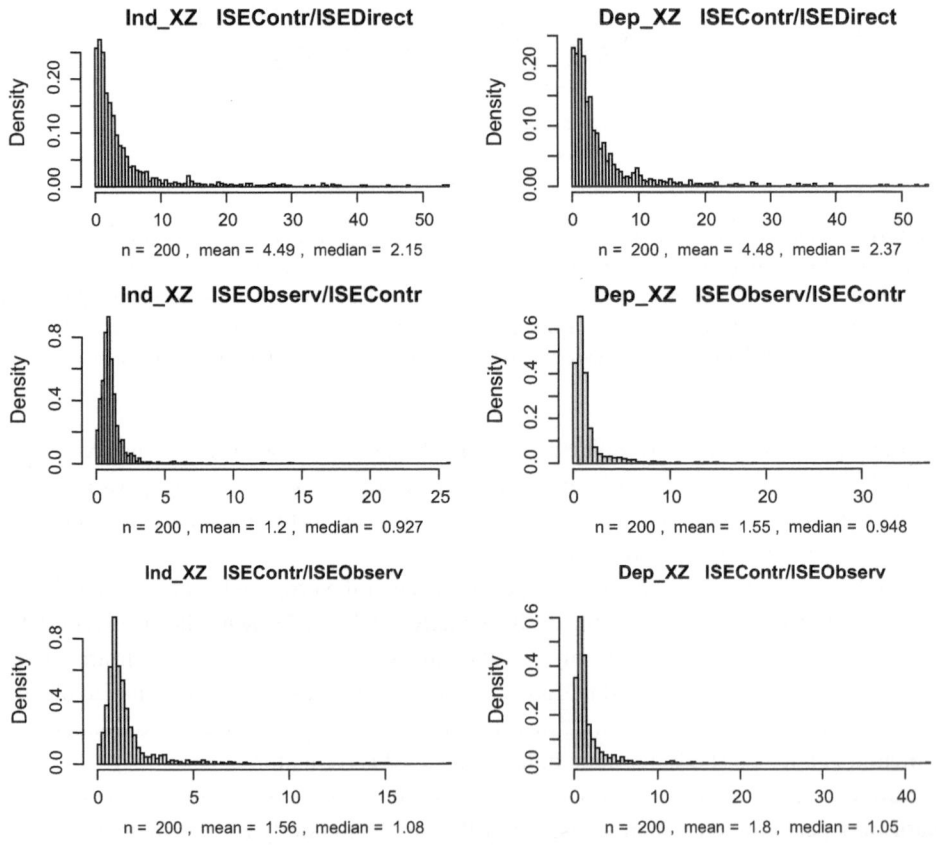

Fig. 2.11 Histograms of ratios between ISEs of estimators. The histograms are based on 1000 simulations, the number of simulations is controlled by the argument *nsim*. Sample mean and sample median of the ratios are shown in the subtitle. [n = c(200, 200), desdenX $=''1 + 0.5 * x''$, desdenZ $=''1 + 0.5 * cos(pi * x)''$, nsim = 1000, corn = 3, cJ0 = 3, CJ1 = 0.8, cTH = 4, cL = 1]

the left column presents results for independent X and Z, the right for dependent. Let us look at diagrams in the left column. The left-top diagram exhibits ratios of ISEContr (the ISE of the CSC regression estimate using the underlying design density $f^{X,Z}$) to ISEDirect (the ISE of regression estimate based on direct observations of (X, T)). As it could be expected, the ratio can be relatively large with the mean 4.49 and median 2.15. Note that for some simulations the direct data estimate may be overwhelmingly better. Now let us look at the left-middle diagram. Here we compare ISE of the data-driven CSC regression estimate (ISEObserv) with ISE of the CSC regression estimate knowing the $f^{X,Z}$ (ISEContr). It could be expected that the ratios ISEObserv/ISEContr would be larger than 1. And indeed, the mean ratio is 1.2. Further, some ratios are very large implying that the used design density estimate is far from being perfect. At the same time, the median ratio is 0.927. This tells us that in a majority of simulations knowing the design density is not needed. This supports and sheds light on the above-made conjecture about why the data-driven estimator may outperform the estimator knowing the underlying design density. The bottom diagram adds extra evidence in support of the conjecture.

Now let us look at results for the experiment with dependent X and Z in the right column of diagrams in Fig. 2.11. The outcomes are similar, and the CSC estimator performs relatively well for this complicated simulation.

The reader is advised to repeat Fig. 2.11 and consider different sample sizes, say 100 and 300. Overall, the outcomes are similar and they support the above-made conclusion, only with larger samples we see less outliers in terms of extremely large ratios and the improvement in relative performance with respect to the oracle-estimator using direct observations of (X, T). This is a natural outcome for larger samples, and it supports the proposed methodology of CSC regression estimation.

Finally, let us return to the real CSC example presented in Fig. 2.8. The reader is advised to look at that figure one more time and try to visualize an underlying regression using the experience gained in analysis of the simulated datasets. The regression E-estimate is shown in Fig. 2.12 by the monotonically decreasing solid line. After the previous training in "reading" CSC scattergrams, the regression looks reasonable. Further, the decreasing regression reflects the underlying physics of anaerobic digestion.

2.4   Multivariate Regression

We begin with discussion of a multivariate anisotropic CSC regression and show that it is possible to match classical results known for the case of direct data discussed in Sect. 1.2.6. An underlying multivariate regression function is $m(\mathbf{x}) = \mathbb{E}\{T | \mathbf{X} = \mathbf{x}\}, \mathbf{x} := (x_1, \ldots, x_k) \in [0, 1]^k$. Its smoothness (the number of derivatives) may be different for each covariate, namely it is assumed that the regression function belongs to an anisotropic Sobolev class

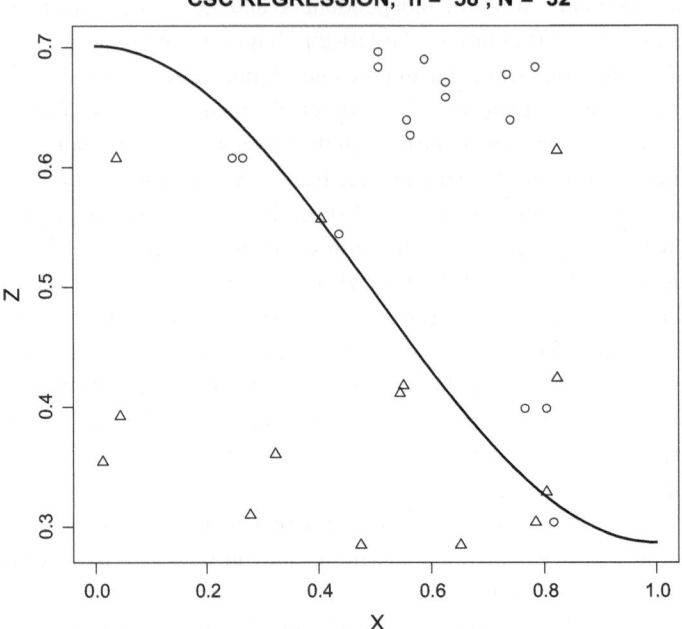

Fig. 2.12 Regression analysis of CSC data for anaerobic digestion of municipal wastewater shown in Fig. 2.8

$$\mathcal{S}_0(\alpha_1, \ldots, \alpha_k, Q) := \left\{ m : m(\mathbf{x}) = \sum_{i_1, \ldots, i_k=0}^{\infty} \theta_{\mathbf{i}} \varphi_{\mathbf{i}}(\mathbf{x}), \ \mathbf{x} \in [0, 1]^k, \right.$$

$$\left. \sum_{i_1, \ldots, i_k=0}^{\infty} \theta_{\mathbf{i}}^2 [1 + \sum_{s=1}^{k} (\pi i_s)^{2\alpha_s}] \le Q < \infty \right\}. \tag{2.4.1}$$

Here $\mathbf{i} := (i_1, \ldots, i_k)$, and $\varphi_{\mathbf{i}}(\mathbf{x}) = \prod_{s=1}^{k} \varphi_{i_s}(x_{i_s})$ are elements of the cosine tensor-product basis on $[0, 1]^k$. Introduce the multivariate effective smoothness $\alpha_* := [\sum_{s=1}^{k} \alpha_s^{-1}]^{-1}$. Consider a direct sample of size n from $(\mathbf{X}, T)$. Then the optimal rate of MISE convergence is $n^{-2\alpha_*/(2\alpha_*+1)}$, and in particular if $\alpha_1 = \ldots = \alpha_k = \alpha$ then the rate is $n^{-2\alpha/(2\alpha+k)}$. Note how dramatically the dimensionality k of the regression slows down the rate.

Let us show that the "direct data" rate $n^{-2\alpha_*/(2\alpha_*+1)}$ is also achievable for a CSC sample of size n from $(\mathbf{X}, Z, \Delta)$, and hence we again can break the curse of CSC in terms of slower rates of convergence for nonparametric estimation of distributions. Further, even for the controlled CSCNO (and similarly for the controlled CSCO due to the symmetry between the subsamples) it is possible to attain the optimal rate as the following theorem shows.

Theorem 2.10 *Consider a controlled CSCNO sample of size n from* $(\Delta'\mathbf{X}, \Delta'Z, \Delta)$ *when the design density* $f^{\mathbf{X},Z}$ *is known and* $\mathbf{X} := (X_1, \ldots, X_k)$ *is supported on* $[0, 1]^k$. *Suppose that* $S^{T|\mathbf{X},Z}(t|\mathbf{x}, z) = S^{T|\mathbf{X}}(t|\mathbf{x})$ *and*

$$\int_0^\infty \Big[\int_{[0,1]^k} \frac{S^{T|\mathbf{X}}(z|\mathbf{x})}{f^{\mathbf{X},Z}(\mathbf{x}, z)} d\mathbf{x} \Big] dz < \infty. \qquad (2.4.2)$$

Introduce a multivariate regression estimator

$$\hat{m}(\mathbf{x}) := \sum_{\mathbf{i} \in \mathcal{J}} \tilde{\theta}_{\mathbf{i}} \varphi_{\mathbf{i}}(\mathbf{x}), \qquad (2.4.3)$$

where $\mathcal{J} := \{0, 1, \ldots, J_1\} \otimes \{0, 1, \ldots, J_2\} \otimes \ldots \otimes \{0, 1, \ldots, J_k\}$, $J_s := 1 + \lceil n^{\frac{\alpha_s^{-1}}{2+\alpha_*^{-1}}} \rceil$,

$$\tilde{\theta}_{\mathbf{i}} := n^{-1} \sum_{l=1}^n \frac{\Delta_l' \varphi_{\mathbf{i}}(\Delta_l' \mathbf{X}_l)}{f^{\mathbf{X},Z}(\Delta_l' \mathbf{X}_l, \Delta_l' Z_l)}. \qquad (2.4.4)$$

Then the regression estimator (2.4.3) is rate optimal and

$$\sup_{m \in S_0(\alpha_1, \ldots, \alpha_k, Q)} \mathbb{E}\Big\{ \int_{[0,1]^k} (\hat{m}(\mathbf{x}) - m(\mathbf{x}))^2 d\mathbf{x} \Big\} \le \zeta n^{-2\alpha_*/(2\alpha_*+1)}. \qquad (2.4.5)$$

We may conclude that even for the case of anisotropic multivariate regression the CSCNO does not slow down rates known in the theory of regression for direct observations.

The same conclusion holds for a controlled CSCO regression and, of course, for a CSC regression.

For small samples the E-estimator, based on the sample mean Fourier coefficient estimator (2.4.4), is recommended. Let us look at how it performs. The top diagram in Fig. 2.13 shows the CSC scattergram, and the caption explains the simulation. It is difficult to realize that there are $n = 100$ observations, it looks like the number of observations is dramatically smaller. This is because we are dealing with the bivariate regression when the predictors are spread over the unit square $[0, 1]^2$. The latter also sheds light on the familiar curse of dimensionality. Further, it is difficult to guess the underlying regression function. Indeed, can you guess the underlying regression? Please keep in mind that Z is not the response, and only knowledge of the indicator $\Delta := I(T \le Z)$ allows one to make a conclusion about the underlying T. Now let us look at how the E-estimator, based on just $n - N = 44$ CSCNO observations, performs. The three bottom diagrams show us three slices of the underlying bivariate regression and its bivariate E-estimate for specific values of y shown in the titles. The solid and dashed lines are the underlying and estimated regressions. First of all, note the shown scales. Overall, as a function in y, the regression $m(x, y)$ increases toward tails and decreases toward the center. The estimates correctly exhibit this character of the regression. The regression profile in x is shown well in the left diagram and worse in the two others.

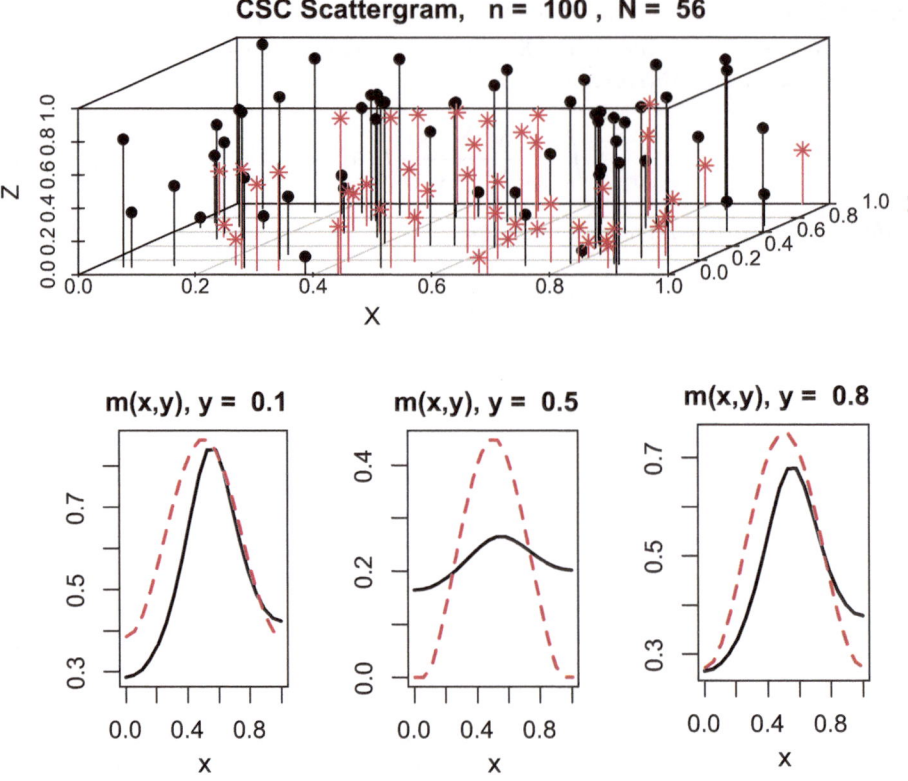

Fig. 2.13 Controlled bivariate regression $m(x, y) = \mathbb{E}\{T|X = x, Y = y\}$ based on CSCNO sub-sample. The monitoring time Z and the covariates X and Y are mutually independent Uniform(0,1) variables. The lifetime of interest T has Beta(1, $1 + f_2(X) + f_3(Y)$) distribution which yields $m(x, y) = 1/[1 + f_2(x) + f_3(y)]$. The used functions f_2 and f_3 are the Bathtub and the Normal corner functions that are controlled by the argument *set.corn*. The dots and the stars show CSCO and CSCNO observations, respectively, and $N = \sum_{l=1}^{n} \Delta_l$. The solid and dashed lines are the underlying and estimated slices of the bivariate regression for fixed values of y controlled by the argument *set.y*. [n = 100, set.corn = c(2, 3), set.y = c(0.1, 0.5, 0.8), cJ0 = 3, CJ1 = 0.8, cTH = 4]

Nonetheless, the all three diagrams correctly indicate the underlying unimodal profile. This particular regression estimate is very good keeping in mind the CSCNO regression and the small sample size. The reader is highly advised to repeat this figure and get used to the binary CSC regression.

Now we are in a position to consider several important and new for us topics in methodology of multivariate regression. They are adaptation to smoothness of continuous covariates, dimension reduction, categorical covariates, and a special setting with necessity of adding an extra covariate that makes T and Z conditionally independent but when this extra covariate is not of interest for the regression.

We begin with adaptation and dimension reduction. In general these are two different topics. The former is to match performance of the oracle who knows smoothness of the regression function. For the setting of Theorem 2.10 the adaptation means that an estimator yields the rate $n^{-2\alpha_*/(2\alpha_*+1)}$ without knowing nuisance parameters $(\alpha_1, \ldots, \alpha_k)$. Dimension reduction is when an estimator matches performance of the oracle who knows that only a subset of covariates defines the regression. To make the presentation shorter, we are considering these two problems together. Namely, the estimator should match the oracle who knows the subset of covariates that define the regression and also knows the corresponding smoothness of regression in those covariates.

It is sufficient to explain the heuristic of adaptation and dimension reduction for the case of a bivariate regression, the general case is considered absolutely similarly. Introduce the tensor-product cosine basis $\{\varphi_j(x_1)\varphi_i(x_2), j, i = 0, 1, \ldots\}$ on $[0, 1]^2$, and write down a bivariate regression as the Fourier series,

$$m(x_1, x_2) = \sum_{j,i=0}^{\infty} \theta_{ji}\varphi_j(x_1)\varphi_i(x_2), \quad (x_1, x_2) \in [0, 1]^2. \tag{2.4.6}$$

Next, we rewrite (2.4.6) as the sum of four terms,

$$m(x_1, x_2) = \theta_{00} + \sum_{j=1}^{\infty} \theta_{j0}\varphi_j(x_1) + \sum_{i=1}^{\infty} \theta_{0i}\varphi_i(x_2) + \sum_{j,i\geq 1} \theta_{ji}\varphi_j(x_1)\varphi_i(x_2). \tag{2.4.7}$$

Following the blockwise adaptation methodology explained in Sect. 2.3, we introduce blocks for Fourier coefficients θ_{j0}, θ_{0i}, and θ_{ji} with indexes $j, i \geq 1$. Blocks B_k of length L_k, defined in Sect. 2.3.2.2, are used for the first and second sums on the right side of (2.4.7). Tensor-product blocks $B_{k_1 k_2} := B_{k_1} \otimes B_{k_2}$ are used for the third sum. Let $\tilde{\theta}_{ji}$ be the Fourier estimator (2.4.4). Then, following (2.3.40), the blockwise-shrinkage oracle-estimator is

$$\tilde{m}^*(x_1, x_2) = \tilde{\theta}_{00} + \sum_{k=1}^{k_n} \frac{\sum_{j\in B_k} \theta_{j0}^2}{\sum_{j\in B_k} [\theta_{j0}^2 + \mathbb{E}\{(\tilde{\theta}_{j0} - \theta_{j0})^2\}]} \left[\sum_{j\in B_k} \tilde{\theta}_{j0}\varphi_j(x_1) \right]$$

$$+ \sum_{k=1}^{k_n} \frac{\sum_{i\in B_k} \theta_{0i}^2}{\sum_{i\in B_k} [\theta_{0i}^2 + \mathbb{E}\{(\tilde{\theta}_{0i} - \theta_{0i})^2\}]} \left[\sum_{i\in B_k} \tilde{\theta}_{0i}\varphi_i(x_2) \right]$$

$$+ \sum_{k_1,k_2=1}^{k_n'} \frac{\sum_{(j,i)\in B_{k_1 k_2}} \theta_{ji}^2}{\sum_{(j,i)\in B_{k_1 k_2}} [\theta_{ji}^2 + \mathbb{E}\{(\tilde{\theta}_{ji} - \theta_{ji})^2\}]} \left[\sum_{(j,i)\in B_{k_1 k_2}} \tilde{\theta}_{ji}\varphi_j(x_1)\varphi_i(x_2) \right]. \tag{2.4.8}$$

Here, similarly to Sect. 2.3, sequences k_n and k'_n are chosen based on the assumption about minimal smoothness of the regression.

Now let us look at the three sums in (2.4.8). If the regression $m(x_1, x_2)$ depends only on x_1, that is $m(x_1, x_2) = m(x_1)$, then the second and third sums are equal to zero. If the regression $m(x_1, x_2)$ depends only on x_2, then the first and third sums are equal to zero. Further, as we already know the oracle's blockwise shrinkage can be mimicked by statistics with accuracy preserving the oracle's rate of the MISE convergence. Accordingly, this special blockwise shrinkage allows us to solve the adaptation and dimension reduction problems.

Next we are considering the case of a *categorical covariate* Y supported on $\{0, 1, \ldots, K - 1\}$.

It is natural to incorporate the Y into a series estimator by using a discrete cosine basis on $\{0, 1, \ldots, K - 1\}$,

$$\xi_0(y) := 1, \ \xi_k(y) := 2^{1/2} \cos(\pi(2y + 1)k/(2K)), k = 1, \ldots, K - 1. \tag{2.4.9}$$

The used inner product for this basis is $\langle g_1, g_2 \rangle = K^{-1} \sum_{y \in \{0,1,\ldots,K-1\}} g_1(y) g_2(y)$.

As an example, consider a regression $m(x, y) = \mathbb{E}\{T | X = x, Y = y\}$ where X is a continuous covariate supported on $[0, 1]$ and Y is categorical. The regression can be written as the Fourier series $m(x, y) = \sum_{j=0}^{\infty} \sum_{k=0}^{K-1} \theta_{jk} \varphi_j(x) \xi_k(y)$, and the Fourier coefficients θ_{jk} can be estimated by the unbiased sample mean Fourier coefficient estimator (compare with (2.4.4))

$$\breve{\theta}_{jk} := K^{-1} n^{-1} \sum_{l=1}^{n} \frac{\Delta'_l \varphi_j(X_l) \xi_k(Y_l)}{f^{X,Y,Z}(X_l, Y_l, Z_l)}. \tag{2.4.10}$$

Then the E-estimator can be used for estimating the regression $m(x, y)$.

Finally, let us consider the following situation. Recall that the developed theory of regression estimation for the CSC is based on the assumption that the lifetime of interest T and the monitoring time Z are conditionally independent given covariate X (the case of conditionally independent CSC). If T and Z are conditionally dependent, then no consistent estimation based on a CSC sample from (X, Z, Δ) is possible. What can be done if the CSC is conditionally dependent? Let us present a possible remedy using the following example. Consider the case of two continuous covariates X_1 and X_2 supported on $[0, 1]^2$. We are interested in regression $m(X_1) := \mathbb{E}\{T | X_1\}$ but need to consider the bivariate regression $m(X_1, X_2) := \mathbb{E}\{T | X_1, X_2\}$ because T and Z are independent only given the pair (X_1, X_2). Then two possible scenarios are discussed.

First, suppose that $m(x_1, x_2) = m(x_1)$. Then we are dealing with the already discussed problem of dimension reduction. In other words, we treat the problem as a bivariate regression and use the above-described blockwise estimator that adapts to the underlying univariate dimensionality.

Second, suppose that the regression $m(x_1, x_2)$ is bivariate but we are interested in the univariate regression $m(x_1) = \mathbb{E}\{T | X_1 = x_1\}$. Assume that the CSC is controlled and hence we know the joint density $f^{X_1, X_2, Z}$. Write,

$$m(x_1) = \int_0^\infty S^{T|X_1}(t|x_1)dt$$

$$= \int_0^\infty \int_0^1 f^{X_2|X_1}(x_2|x_1) S^{T|X_1,X_2}(t|x_1, x_2) dx_2 dt. \qquad (2.4.11)$$

Fourier coefficients of the univariate regression of interest can be written as

$$\theta_j := \int_0^1 m(x_1)\varphi_j(x_1)dx_1$$

$$= \int_0^\infty \int_{[0,1]^2} f^{X_2|X_1}(x_2|x_1) S^{T|X_1,X_2}(t|x_1, x_2)\varphi_j(x_1) dx_1 dx_2 dt. \qquad (2.4.12)$$

Note that $S^{T|X_1,X_2}(t|x_1, x_2) = f^{X_1,X_2,Z,\Delta'}(x_1, x_2, t, 1)/f^{X_1,X_2,Z}(x_1, x_2, t)$ and continue (2.4.12),

$$\theta_j = \mathbb{E}\{\frac{\Delta' f^{X_2|X_1}(X_2|X_1)\varphi_j(X_1)}{f^{X_1,X_2,Z}(X_1, X_2, Z)}\} = \mathbb{E}\left\{\frac{\Delta'\varphi_j(X_1)}{f^{X_1}(X_1)f^{Z|X_1,X_2}(Z|X_1, X_2)}\right\}.$$

This formula yields the unbiased sample mean Fourier coefficient estimator

$$\tilde{\theta}_j := n^{-1}\sum_{l=1}^n \frac{\Delta'_l\varphi_j(X_{1l})}{f^{X_1}(X_{1l})f^{Z|X_1,X_2}(Z_l|X_{1l}, X_{2l})}. \qquad (2.4.13)$$

Finally, the E-estimator can be used to calculate the regression estimator of $m(x_1)$.

The described remedy is called the "increase-decrease dimensionality". The increase step is to find extra covariates that make T and Z conditionally independent. This step makes consistent estimation possible but in general slows down rate of convergence due to the increased dimensionality. The second step is to decrease the dimensionality (or we may say restore the original dimensionality) and consider only the covariates of interest.

2.5 Several More Topics

This section is devoted to a brief discussion of several problems related to the above-discussed distribution and regression estimation for the CSC.

2.5.1 Conditional Survival Function

Following the above-presented setting of a multivariate regression, let us consider estimation of a conditional survival function $S^{T|X}(t|x) := \mathbb{P}(T > t|X = x)$. Note that now we are estimating a $(1 + k)$-dimensional function in t and x. It is known from Sect. 2.2 that estimation of a survival function $S^T(t)$ based on a CSC sample from (Z, Δ) is ill-posed, no

longer the classical rate n^{-1} is achievable, and for α-fold differentiable survival function
the optimal rate of MISE convergence is $n^{-2\alpha/(2\alpha+1)}$. As we will see shortly, estimation of
a conditional survival function is also ill-posed with respect to its direct data counterpart. To
make presentation of the following proposition shorter, let us assume that T is supported on
$[0, 1]$ and that $S^{T|\mathbf{X}}(t|\mathbf{x})$, as a $(1 + k)$-dimensional function in $(t, \mathbf{x})$, belongs to a Sobolev
class $S_0(\alpha_1, \ldots, \alpha_{k+1}, Q)$ defined in (2.4.1).

Theorem 2.11 *Suppose that assumptions of Theorem 2.10 hold and $S^T(1) = 0$. Introduce
the series estimator*

$$\hat{S}^{T|\mathbf{X}}(t|\mathbf{x}) := \sum_{\mathbf{i} \in \mathcal{J}} \hat{\theta}_{\mathbf{i}} \varphi_{\mathbf{i}}(t, \mathbf{x}). \tag{2.5.1}$$

Here $\mathbf{i} := (i_1, \ldots, i_{k+1})$, $\mathcal{J} := \{0, 1, \ldots, J_1\} \otimes \{0, 1, \ldots, J_2\} \otimes \ldots \otimes \{0, 1, \ldots, J_{k+1}\}$,
$J_s := 1 + \lceil n^{\frac{\alpha_s^{-1}}{2+\alpha_*^{-1}}} \rceil$, $\alpha_* = [\sum_{s=1}^{k+1} \alpha_s^{-1}]^{-1}$, $\varphi_{\mathbf{i}}(t, \mathbf{x}) = \varphi_1(t) \prod_{s=2}^{k+1} \varphi_{i_s}(x_{i_s})$, *and*

$$\hat{\theta}_{\mathbf{i}} := n^{-1} \sum_{l=1}^{n} \frac{\Delta'_l \varphi_{\mathbf{i}}(Z_l, \mathbf{X}_l)}{f^{\mathbf{X}, Z}(\mathbf{X}_l, Z_l)}. \tag{2.5.2}$$

Then the estimator (2.5.1) is rate optimal and

$$\sup_{S^{T|\mathbf{X}} \in S_0(\alpha_1, \ldots, \alpha_{k+1}, Q)} \mathbb{E}\left\{ \int_{[0,1]^{k+1}} (\hat{S}^{T|\mathbf{X}}(t|\mathbf{x}) - S^{T|\mathbf{X}}(t|\mathbf{x}))^2 dt d\mathbf{x} \right\}$$

$$\leq \zeta n^{-2\alpha_*/(2\alpha_*+1)}. \tag{2.5.3}$$

To get a feeling of the rate and compare it with a regression setting of the previous
subsection, let us assume that all α_s are the same and equal to α, that is we are considering
Sobolev classes of $S^{T|\mathbf{X}}(t|\mathbf{x})$ and $m(\mathbf{x})$ that are α-fold differentiable with respect to each
variable. Then the optimal rate for regression is $n^{-2\alpha/(2\alpha+k)}$ versus a slower $n^{-2\alpha/(2\alpha+k+1)}$
for the conditional survival function. The difference in one dimension is explained by the
integral $m(\mathbf{x}) = \int_0^\infty S^{T|\mathbf{X}}(t|\mathbf{x})dt$. On the other hand, because $S^{T|\mathbf{X}}(t|\mathbf{x}) = \mathbb{E}\{I(T > t)|\mathbf{X} = \mathbf{x}\}$, for direct data the conditional survival is estimated with the classical rate $n^{-2\alpha/(2\alpha+k)}$.

2.5.2   Conditional Linear Functionals

We begin with several motivating examples that explain why conditional linear functionals of $S^{T|X}$ are of interest.

Recall formula (2.3.2) for the regression,

$$m(x) := \mathbb{E}\{T|X = x\} = \int_0^\infty S^{T|X}(t|x)dt. \qquad (2.5.4)$$

We also have the following formula for the regression,

$$m(x) = \int_0^\infty t f^{T|X}(t|x)dt. \qquad (2.5.5)$$

As we see, the regression can be considered as a *linear functional* of the conditional survival or of the conditional density. Similarly, $\int_0^\infty t^k f^{T|X}(t|x)dt = \int_0^\infty kt^{k-1}S^{T|X}(t|x)dt$ whenever the conditional kth moment exists.

The regression and the conditional moments are important characteristics of the lifetime of interest. Due to the dual formulas based on the conditional density and the conditional survival, which one is better for estimation based on CSC data? As we will see shortly, in general it is preferable to work with linear functionals of $S^{T|X}$ because they may be more accurately estimated.

Introduce a known and continuous on $[0, \infty)$ function ψ, and define a *restricted linear functional*

$$M_r(x) := M_r(x, \psi, S^{T|X}) := \int_0^r \psi(t)S^{T|X}(t|x)dt. \qquad (2.5.6)$$

Note that $\psi(t) = 1$ yields the restricted regression μ_r, while $\psi(t) = kt^{k-1}$, $k = 2, 3, \ldots$ are used to evaluate higher moments.

To make the following theorems shorter, we are considering CSC and CSCNO data because due to the symmetry the CSCO is considered similarly to CSCNO only with Δ' being replaced by Δ and the survival being replaced by the cdf. Further, in Assumption 2.5 we replace μ_r and $\mu_{r0}(x)$ by M_r and $M_{r0}(x) := \int_0^r \psi(t)S_0^{T|X}(t|x)dt$, respectively. To stress this change, we denote the corresponding function class (2.3.7) as $\mathcal{F}_n(S_0^{T|X}, \alpha, Q, r, \psi)$. Recall that $\Delta' := 1 - \Delta$.

Theorem 2.12 *Suppose that Assumptions 2.3–2.7 hold, and let a CSC sample of size n from* (X, Z, Δ) *be given. Then*

$$\inf_{\tilde{M}_r^*} \sup_{S^{T|X} \in \mathcal{F}_n(S_0^{T|X}, \alpha, Q, r, \psi)} \left\{ [n/d'(S^{T|X}, f^{X,Z}, r, \psi)]^{2\alpha/(2\alpha+1)} \right.$$

$$\left. \times \mathbb{E}_{S^{T|X}} \left\{ \int_0^1 (\tilde{M}_r^*(x) - M_r(x))^2 dx \right\} \right\} \geq P(\alpha, Q)(1 + o_n(1)). \qquad (2.5.7)$$

Here P is defined in (2.3.14) and

$$d'(S^{T|X}, f^{X,Z}, r, \psi) := \int_{\mathcal{R}_r} \frac{\psi^2(t)(1 - S^{T|X}(t|x))S^{T|X}(t|x)}{f^{X,Z}(x, t)} dt\, dx. \qquad (2.5.8)$$

Now suppose that a CSCNO sample of size n from $(\Delta'X, \Delta'Z, \Delta)$ *is given. Then*

$$\inf_{\tilde{M}_r^*} \sup_{S^{T|X} \in \mathcal{F}_n(S_0^{T|X}, \alpha, Q, r, \psi)} \left\{ [n/d_*'(S^{T|X}, f^{X,Z}, r, \psi)]^{2\alpha/(2\alpha+1)} \right.$$

$$\left. \times \mathbb{E}_{S^{T|X}} \left\{ \int_0^1 (\tilde{M}_r^*(x) - M_r(x))^2 dx \right\} \right\} \geq P(\alpha, Q)(1 + o_n(1)). \qquad (2.5.9)$$

Here

$$d_*'(S^{T|X}, f^{X,Z}, r, \psi) := \int_{\mathcal{R}_r} \frac{\psi^2(t)S^{T|X}(t|x)}{f^{X,Z}(x, t)} dt\, dx. \qquad (2.5.10)$$

In (2.5.7) and (2.5.9) the infimum is taken over all oracle-estimators that know the corresponding sample, $\mathcal{F}_n(S_0^{T|X}, \alpha, Q, r, \psi)$, $f^{X,Z}$ *and* ψ.

Let us present an oracle-estimator that attains the lower bound (2.5.9) and whose MISE is within a constant factor from the lower bound (2.5.7).

Theorem 2.13 *Let Assumptions 2.3 and 2.7 hold. Consider the oracle-estimator based on a CSCNO sample of size n from* $(\Delta'X, \Delta'Z, \Delta)$,

$$\bar{M}_r^*(x) := \sum_{j=0}^{J_n} \bar{\theta}_j I(\bar{\theta}_j^2 > c_{TH}\mathbb{V}\{\bar{\theta}_j\})\varphi_j(x) + \sum_{j=J_n+1}^{J_n^*} (1 - (j/J_n^*)^\alpha)\bar{\theta}_j\varphi_j(x). \qquad (2.5.11)$$

Here $J_n^* := J_n + \lceil [(n/d_*')Q\pi^{-2\alpha}(\alpha + 1)(2\alpha + 1)/\alpha]^{1/(2\alpha+1)} \rceil$, *and the Fourier coefficient estimator is*

$$\bar{\theta}_j := n^{-1} \sum_{l=1}^n \frac{\Delta' I(\Delta_l'Z_l \leq r)\psi(\Delta_l'Z_l)\varphi_j(\Delta_l'Z_l)}{f^{X,Z}(\Delta_l'X_l, \Delta_l'Z_l)}. \qquad (2.5.12)$$

The MISE of this oracle-estimator attains the lower bound (2.5.9).

We can also use the blockwise-shrinkage methodology of Section 2.3 to prove that the adaptive estimator is also efficient.

Now let us look at linear functionals of $f^{T|X}$. Set $\Psi(t) := \int_0^t \psi(u)du$, and write using integration by parts,

$$M_r(x, \Psi, f^{T|X}) = \int_0^r \Psi(t) f^{T|X}(t|x)dt$$

$$= \left[\int_0^r \psi(t) S^{T|X}(t|x)dt\right] - [\Psi(r)S^{T|X}(r|x)]$$

$$= M_r(x, \psi, S^{T|X}) - [\Psi(r)S^{T|X}(r|x)]. \tag{2.5.13}$$

Let us look at the two terms on the right side of (2.5.13). Suppose that $\Psi(r) \neq 0$ to avoid triviality. The first term is the already studied linear functional of $S^{T|X}$, and we know that it can be estimated with the univariate rate defined by smoothness of the conditional survival in x. Let b be the endpoint of the support of T. Unless $S^{T|X}(r|x) = 0$ due to $r \geq b$, the second term on the right side of (2.5.13) is the bivariate function in (r, x), and it is estimated with a slower bivariate rate.

The latter is not the only complication in estimating linear functionals of $f^{T|X}$. First, recall that adaptive nonparametric pointwise estimation triggers the Lepski's logarithmic penalty for the rate of convergence. Second, an intriguing outcome may occur if $b = b(x)$, that is when the endpoint of the support of T depends on X. In this case for some x we may have a univariate rate of estimating $M_r(x, \Psi, f^{X|T})$ and for others a bivariate rate.

We can make the following conclusion. The survival analysis uses the above-discussed linear functionals as interpretable and meaningful survival metrics. Accordingly, whenever it is possible to choose between linear functionals of $S^{T|X}$ or $f^{T|X}$, for CSC it is prudent to choose the former.

2.5.3   Fixed Size Case-CSC and Control-CSC

These are sampling models closely related to already studied ones only now the number of available CSCO observations, called the cases, and the number of available CSCNO observations, called the controls, are a priori fixed rather than being random with Binomial distribution. In other words, the sampling is Negative Binomial (also called Pascal) rather than Binomial. For instance, consider the fixed size case-CSC. Let n_1 be a priori fixed number of cases. Then the sampling from $(Z, \Delta) := (Z, I(T \leq Z))$ stops when n_1 cases $Y_1, \ldots, Y_{n_1}$ are obtained, and the total random number N_1 of underlying CSC observations is unknown. We only know that $\sum_{l=1}^{N_1} \Delta_l = n_1$, and the distribution of N_1 is Negative Binomial (Pascal) with parameters $(n_1, \mathbb{P}(\Delta = 1))$. The fixed size control-CSC is defined identically only here n_0 controls $X_1, \ldots, X_{n_0}$ are collected.

Note that the sample $Y_1, \ldots, Y_{n_1}$ of cases may be considered as a sample of size n_1 from the lifetime Y with the distribution equal to the conditional distribution of Z given $\Delta = 1$. Accordingly, using Assumption 2.2.1 and formula (2.1.1) we get for the case-CSC,

$$f^Y(y) = f^{Z|\Delta}(y|1) = \frac{f^Z(y)F^T(y)}{\mathbb{P}(\Delta = 1)}, \qquad (2.5.14)$$

and for the control-CSC we get

$$f^X(x) = f^{Z|\Delta}(x|0) = \frac{f^Z(x)S^T(x)}{\mathbb{P}(\Delta = 0)}. \qquad (2.5.15)$$

If we compare (2.5.14) with (2.1.1), then it becomes clear that, in terms of the theory of estimating the distribution of T, the key difference between the CSCO and the fixed size case-CSC is that the former allows us to estimate the probability $\mathbb{P}(\Delta = 1)$ by the sample mean estimate $n^{-1} \sum_{l=1}^{n} \Delta_l$, while the latter, if only the sample $(Y_1, \ldots, Y_{n_1})$ is available, precludes us from estimating $\mathbb{P}(\Delta = 1)$ and hence from consistent estimation of the distribution of Y.

If $\mathbb{P}(\Delta = 1)$ is known, then modification of the corresponding CSCO and CSCNO efficient estimators is straightforward. For instance, consider estimation of the cdf F^T for the fixed size case-CSC. Assume that T is supported on $[0, 1]$. Then we may use the sample mean Fourier coefficient estimator

$$\tilde{\theta}_j := \mathbb{P}(\Delta = 1) \Big[n_1^{-1} \sum_{l=1}^{n_1} \frac{\varphi_j(Y_l)}{f^Z(Y_l)} \Big]. \qquad (2.5.16)$$

This is unbiased estimator of the Fourier coefficient $\theta_j := \int_0^1 \varphi_j(t) F^T(t) dt$.

The probability $\mathbb{P}(\Delta = 1)$ is typically unknown, and then in general an additional information is needed for consistent estimation of the distribution of T. Nonetheless, the Fourier coefficient estimator (2.5.16) sheds light on an attractive possibility to estimate the so-called shape of a curve of interest. Indeed, if we check formula (2.5.16), then we may notice that all Fourier coefficients are proportional to the probability $\mathbb{P}(\Delta = 1)$. By definition, the *shape* of a curve is the underlying curve multiplied by some factor whose value is not of interest to us. Analysis of the shape may be useful if we are interested in modes, symmetry, or a relative rate of growth. As a result, even if $\mathbb{P}(\Delta = 1)$ is unknown, we can estimate shape of the curve. Of course, similarly to the previously discussed CSCO and CSCNO, we also need to know or be able to estimate the underlying density f^Z.

2.5.4   Doubly Current Status Censoring

In the doubly current status censoring (DCSC) problem we are interested in the distribution of the lifetime of interest T between two events that occur sequentially and not observed directly. Instead we observe a sample from

$$(Z, \Delta) := (Z, I(Y \leq Z)), \text{ where } Y = T + W, \tag{2.5.17}$$

and W is a nuisance random lifetime. Examples of the DCSC are the time T to divorce after marriage at age W, or what time T (how long) it takes to get sick after exposure at time W to a virus. To shed light on DCSC, consider the following formula,

$$f^{Z,\Delta}(z, 1) = f^Z(z)\mathbb{P}(T + W \leq z | Z = z)$$

$$= f^Z(z) \int_0^z F^{T|Z,W}(z - w|z, w) dF^{W|Z}(w|z). \tag{2.5.18}$$

This formula explains complexity of DCSC, and it also sheds light on a possible solution for some settings studied in the DCSC literature. For instance, consider a DCSC model where T and $Z - W$ are mutually independent and W is observed. In this case the induced monitoring time of T is $Z - W$, and we have a sample from $(Z - W, I(T \leq Z - W))$. As a result, the DCSC problem becomes the already considered CSC problem with $Z - W$ being the induced (and observable) monitoring time. If W is not observed but its distribution is known and the random variables W, T and Z are mutually independent, then (2.5.18) implies that we need to solve the deconvolution problem. It is known that this deconvolution problem is ill-posed even if $Y = T + W$ is observed directly. The problem becomes even more challenging when the assumption of independence is relaxed and no longer nuisance distributions are known. Then an extra information is needed for consistent estimation. Another often considered DCSC model is when a sample from $(\Delta^*, \Delta^{**}, Z)$ is available where $\Delta^* := I(W \leq Z)$ and $\Delta^{**} := I(Y \leq Z)$.

It is an open, challenging and practically important problem to develop theory and methodology of efficient estimation for DCSC data.

2.6   Exercises

The asterisk denotes a more difficult exercise.

2.1.1 What is the current status censoring (CSC)? How does it modify the underlying direct sample from the lifetime of interest?

2.1.2 What is the monitoring time Z? What is the status Δ of the event of interest? Explain and write down a corresponding formula for the status.

2.1.3 CSC creates two complementary subsamples CSCO and CSCNO. Define them and explain.

2.1.4 Is the lifetime of interest observed under CSC?

2.1.5* Can the CSCO and CSCNO be explained as a missing CSC? If the answer is "yes", then is the missing mechanism MCAR, MAR or MNAR?

2.1.6 What is the problem of aggregating CSCO and CSCNO observations?

2.1.7 Consider Examples 1–4 and explain why the studied sampling mechanisms yield CSC.

2.1.8 Definitions of dependent and independent CSC. Complement the definitions by discussion of corresponding examples.

2.1.9* Verify formula (2.1.1).

2.1.10* Verify formula (2.1.2).

2.1.11 Why does formula (2.1.2) imply that the available monitoring times in a CSCO subsample or CSCNO subsample are biased with respect to the underlying monitoring times?

2.1.12* Suppose that the density f^Z of the monitoring time is unknown. Can either CSCNO or CSCO subsample be used for consistent estimation of f^Z?

2.1.13 Why is it possible for many applied problems to assume that the lifetime of interest T is supported on [0, 1]? Hint: Think about a bounded lifetime and then propose a corresponding transformation.

2.1.14 Explain the diagrams in Fig. 2.1.

2.1.15 Repeat Fig. 2.1 using different sample sizes. Explain how the size n affects the CSC, CSCO and CSCNO histograms and the corresponding estimates.

2.1.16 Use Fig. 2.1 and find optimal arguments of the E-estimator.

2.1.17 What is the difference between Δ and Δ'?

2.2.1 What is the relationship between the cumulative distribution function, the survival function and the density?

2.2.2 Explain all elements of the local Sobolev class (2.2.1).

2.2.3 What is a coefficient of difficulty? What do coefficients of difficulty (2.2.3) tell us about CSC, CSCO and CSCNO?

2.2.4 Explain the lower bound (2.2.4) and the underlying assumptions.

2.2.5* The lower bound (2.2.4) holds for oracle-estimators. Can it be used as a lower bound for estimators? Prove your answer. Hint: It is insufficient to say that the oracle knows more than the statistician. A proof should be presented. For instance, begin with the assumption that the lower bound for an estimator is smaller than the right side of (2.2.4).

2.2.6* For direct observations of T the optimal rate of the estimating F^T is the parametric n^{-1}. What is the rate for CSC? Give example of another nonparametric estimation problem with the same rate of the MISE convergence.

2.2.7 Consider density estimation problem. What are the optimal rates for direct and CSC samples? Describe complexity of the CSC problem.

2.2.8* Verify (2.2.5)–(2.2.7).

2.2.9 Explain the Fourier coefficient estimators (2.2.9) and (2.2.10).

2.2.10* Calculate variances of Fourier coefficient estimators (2.2.9) and (2.2.10).

2.2.11* Prove Lemma 2.1.

2.2.12* Explain the underlying idea of the Fourier coefficient estimator (2.2.12). Hint: Recall the corresponding discussion about efficient estimation for nonparametric regression. Use (2.2.13).

2.2.13 Explain how the estimator (2.2.12) aggregates CSCO and CSCNO observations.

2.2.14 Explain the estimator (2.2.16). Is it unbiased?

2.2.15 Explain the estimator (2.2.17). Is it unbiased?

2.2.16* Evaluate variances of the estimators (2.2.16) and (2.2.17).

2.2.17 Explain how the estimator (2.2.19) is constructed.

2.2.18* Propose an estimator of θ_j based on a linear unbiased aggregation of estimators (2.2.16) and (2.2.17).

2.2.19 Let the distribution of Z be unknown. In this case consistent distribution estimation, based on CSCO or CSCNO, is impossible. Suggest a remedy and explain it. Hint: Use Theorem 2.3.

2.2.20 Describe results shown in Fig. 2.2.

2.2.21 Use Fig. 2.3 to get experience in CSC distribution estimation. Consider several different sample sizes and try to find better parameters of the E-estimator.

2.2.22* Explain how the estimates, shown in Fig. 2.4, are constructed. Then repeat the figure using several different arguments nE. Do the results support the theory or the sample sizes are too small? Try to find minimal sizes of the extra samples yielding satisfactory outcomes.

2.2.23 Present an example of dependent CSC.

2.2.24 Explain formula (2.2.20).

2.2.25 What can be estimated for dependent CSC?

2.2.26 Explain results shown in Fig. 2.5.

2.2.27 Repeat Fig. 2.6 for two different corner functions and comment on the results.

2.2.28 Verify formula (2.2.22).

2.2.29 Prove (2.2.23). Hint: Review the notion of conditional probability.

2.2.30 Explain the underlying idea of estimator (2.2.25). Hint: Begin with verification of formula (2.2.24).

2.2.31 Repeat Fig. 2.7 five times and comment on the outcomes.

2.2.32* Have a guess about the underlying simulation used for Fig. 2.7. Hint: Check your guess using the R function ch2(Fig=7).

2.3.1 Explain the classical regression estimation problem for directly observed and CSC responses.

2.3.2 Can you visualize the regression curve in Fig. 2.8?

2.3.3 Define the conditional survival function.

2.3.4 What is the difference between regression and restricted regression? Why is it worthwhile to study restricted regression for CSC responses?

2.3.5 What is the difference between the status Δ and Δ'?

2.3.6* Explain the Assumptions 2.3–2.5.

2.3.7 What is the optimal rate of the MISE convergence for nonparametric regression? Compare with optimal rate for the survival function estimation.

2.3.8 Write down the coefficient of difficulty (2.3.15) and explain its behavior as $r \to \infty$. What is a sufficient condition for consistent estimation of the regression?

2.3.9* Use Theorem 2.5 and explain how CSCO and CSCNO affect regression estimation. Then compare with CSC regression.

2.3.10* Explain why it is of interest to study both local and global Sobolev function classes.

2.3.11 Prove inequality (2.3.20).

2.3.12* Prove inequality (2.3.23). Hint: Use definition of the Sobolev function class.

2.3.13 Explain why (2.3.24) does not follow from (2.3.23).

2.3.14* Verify inequality (2.3.24).

2.3.15 Verify relations (2.3.28).

2.3.16 Explain the underlying idea of the Fourier coefficient estimator (2.3.29).

2.3.17 Prove relations (2.3.30).

2.3.18 Verify (2.3.31).

2.3.19 Explain all steps in establishing (2.3.32).

2.3.20 Why is the Fourier coefficient estimator (2.3.33) feasible for CSCO?

2.3.21 Verify relations (2.3.34).

2.3.22* Explain the underlying idea of the Fourier coefficient estimator (2.3.37).

2.3.23 Can the relation (2.3.38) be used as a justification for the aggregation (2.3.37) of CSCO and CSCNO observations?

2.3.24* Propose an unbiased linear aggregation of Fourier coefficient estimators, based on CSCO and CSCNO subsamples, for CSC.

2.3.25* Explain the underlying idea of blockwise adaptation to parameters (α, Q) of the Sobolev class, that is adaptation to unknown smoothness of the regression function. What is the role of Sobolev functionals?

2.3.26 Definition of restricted regression. Explain why this notion is introduced for CSC responses.

2.3.27 What is the CSCNO statistical model? Explain it via a sampling model and a missing mechanism.

2.3.28* Present a justification for the Fourier coefficient estimator (2.3.42).

2.3.28* Explain the underlying motivation of statistics (2.3.43) and (2.3.44).

2.3.29 How is the cutoff k_n, used in (2.3.45), chosen?

2.3.30 Explain the low-frequency and high-frequency components of the restricted regression estimator (2.3.45).

2.3.31* Propose the analog of Theorem 2.6 for CSCO.

2.3.32 Explain why the efficient estimator for CSCNO is not efficient for CSC.

2.3.33 Why do we estimate the conditional survival function $S^{T|X}$ for efficient CSC regression estimation? Hint: There is a need to aggregate CSCO and CSCNO observations.

2.3.34* Find the mean and variance of the Fourier coefficient estimator (2.3.48). Make the corresponding conclusions about statistical properties of the estimator.

2.3.35 Why is statistic (2.3.47) called U-statistic? What is its kernel?

2.3.36* Calculate expectation of the statistic (2.3.49). Then explain its role in the blockwise-shrinkage adaptation.

2.3.37 Explain low-frequency and high-frequency components of the nonparametric estimator (2.3.51).

2.3.38 What is the theoretical conclusion of Theorem 2.7? Hint: Think about efficiency and compare the lower and upper bounds.

2.3.39 Describe complexity of regression with unbounded lifetime of interest.

2.3.40 Explain when the condition (2.3.53) may hold.

2.3.41* Calculate the mean and variance of the Fourier coefficient estimator (2.3.54).

2.3.42* Calculate the MISE of estimator (2.3.55). Then establish validity of Theorem 2.8.

2.3.43 Formulate and then comment on the assumptions of Theorem 2.8.

2.3.44 Verify the assertion of Corollary 2.1.

2.3.45 Explain the heuristic of Theorem 2.9. Hint: Recall the underlying idea of the blockwise-shrinkage adaptation.

2.3.46 Compare rates of the MISE convergence for the density and regression estimation for CSC data.

2.3.47 Definition of an observational study. Present several examples when the design density is unknown.

2.3.48 What is the restriction on smoothness under Assumption 2.8?

2.3.49 Explain the density estimator (2.3.59).

2.3.50 Explain the assumption $\alpha \geq \alpha_0 = 2$ in Theorem 2.9.

2.3.51 Repeat Fig. 2.9 at least five times and present your analysis of the estimator.

2.3.52 Consider different sample sizes and corner functions in Fig. 2.9. Then suggest optimal arguments of the E-estimator.

2.3.53 Repeat the previous assignment only now using Fig. 2.10.

2.3.54 Repeat the previous assignment only now using Fig. 2.11.

2.3.55 Explain the data and the E-estimate shown in Fig. 2.12.

2.4.1 Discuss the assumption (2.4.2). Present an example when it does not hold.

2.4.2* Prove that the Fourier coefficient estimator (2.4.4) is unbiased. What is its variance?

2.4.3 Consider a multivariate regression. What is the difference, if any, between adaptation and dimension reduction?

2.4.4 Can the blockwise-shrinkage adaptation perform dimension reduction? If the answer is "yes", then explain how.

2.4.5 Explain all components of the regression estimator (2.4.8).

2.4.6 How can a categorical predictor be incorporated into a series nonparametric regression?

2.4.7 Is the Fourier coefficient estimator (2.4.10) unbiased?

2.4.8* Explain how an extra covariate can provide conditional independence between Z and T. Hint: Begin with the necessity of the independence.

2.4.9* Propose aggregation of CSCO and CSCNO observations for bivariate regression.

2.4.10 Repeat Fig. 2.13 ten times and evaluate quality of estimation.

2.4.11* Using Fig. 2.13, propose optimal parameters of the estimator.

2.4.12 Choose a different set of corner functions and analyze performance of the estimator using Fig. 2.13.

2.5.1 Definition of the conditional survival function.

2.5.2 Does conditional survival function or regression better characterize the relationship between the response and predictor?

2.5.3 What is the conclusion of Theorem 2.11?

2.5.4* Find the mean and variance of the Fourier estimator (2.5.2).

2.5.5 Present examples of conditional linear functions.

2.5.6 Explain the assumption and the assertion of Theorem 2.12.

2.5.7 Explain the motivation behind the estimators (2.5.11) and (2.5.12).

2.5.8. Verify (2.5.13).

2.5.9. Why does (2.5.13) imply worse quality of estimating linear functionals of $f^{T|X}$ than linear functionals of $S^{T|X}$?

2.5.10 You need to compare two populations using either linear functionals of conditional densities or linear functionals of conditional survivals. What is your choice? Present an example.

2.5.11 Why are linear functionals of interest in statistical inference? Hint: Think about a test to compare two populations.

2.5.12 What is fixed size case-CSC?

2.5.13 What is fixed size control-CSC?

2.5.14 Verify formula (2.5.14).

2.5.15 Find the mean and variance of estimator (2.5.16).

2.5.16 Explain the doubly CSC sampling model.

2.5.17 Verify and then explain (2.5.18).

2.5.18 When does DCSC become CSC?

2.5.19* Propose a consistent estimator of S^T for DCSC.

2.7   Notes

We begin with a brief summary of results devoted to optimal rates for direct and CSC samples of size n. Then notes for the sections will be presented.

To discuss rates, assume that a function of interest has α derivatives in each variable. We begin with estimation of a population mean $\mu := \mathbb{E}\{T\}$ and a survival function $S^T(t) := \mathbb{E}\{I(T > t)\}$. Note that the mean is a parameter and the survival function is a univariate function. Nonetheless, for direct observations of T the rate of their estimation is the same n^{-1} because the estimands are expectations of the observable variables T and $I(T > t)$, respectively. For CSC the situation changes, and while the population mean is still estimated with the parametric rate n^{-1}, the survival function S^T is estimated with the rate $n^{-2\alpha/(2\alpha+1)}$ which is the same as for estimation of an α-fold differentiable univariate regression $m(x) := \mathbb{E}\{T|X = x\}$ based on direct observations of the pair (X, T). Relation $\mu = \int_0^\infty S^T(t)dt$

sheds light on why the parametric rate for μ is preserved, and the fact that the indicator $I(T > t)$ is no longer directly observed explains "return to normality" in estimation of a univariate survival function $S^T(t)$. An insightful theory of estimating functionals can be found in van der Vaart (1991).

Now let us review results on nonparametric estimation of a k-variate regression function $m(\mathbf{x}) := \mathbb{E}\{T|\mathbf{X} = \mathbf{x}\}$ and a corresponding conditional survival function $S^{T|\mathbf{X}}(t|\mathbf{x}) := \mathbb{E}\{I(T > t)|\mathbf{X} = \mathbf{x}\}$, $\mathbf{x} := (x_1, \ldots, x_k)$. The regression function is k-variate, the conditional survival function is $(k + 1)$-variate, but for direct observations they are estimated with the same optimal k-dimensional rate $n^{-2\alpha/(2\alpha+k)}$. The fact that conditional survival is the conditional expectation of the observed indicator $I(T > t)$ explains the result. For CSC the outcome changes. Because the indicator $I(T > t)$ is no longer observed, for the survival function the rate slows down to the classical $(k + 1)$-dimensional $n^{-2\alpha/(2\alpha+k+1)}$, and note that the rate "fits" the dimensionality of $S^{T|\mathbf{X}}$. Formula $m(\mathbf{x}) = \int_0^\infty S^{T|\mathbf{X}}(t|\mathbf{x})dt$ explains the faster rate $n^{-2\alpha/(2\alpha+k)}$ for CSC regression, and note that the integral effectively performs one dimension reduction with respect to the conditional survival. Let us also stress that while there exists the asymptotic equivalence between density and regression settings for direct data, see Efromovich (1999a), there is no such equivalence for CSC data. In other words, the CSC limits the classical theory of asymptotic equivalence between density and regression problems.

These are interesting and practically important phenomena of CSC. Let us also add that presented in the chapter sharp constants shed additional light on CSC via explanation of how an underlying estimand, together with nuisance functions, affect efficient estimation.

Now we are in a position to comment about the sections.

2.1 Current status censoring (CSC), also known as "case I" interval censoring, is a well-known problem in survival analysis. The effect of CSC on nonparametric estimation is dramatically more severe than the one caused by a classical right-censoring because CSC slows down rates of convergence for classical risks, may make an estimation problem ill-posed, and for many settings it even prevents consistent estimation unless an additional information is provided. As a result, it is prudent to know what quality of estimation may be achieved for CSC data and what, if needed, additional information should be obtained for consistent estimation. A nice exposition can be found in books Sun (2007), Chen, Sun and Peace (2012), Sun and Zhao (2013), Groeneboom and Jongbloed (2014), Klein et al. (2014), and James et al. (2023). Interesting examples and a discussion can be also found in Ma and Kosorok (2005), Becker, Braun and White (2017), Li, Wang and Sun (2017), van Es and Graafland (2017), Diao and Yuan (2018), Groeneboom and Hendrickx (2018), Li et al. (2019), Malov (2019), Cui et al. (2023), Cui, Hanning and Kosorok (2024), Koley and Dewanji (2024), Zhou and Wong (2024) where further references may be found. That literature is devoted primarily to biostatistical applications. Another cluster of CSC applications is in econometrics when the interest is in developing choice models for individual and household behavior (Nobel Prize in Economic Sciences in 2000), see Cosslett (1983), Manski (1988), Manski and McFadden (1981), Windmeier (1995), Horowitz and Savin (2001), Train (2009).

Discussion of R packages can be found in Anderson-Bergman (2017).

Let us mention several interesting and practically important extensions of the considered CSC setting. There is a more general sampling with *interval censoring*. The simplest is the interval censoring case II when one observes a sample from a triplet (L, U, Δ). Here $L < U$ are two monitoring times and the status $\Delta = -1$ if $X \leq L$, $\Delta = 0$ if $L < X \leq U$, and $\Delta = 1$ otherwise, see Wellner (1995), Birge (1999), Becker, Braun and White (2017), Gomez et al. (2009) and Gill and Levit (1995). No results, matching the presented ones for the CSC, are known for the interval censoring. Missingness is a another typical complications in CSC. A thorough discussion of different missing mechanisms in survival analysis can be found in Efromovich (2018). The setting with dependent lifetime of interest and monitoring time (dependent CSC) is referred to as "intriguing" in Jewell and van der Laan (2004a), and see also Ma, Hu and Sun (2015), Li, Wang and Sun (2017) where further references may be found. Let us also mention the *cure models*. It often happens that a certain fraction of events of interest never occur. For instance, some patients after cancer surgery are completely cured. Then the corresponding lifetimes of interest are considered as infinite. Survival models that take this feature into account are commonly referred to as cure models, see comprehensive reviews in Peng and Taylor (2014) and Amico and Van Keilegom (2018).

Finally, let us stress that a CSC sampling may be dramatically simpler and cheaper than the direct sampling of a lifetime of interest, but it may make the data ill-posed. Accordingly, a prudent cost-effectiveness analysis is the must.

2.2 The CSC literature on distribution estimation is primarily devoted to the univariate cumulative distribution function (cdf). In van de Geer (1993) it is established that, under a mild assumption on differentiability of the cdf, it may be estimated pointwise with the optimal rate $n^{-2/3}$ by a nonparametric maximum likelihood estimator. Minimal assumptions, for this rate to be optimal, can be found in Gill and Levit (1995). Note that the rate is dramatically slower than n^{-1} for the case of direct observations. Another seminal paper, devoted to estimation under minimal assumptions, is Birge (1999) where a piecewise constant (histogram-type) estimator of cdf is proposed. Yang (2000) studied locally linear smoothers. Spline methods were introduced in Kooperberg and Stone (1992). A novel kernel method was proposed in Groeneboom and Ketelaars (2011). Chagny (2015) explored warped adaptation for a kernel estimator which was motivated by Goldenshluger–Lepskii procedure which yielded the squared-bias and variance trade-off. Log-concave constraint was proposed in Anderson-Bergman and Yu (2016). Bootstrapped confidence bands are developed in Gro and Hend (2017). A number of interesting and thought-provoking papers are devoted to orthogonal series estimation. In Brunel and Comte (2009) a rigorous analysis of a so-called quotient estimator is performed. The underlying idea is to write the cumulative distribution function of interest as a ratio of two densities of directly observed random variables, recall (2.2.20). Then each density is estimated via a series projection estimator with cutoffs chosen via minimization of a penalized contrast function. It is shown that the adaptive estimator attains optimal nonparametric rates whenever smoothnesses of the two densities are the same. A regression-type estimator was also explored. Further development and literature review on series estimation can be found in Bouaziz, Brunel and Comte (2019) where both compact and non compactly supported bases are considered. Plancade (2013) used penalization for a projection series estimator of a conditional cumulative distribution function.

Density estimation problem for CSC data is less explored. Let us mention Braun, Duchesne and Stafford (2005) where the authors consider a kernel density estimation via solving iterative equations, nonparametric maximum likelihood estimation, and also via a local EM approach. The latter requires an explicit solution of the local likelihood equations which is done via the symbolic Newton?Raphson algorithm. Convergence of proposed algorithms is studied. In Becker, Braun and White (2017) data sharpening is proposed to increase robustness of a kernel density estimator to bandwidth misspecification and measurement errors. Groeneboom, Jongbloed and Witte (2010) used a maximum smoothed likelihood approach and a smoothing the (discrete) MLE of the distribution function approach. In particular, under assumption that the cumulative distribution function is three-fold differentiable, the density is estimated with the rate $n^{-4/7}$. Asymptotic distribution of the estimate was further explored in Gro and Ket (2011). In Van Es and Graafland (2017) a kernel estimator is proposed under an assumption that the density of interest is twice differentiable and the density of monitoring time is three-fold differentiable. A smart procedure of data transformation is used to convert the problem into deconvolution. Then the optimal rate $n^{-4/7}$ is achieved, and expansions of the expectation and variance as well as asymptotic normality are derived. This paper also contains a nice literature review.

Presented in the section results are based on Efromovich (2021a, 2021b; 2022b) where discussion of different aggregation methods and the practical examples can be found.

An interesting topic for future research is analysis of multivariate CSC, see a discussion in Wu and Zhang (2012).

2.3 CSC regression is a well-known problem in survival analysis, see a discussion in books Jewell and van der Laan (2002), Chen, Sun and Peace (2012), Sun and Zhao (2013), Groeneboom and Jongbloed (2014), Klein et al. (2014), Efromovich (2018), as well as reviews in Grummer-Strawn (1993), Sun (2005), Becker, Braun and White (2017), Li, Wang and Sun (2017), van Es and Graafland (2017), Diao and Yuan (2018), Li et al. (2019), Malov (2019). Differentiable functionals were theoretically studied in Van der Vaart (1991), and the theory pointed upon regular and irregular convergence rates for CSC observations. As we have seen, similar phenomena exist for adaptive estimation of nonparametric curves under MISE criterion. Proportional and additive hazards are popular models discussed in Huang (1996) and Feng and Chen (2018). The accelerated failure time model is explored in Rabinowitz, Tsiatis and Aragon (1995). The proportional odds regression is studied in Rosinni and Tsiatis (1996). Additive hazard regression is discussed in Lin, Oakes and Ying (1998). EM methodology is explored in McMahan, Wang and Tebbs (2013). Theory and methodology for linear regression can be found in Li and Zhang (1998), Murphy, van der Vaart and Wellner (1999) and Shen (2000). Hazard regression is explored in Cai and Betensky (2003). An estimator of regression parameters in the accelerated failure time model by inverting a Wald-type test for testing a null proportional hazards model is proposed in Tian and Cai (2006). Study of a semiparametric probit model and its applications can be found in Liu and Qin (2018) and Du, Hu and Sun (2019). Model with varying-coefficient partially linear proportional odds is investigated in Lu, Wu and Lu (2019). Theory of semiparametric linear regression is developed in Groeneboom and Hendrickx (2018) where asymptotically normal estimate is proposed, see also Xu et al. (2019). Nonparametric regression of the status on the predictor is explored in Honda (2004) where a modified maximum rank correlation estimator is proposed. There is a relatively large literature devoted to sieve maximum likelihood regression, see a discussion and reviews in Xue, Lam and Li (2004), Ma, Hu and Sun (2015), Zhou, Hu and Sun (2017). Review of recent results in linear quantile regression can be found in Choi et al. (2024).

The first use of the restricted mean survival time (RMST) for analysis of tumourless life was in Irwin (1949), and the enlightening discussion of the approach can be found in Susarla and Van Ryzin (1980). In Karrison (1987) estimation of RMST for right-censored data with available covariates is investigated, and this paper pioneered the methodology of restricted regression. In particular, that paper presents an interesting discussion of advantages of the restricted regression with respect to other regression models and the Cox's model in particular. For now there is a relatively large literature devoted to these functionals, and reviews can be found in Huang and Wellner (1995), Qui et al. (2019), Zhang, Wu and Yin (2020), Gardiner (2021), Han and Jung (2022), and Zhong and Schabel (2022). It is an open and interesting problem to consider a restricted regression with the restriction r being a function of the predictor.

To shed light on Assumption 2.3, recall that it is stressed in the review Jewell and van der Laan (2004) that "... the monitoring time is almost always assumed independent of the lifetime of interest." Also recall that independence between X and Z is the necessary assumption for consistent estimation of the distribution of T based on a sample from (Z, Δ), see Efromovich (2021a).

The focal point of the blockwise-shrinkage adaptation is the classical Sobolev functional $\Theta := L^{-1} \sum_{j \in B} \theta_j^2$. The theory of estimating Sobolev functionals and the blockwise adaptation is well developed and its discussion can be found in Efromovich (1985, 1999a, 2018), Hoffmann and Lepski (2002). Of course, there is a number of other procedures for adaptation proposed in the literature, but they are primarily concerned with rate optimal adaptation. The interested reader can find reviews in Efromovich (1996a,1999a), Hoffmann and Lepski (2002), Wasserman (2006), Horowitz (2009), Lepski (2022).

The presented theory and some practical examples are based on Efromovich (2024c).

An interesting topic for future research is to consider left truncated CSC data, see Efromovich (2018) and Lu et al. (2023).

2.4 A discussion of multivariate Sobolev classes can be found in Nikolskii (1975), Efromovich (1999a) and Hoffmann and Lepski (2002). For the case of direct data, the theory of efficient estimation for isotropic Sobolev classes is due to Efromovich (2000b). Rate optimal estimation for anisotropic Sobolev classes is pioneered in Hoffmann and Lepski (2002), and efficient estimation for bivariate anisotropic Sobolev classes with mixed variables in Efromovich (2011c). The presented results are based on Efromovich (2024e). An interesting new topic is efficient estimation for multivariate CSC, see Liu and Qin (2018).

2.5 A nice discussion of the case-CSC and control-CSC may be found in Jewell and van der Laan (2004a, 2004b), Vandenbroucke and Pearce (2012), Chan (2013), Keogh and Cox (2014), Klein et al. (2014), Breslow and Crowley (1974), Hsu, Gorfine and Zucker (2018). Interesting practical examples and an excellent discussion of the Doubly CSC can be found in Rabinowitz and Jewell (1996), van der Laan, Bickel and Jewell (1997), Jewel and van der Laan (2004a), Wang, Tong and Sun (2018), Li et al. (2019) and Malov (2019). Statistical methods used for solving deconvolution problems are discussed in the book Meister (2009). This is an important and open topic to propose efficient estimators for the Doubly CSC.

Right-Censoring

3

In survival analysis it is often the case that a lifetime of interest is right-censored by a random censoring lifetime. Under right-censoring, if the censoring lifetime is larger than the lifetime of interest then the lifetime of interest is observed, and otherwise the censoring lifetime is observed. Accordingly, a right-censored sample consists of two complementary subsamples containing observations of the lifetime of interest and observations of the censoring lifetime. One of the aims of this chapter is to shed light on how these subsamples may be used for efficient estimation, and then get experience in statistical analysis of small samples with high rate of censoring.

The content of the chapter is as follows. Introduction to the right-censoring can be found in Sect. 3.1. The classical problem of survival function estimation is considered in Sect. 3.2. Here the Kaplan-Meier and the method of moments estimators are discussed. Section 3.3 is devoted to density estimation.

Hazard rate is another classical characteristic of a lifetime, and its nonparametric estimation is explained in Sect. 3.4. Nonparametric regression with right-censored response or right-censored predictor is explored in Sect. 3.5. Nonparametric estimation of conditional distributions is considered in Sect. 3.6. Interesting and practically important topic of estimating bivariate survival function and bivariate density is studied in Sect. 3.7.

A classical in survival analysis problem of estimating the mean residual life and the bivariate mean residual life is considered in Sect. 3.8. Exercises are placed in Sect. 3.9, and the literature review and topics for future research can be found in the Notes.

© The Author(s), under exclusive license to Springer Nature Switzerland AG 2026   115
S. Efromovich, *Survival Analysis*, Synthesis Lectures on Mathematics & Statistics,
https://doi.org/10.1007/978-3-031-82814-0_3

3.1   Uncensored and Censored Observations in Right-Censored Data

Consider a continuous random lifetime of interest T which is right-censored (RC) by a continuous random censoring lifetime C. Then, instead of a direct sample from T, we get a sample from pair

$$(V, \Delta) := (\min(T, C), I(T \leq C)), \tag{3.1.1}$$

where $I(\cdot)$ is the indicator function and Δ is called the *indicator* of censoring. Accordingly, the right-censoring partitions data into *uncensored observations* when we observe realizations of T and *censored observations* when we observe realizations of C. We may also say that there are two complementary subsamples of uncensored and censored observations.

Uncensored observations allow us to observe the lifetime of interest, while censored observations are realizations of the censoring lifetime. This creates a dramatic difference between the two complementary subsamples and explains the traditional opinion about dominance of uncensored observations. Another general remark is that the censored subsample is identical to the CSCNO subsample studied in Chap. 2 if the C is considered as the monitoring time.

Let us also note that the following symmetry in the right-censoring. While C right-censors T, the T right-censors C and then the only change in the model is that $\Delta' := 1 - \Delta$ is used in place of Δ. This is a useful fact to know. Indeed, suppose that a good estimator of the survival function S^T of the lifetime T is proposed, then replace observations of Δ by observations of Δ' and get an estimator of S^C.

The main assumption about right-censoring, used in this chapter, is as follows.

Assumption 3.1 The lifetime of interest T and the censoring lifetime C are nonnegative and independent continuous random variables.

Given this assumption, we can present the following useful formula that sheds light on the right-censoring. Recall notations f^X and S^X for the density and survival function of a continuous random variable X. Using these notations, the joint mixed density (likelihood) of (V, Δ) can be written as

$$f^{V,\Delta}(t, \delta) = [f^T(t) S^C(t)]^\delta [f^C(t) S^T(t)]^{1-\delta}, \quad (t, \delta) \in [0, \infty) \times \{0, 1\}. \tag{3.1.2}$$

Note the symmetry in the formula with respect to T and C, it reflects the fact that C right-censors T and T right-censors C.

The survival analysis literature treats the uncensored subsample as the dominant one. For instance, the classical Kaplan-Meier estimator of the survival function has jumps only at uncensored observations of T, and moreover the pioneering paper Kaplan and Meier (1958) refers to censored observations as the "loss". Further, a large portion of statistical literature treats censored observations as "missing", and then recommends the Buckley-James imputation of censored observations by statistics based on uncensored observations.

The aim of the chapter is twofold. First, to develop the theory and methodology of efficient estimation for a large spectrum of classical survival analysis problems including estimation of the density, hazard rate and regression. Second, it is of interest to understand how uncensored and censored subsamples may be used for efficient estimation, and when and how these supplementary subsamples should be aggregated. Of course, in general both these subsamples are needed for consistent estimation. Accordingly, the oracle's approach is used. The oracle knows distribution of the censoring variable C and may use uncensored and censored observations separately to answer the raised question. Then the recommended oracle's estimators are mimicked by corresponding data-driven nonparametric estimators based on an estimated distribution of C. At the same time, if the distribution of C is known, then the oracle's approach can be used directly.

Let us stress that it is of a special interest to consider problems of density and regression estimation together because, as we will see shortly, conclusions of the asymptotic theory for these problems are different. Namely, for density no aggregation is needed for asymptotically efficient estimation while for regression the aggregation is always beneficial. At the same time, conclusions of numerical studies for small samples and high rates of censoring coalesce in terms of feasibility of aggregation for both density and regression estimation problems.

Finally, let us one more time recall the used terminology. With some obvious but not confusing abuse of the notions, a sample from (V, Δ) is called right-censored (RC), its subsample with the unit indicator of censoring $\Delta = 1$ is called uncensored subsample and it contains uncensored observations because T is observed, and the complementary subsample with the zero indicator of censoring $\Delta = 0$ is called censored subsample and it contains censored observations because censoring variable C is observed instead of the lifetime of interest T. As a corresponding example, the term censored-data estimator means that the estimator is based on censored observations, uncensored-data estimator is based on uncensored observations, and aggregated estimator is based on right-censored data (the RC sample).

3.2   Estimation of Survival Function

We begin with the classical case of a directly observed sample $T_1, \ldots, T_n$ of size n from T. The problem is to estimate the underlying survival function $S^T(t) := \mathbb{P}(T > t)$. We can write

$$S^T(t) := \mathbb{P}(T > t) = \mathbb{E}\{I(T > t)\}. \tag{3.2.1}$$

Accordingly, the survival function S^T is the expectation of the indicator function $I(T > t)$, and it may be estimated by the sample mean (method of moments) estimator

$$\tilde{S}^T(t) := n^{-1} \sum_{l=1}^{n} I(T_l > t). \qquad (3.2.2)$$

This estimator, called the empirical survival function, is a stepwise function with equal jumps n^{-1} at each observation. Note how simple and intuitively clear the empirical survival function is. Further, its statistical analysis is also relatively simple. For instance, its variance is

$$\mathbb{V}\{\tilde{S}^T(t)\} = n^{-1} S^T(t)[1 - S^T(t)], \qquad (3.2.3)$$

the estimator is efficient, and also recall several other useful properties discussed in Sect. 1.2.1. To get a feeling of the problem and of the empirical survival function, the top diagrams in Fig. 3.1 exhibit two samples from T by the circles, the underlying survival function by the dashed line, and the empirical survival function by the stepwise solid line.

The situation changes rather dramatically if the lifetime of interest is right-censored, and we observe a sample $(V_1, \Delta_1), \ldots, (V_n, \Delta_n)$ from (V, Δ) defined in (3.1.1). There is no way to straightforwardly expand the empirical survival function from direct data to right-censored data. It is fair to say that the modern survival analysis of censored data, and the distribution estimation in particular, are based on the pathbreaking product-limit methodology of Kaplan and Meier (1958) for nonparametric estimation of survival function by a stepwise function with not-equal steps at uncensored lifetimes. The Kaplan-Meier estimator is defined as follows,

$$\tilde{S}^T_{KM}(t) := 1 \ \text{ for } \ t < V_{(1)}, \quad \tilde{S}^T_{KM}(t) := 0 \ \text{ for } \ t > V_{(n)},$$

$$\tilde{S}^T_{KM}(t) := \prod_{i=1}^{l-1} [(n-i)/(n-i+1)]^{\Delta_{(i)}} \ \text{ for } \ V_{(l-1)} < t \leq V_{(l)}. \qquad (3.2.4)$$

Here $(V_{(l)}, \Delta_{(l)})$ are ordered V_l's with their corresponding $\Delta_l, l = 1, \ldots, n$.

Let us also recall that the right-censoring is symmetric with respect to T and C, that is C right-censors T and T right-censors C. Accordingly, the Kaplan–Meier estimator for S^C is defined by formula (3.2.4) with Δ being replaced by $\Delta' := 1 - \Delta$.

This is not a simple task to explain the Kaplan–Meier estimator, and Fig. 3.1 may help us. Two simulated samples are considered in its two columns, see the caption. The top diagrams show the underlying samples, the middle diagrams show the right-censored samples. Let us look at the left-middle diagram with the underlying sample of size $n = 20$ and just $N = 13$ uncensored observations. This small sample size allows us to shed light on how the Kaplan-Meier estimator is constructed. The estimator is stepwise and has jumps at uncensored lifetimes $\Delta_l V_l > 0$. Censored observations affect only the size of each jump. The last step at $V_{(n)} := \max_l(V_l)$ to zero is just a convention of (3.2.4), and some statistical softwares do not

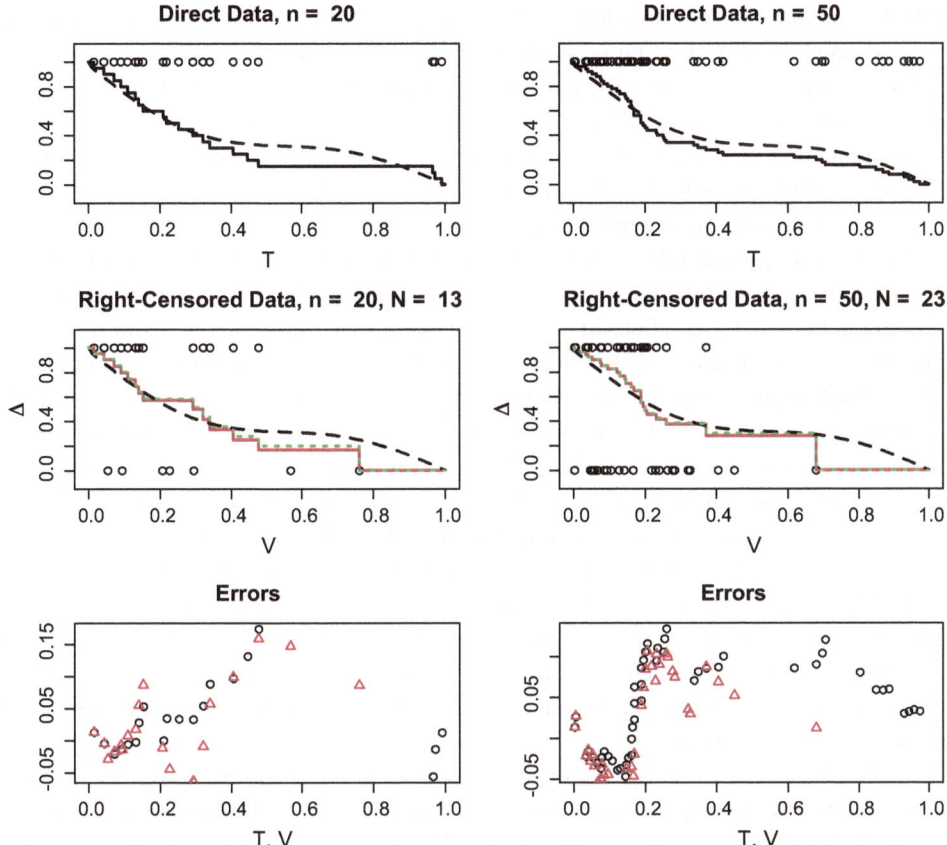

Fig. 3.1 Two simulated right-censored samples. A top diagram shows by the circles the underlying observations of T available to the oracle, that is directly observed lifetimes of interest. The dashed line is the underlying survival function, the stepwise solid line is the empirical survival function. A middle diagram shows right-censored data by the circles, $N := \sum_{l=1}^{n} \Delta_l$ is the number of uncensored observations. The dashed line is the underlying survival function, the stepwise solid line is the Kaplan-Meier estimate, and the stepwise dotted line is the proposed method of moments estimate (3.2.6) of the survival function. In the left-middle diagram the used f^C is the Uniform, in the right-middle diagram $f^C(t) = (1 + 0.9\cos(\pi t))I(t \in [0, 1])$. The bottom diagrams show by the circles the errors of the empirical survival function at T_l, and by the triangles the errors of the Kaplan-Meier estimator at V_l. {The choice of sample sizes is controlled by the argument n, the choice of underlying S^T by the argument $corn$, and the censoring density f^C for the right column's experiment is defined by $denC$.} [n = c(20, 50), corn = c(2, 2), denC =''1 + 0.9 * cos(pi * x)'']

show the last jump to zero. In other words, the Kaplan-Meier estimator and its underlying theory stops at the largest uncensored observation, that is at $\max_l (\Delta_l V_l)$.

Three remarks are due about the left-middle diagram. First, note that there are no uncensored observations to the right of 0.5. Accordingly, the Kaplan-Meier gives us no information about distribution of T for the right half of the support of T. Second, note that among 20 observations there are only $N = 13$ uncensored ones. Finally, the two largest observations of C cover a relatively large portion of the support.

Now let us increase sample size to $n = 50$ and look at a new simulation in the right column. The empirical survival function, based on directly observed lifetimes, is much better due to the larger number of observations in the right tail, see the right-top diagram. Now let us look at the RC data shown in the right-middle diagram. There are more uncensored observations in the left tail, but for this particular simulation there is no dramatic improvement in the right tail. This is due to the more challenging distribution of C which is skewed to the right. Note that there is no way to understand the underlying distribution of T for approximately 60% of its support. Interestingly, the range of V is even smaller than in the left-middle diagram. What we observe is the *curse of right tail* caused by the right-censoring.

The bottom diagrams shed light on accuracy of estimation by the classical empirical survival function for the case of direct observations and the Kaplan-Meier estimator for right-censored observations. These diagrams are useful for getting used to the quality of estimation. It is recommended to repeat this figure, using different parameters, and get first-hand experience in the quality of estimating the survival function using direct and right-censored observations.

The Kaplan-Meier estimator is efficient, and it is an excellent estimator. At the same time, it is difficult for statistical analysis and requires using special statistical techniques of the theory of point processes and martingales. This is why in what follows we will use a different estimator which, similarly to the empirical survival function (3.2.2), is based on the method of moments.

The underlying idea of the proposed method of moments estimator of S^T is as follows. We can write using Assumption 3.1 and formulas (1.1.9) and (3.1.2),

$$S^T(t) = \exp\{-H^T(t)\} = \exp\left\{ - \int_0^t \frac{f^T(v)}{S^T(v)} dv \right\} = \exp\left\{ - \int_0^t \frac{f^{V,\Delta}(v,1)}{S^T(v) S^C(v)} dv \right\}$$

$$= \exp\left\{ - \int_0^t \frac{f^{V,\Delta}(v,1)}{S^V(v)} dv \right\} = \exp\left\{ - \mathbb{E}\{ \frac{\Delta I(V \leq t)}{S^V(V)} \} \right\}. \tag{3.2.5}$$

Recall that H^T is the cumulative hazard function of T.

Formula (3.2.5) yields the following method of moments estimator of survival function,

$$\hat{S}^T(t) := \exp\{ -n^{-1} \sum_{l=1}^{n} \Delta_l I(V_l \leq t)/\hat{S}^V(V_l) \}, \tag{3.2.6}$$

where

$$\hat{S}^V(t) := n^{-1} \sum_{l=1}^{n} I(V_l \geq t) \tag{3.2.7}$$

is the empirical survival estimator of S^V for a continuous random variable V. Note that $\hat{S}^V(V_l) \geq n^{-1}$, and hence it can be used in the denominators of (3.2.6).

Due to the above-explained symmetry between T and C in the right-censoring, the following method of moments estimator of S^C may be proposed via replacing in (3.2.6) the indicator Δ by $1 - \Delta$,

$$\hat{S}^C(t) := \exp\{-n^{-1} \sum_{l=1}^{n} (1 - \Delta_l) I(V_l \leq t)/\hat{S}^V(V_l)\}. \tag{3.2.8}$$

Let us present several useful properties of the proposed method of moments estimator of the survival function whose proofs are based on classical properties of Bernoulli sums and the delta method.

Theorem 3.1 *Consider estimation of a survival function S, that may be either S^T or S^C, by a corresponding method of moments estimator (3.2.6) or (3.2.8). Suppose that Assumption 3.1 holds and $S^V(r) > 0$, $r > 0$. Then there exist finite positive constants ζ_*, ζ, and a sequence of finite constants ζ_k such that for any $l = 1, \ldots, n$, $z \in [0, r]$, positive v and positive integer k,*

$$|\mathbb{E}\{[\hat{S}(V_l) - S(V_l)]|V_l = z\}| \leq \zeta_* n^{-1}, \tag{3.2.9}$$

$$\mathbb{P}(|\hat{S}(V_l) - S(V_l)| > v|V_l = z\} \leq \zeta n e^{-nv^2/B}, \tag{3.2.10}$$

$$\mathbb{E}\{[\hat{S}(V_l) - S(V_l)]^{2k}|V_l = z\} \leq \zeta_k n^{-k}. \tag{3.2.11}$$

Note the remarkable symmetry between T and C in the assumptions and in the properties of the estimators.

Now we can return to the middle diagrams in Fig. 3.1. The stepwise dotted line shows us the method of moments estimate $\hat{S}^T$. As we see, it is close to the Kaplan-Meier estimate. Overall, the both estimators may be used in applications, but the method of moments estimator is simpler for theoretical analysis.

3.3   Density Estimation

This section is devoted to estimating density f^T of the lifetime of interest T based on a right-censored sample of size n from the pair $(V, \Delta) := (\min(T, C), I(T \leq C))$ where C is the continuous random censoring lifetime. Note that due to the symmetry between T and C in the right-censoring model, by developing an estimator of f^T we also will know how to estimate f^C. As we will see shortly, estimation of the nuisance density f^C is an important problem in survival analysis.

We begin with the theory of density estimation when the distribution of censoring life-time C is known to the oracle. Under this assumption, it is shown that both censored and uncensored observations may be used for consistent estimation of the density f^T, but cor-responding rates of the MISE convergence are dramatically different. Namely, the rate for censored observations is slower than for its uncensored counterpart. In other words, the censored subsample is ill-posed with respect to the uncensored one. Accordingly, the aggre-gation of the two subsamples becomes a challenging problem which may benefit samples with high rate of censoring.

Finally, a note about notations. Similarly to the previous chapter, we set $\Delta' := 1 - \Delta$. Recall that $\{\varphi_0(t) := 1, \varphi_j(t) := 2^{1/2}\cos(\pi jt), j = 1, 2, \ldots\}$ is the cosine orthonormal basis on $[0, 1]$, $\{\psi_j(t) := 2^{1/2}\sin(\pi jt), j = 1, 2, \ldots\}$ is the sine orthonormal basis on $[0, 1]$, $J_n := \lceil c_{J0} + c_{J1}\ln(n)\rceil$ where c_{J0} and c_{J1} are constants and $\lceil x \rceil$ is the smallest integer not smaller than x. In what follows we use subscripts u and c to stress that statistics are based on uncensored and censored subsamples, respectively.

3.3.1 Theory

Let us recall the right-censored (RC) model and used terminology. Estimation of f^T is based on a sample $(V_1, \Delta_1), \ldots, (V_n, \Delta_n)$ from $(V, \Delta) := (\min(T, C), I(T \leq C))$. The censoring lifetime C partitions the sample into two subsamples where observations of either T or C are available. We refer to the two subsamples as uncensored and censored subsamples, and refer to the corresponding observations as uncensored and censored. The indicator Δ points upon the type of an observation, namely $\Delta = 1$ implies uncensored observation $V = T$ while $\Delta = 0$ (or equivalently $\Delta' = 1$) implies censored observation $V = C$. Note that the number $N := \sum_{l=1}^n \Delta_l$ of uncensored observations has Binomial$(\mathbb{P}(\Delta = 1), n)$ distribution. The oracle knows the data and density f^C of the continuous random censoring lifetime C, and accordingly can estimate the density of interest f^T using either censored or uncensored observations. The latter is the reason why we are using the oracle's help in developing the theory of estimation based on uncensored, censored and right-censored (we may also say aggregated) observations.

The main aim of this subsection is to explain what the oracle can and cannot achieve by using uncensored and censored observations for estimating density f^T of a bounded lifetime of interest T, and then explain how to mimic the oracle. To simplify formulas, let us assume that T is supported on $[0, 1]$ and the oracle estimates f^T over the support. The oracle knows the nuisance density f^C.

Similarly to Chap. 2, we introduce shrinking local Sobolev classes of α-fold differentiable densities supported on $[0, 1]$,

$$\mathcal{F}_n := \mathcal{F}_n(f_0, \alpha, Q) := \Big\{ f : f(t) = f_0(t) + g(t)I(t \in [0, 1]),$$

$$g \in \mathcal{S}_1(\alpha, Q), \ |g(t)| \leq \min_{t \in [0,1]} f_0(t)/\ln(\ln(n + 20)), \ t \in [0, 1]\Big\}. \quad (3.3.1)$$

Here the anchor density f_0 is supported, continuous and positive on $[0, 1]$, and

$$S_k(\alpha, Q) := \{g : g(t) = \sum_{j=k}^{\infty} \theta_j \varphi_j(t), \ \sum_{j=k}^{\infty} (1 + (\pi j)^{2\alpha}) \theta_j^2 \leq Q, \ t \in [0, 1]\}. \quad (3.3.2)$$

Recall that $\mathcal{S}_0(\alpha, Q)$ is called the global Sobolev class, and in (3.3.1) we use the sequence $\ln(\ln(n + 20))$ because it is larger than 1 for all n. As we will see shortly in Theorem 3.2, we need to use the local shrinking Sobolev class (3.3.1) because sharp constant of the MISE convergence depends on an underlying density. Let us also stress that f_0 is not necessarily the underlying density of interest, it simply anchors all underlying densities f^T in its vanishing in L_∞-norm vicinity.

Set

$$P_u := \frac{Q^{1/(2\alpha+1)} \alpha^{2\alpha/(2\alpha+1)} (2\alpha + 1)^{1/(2\alpha+1)}}{[\pi(\alpha + 1)]^{2\alpha/(2\alpha+1)}}, \quad (3.3.3)$$

$$P_c := [Q(2\alpha + 3)]^{3/(2\alpha+3)} (1/3) [\alpha/(\pi(\alpha + 3))]^{2\alpha/(2\alpha+3)}, \quad (3.3.4)$$

and

$$d_u := \int_0^1 \frac{f^T(t)}{S^C(t)} dt, \quad d_c := \int_0^1 \frac{S^T(t)}{f^C(t)} dt. \quad (3.3.5)$$

Theorem 3.2 *Consider a right-censored sample of size n from $(V, \Delta) := (\min(T, C), I(T \leq C))$. The problem is to estimate density f^T of the lifetime of interest T over its support $[0, 1]$ under the MISE criterion. Let Assumption 3.1 hold, density f^C is positive and continuous on $[0, 1]$, $S^C(1) > 0$, and the oracle knows the right-censored sample, density f^C and the function class $\mathcal{F}_n$ defined in (3.3.1). Then:*
(1) The following lower bound for the MISE holds,

$$\inf_{\tilde{f}^*} \sup_{f^T \in \mathcal{F}_n} \mathbb{E}\{(n/d_u)^{2\alpha/(2\alpha+1)} \int_0^1 (\tilde{f}^*(t) - f^T(t))^2 dt\} \geq P_u(1 + o_n(1)), \quad (3.3.6)$$

where the infimum is over all possible oracle-estimators $\tilde{f}^*$.
(2) The lower bound (3.3.6) is sharp and attainable by an oracle-estimator $\tilde{f}_u^*$ based solely on uncensored observations.
(3) If the oracle uses only censored observations and $\alpha \geq 2$, then the following lower bound for the MISE holds,

$$\inf_{\tilde{f}_c^*} \sup_{f^T \in \mathcal{F}_n} \mathbb{E}\{(n/d_c)^{2\alpha/(2\alpha+3)} \int_0^1 (\tilde{f}_c^*(t) - f^T(t))^2 dt\} \geq P_c(1 + o_n(1)), \quad (3.3.7)$$

and the lower bound is sharp. Accordingly, censored observations are ill-posed with respect to uncensored observations, and using only them slows down rate of the MISE convergence from $n^{-2\alpha/(2\alpha+1)}$ to $n^{-2\alpha/(2\alpha+3)}$.

(iii) While for uncensored observations parametric Fisher informations for Fourier coefficients $\theta_j := \int_0^1 f^T(t)\varphi_j(t)dt$ are bounded below from zero for all j, for censored observations the Fisher informations vanish with rate j^{-2}.

Let us comment on the results presented in Theorem 3.2. It follows from (3.3.6) that the same lower bound holds for oracle-estimators based solely on uncensored observations. This and part (2) of Theorem 3.2 support the Kaplan-Meier paradigm about dominance of uncensored observations. Indeed, using uncensored observations we can attain the rate $n^{-2\alpha/(2\alpha+1)}$ known for the case of direct observations of T. On the other hand, even the oracle cannot attain this rate using censored observations. Moreover, using censored observations the optimal rate slows down to $n^{-2\alpha/(2\alpha+3)}$. That rate, according to Subsection 1.2.6, is the optimal rate for estimation of a trivariate density $f(x_1, x_2, x_3)$ based on direct observations from (X_1, X_2, X_3). In other words, censored observations are ill-posed with respect to uncensored ones. This is the bad news. The good news is the third part of Theorem 3.2. It shows that the ill-posedness increases as j increases but may be only moderate for low frequencies. This opens a window of opportunity for aggregation of uncensored and censored observations on low frequencies, and the aggregation may be of interest for right-censored data with high rates of censoring.

3.3.2 Estimation Based on Uncensored and Censored Observations

The aim is to construct estimators that attain the sharp lower bounds (3.3.6) and (3.3.7). Our first estimator is based on uncensored observations, and recall that we are using subscript u to highlight that. First, let us explain the heuristic. Using (3.1.2) we can write,

$$\theta_j = \int_0^1 f^T(t)\varphi_j(t)dt = \mathbb{E}\{\frac{\Delta\varphi_j(V)}{S^C(V)}\}. \tag{3.3.8}$$

Using $\theta_0 := \int_0^1 f^T(t)dt = 1$ and (3.3.8) we propose the following sample mean Fourier coefficient oracle-estimator,

$$\tilde{\theta}_{u0}^* := 1, \quad \tilde{\theta}_{uj}^* := n^{-1}\sum_{l=1}^{n}\Delta_l\frac{\varphi_j(V_l)}{S^C(V_l)}, \quad j \geq 1. \tag{3.3.9}$$

In its turn, the Fourier coefficient oracle-estimator yields the following density oracle-estimator based on uncensored observations,

$$\tilde{f}_u^*(t) := \sum_{j=0}^{J_n} \tilde{\theta}_{uj}^* I((\tilde{\theta}_{uj}^*)^2 > c_{TH} \mathbb{V}\{\tilde{\theta}_{uj}\}) \varphi_j(t)$$

$$+ \sum_{j=J_n+1}^{J_{un}^*} (1 - (j/J_{un}^*)^\alpha) \tilde{\theta}_{uj}^* \varphi_j(t). \tag{3.3.10}$$

Here $J_n := \lceil c_{J0} + c_{J1} \ln(n) \rceil$, c_{TH} is a positive constant, d_u is defined in (3.3.5), and

$$J_{un}^* := J_n + \lceil [(n/d_u) Q \pi^{-2\alpha} (\alpha+1)(2\alpha+1)/\alpha]^{1/(2\alpha+1)} \rceil. \tag{3.3.11}$$

Note that in (3.3.10) the classical hard thresholding is used on low frequencies and the shrinkage on high frequencies. As we will see shortly, this simple oracle-estimator is efficient and attains the lower bound of Theorem 3.2. Further, its low-frequency component (the first sum) is the underlying motivation of the E-estimator used for small samples. The E-estimator is defined in Sect. 1.2.3 and will be reviewed shortly.

Now consider the censored subsample and explain how the Fourier coefficients θ_j of f^T may be estimated by the sample mean approach. Recall our notation $\psi_j(t) := 2^{1/2} \sin(\pi j t)$, $j = 1, 2, \ldots$ for elements of the sine basis on $[0, 1]$. Using integration by parts and the assumed $S^T(0) = 1$ and $S^T(1) = 0$ we write for $j \geq 1$,

$$\theta_j := \int_0^1 f^T(t) \varphi_j(t) dt = -S^T(t) \varphi_j(t)|_0^1 - (\pi j) \int_0^1 S^T(t) \psi_j(t) dt$$

$$= 2^{1/2} - (\pi j) \int_0^1 S^T(t) \psi_j(t) dt. \tag{3.3.12}$$

With the help of (3.1.2) we can continue (recall the notation $\Delta' := 1 - \Delta$),

$$\theta_j = 2^{1/2} - (\pi j) \mathbb{E}\{\frac{\Delta' \psi_j(V)}{f^C(V)}\}, \quad j \geq 1. \tag{3.3.13}$$

If we compare (3.3.13) with (3.3.8), then the main difference is in the factor πj used for censored observations. This factor sheds light on the ill-posedness of censored observations, and also explains why for low frequencies we may see only its onset. Another important difference is in the natural nuisance functions. Recall that the oracle uses data and the natural nuisance function for efficient estimation, and that function is the oracle's hint about optimal estimation. Oracle-estimators (3.3.8) and (3.3.13) show that for uncensored observations the natural nuisance function is the survival function S^C, while for censored observations the natural nuisance function is the density f^C which may be less accurately estimated than the survival function S^C. The latter is another difficulty in dealing with censored observations that will be addressed shortly.

Using (3.3.13) we can suggest the sample mean Fourier coefficient oracle-estimator for censored observations. Recall that subscript c highlights the fact that a statistic is based on censored observations. Set

$$\tilde{\theta}_{c0}^* := 1, \quad \tilde{\theta}_{cj}^* := 2^{1/2} - n^{-1}(\pi j) \sum_{l=1}^{n} \Delta_l' \frac{\psi_j(V_l)}{f^C(V_l)}, \quad j \geq 1. \tag{3.3.14}$$

Finally, the density oracle-estimator, based on censored observations, is defined as

$$\tilde{f}_c^*(t) := \sum_{j=0}^{J_n} \tilde{\theta}_{cj}^* I((\tilde{\theta}_{cj}^*)^2 > c_{TH} \mathbb{V}\{\tilde{\theta}_{cj}^*\}) \varphi_j(t)$$

$$+ \sum_{j=J_n+1}^{J_{cn}^*} (1 - (j/J_{cn}^*)^\alpha) \tilde{\theta}_{cj}^* \varphi_j(t), \tag{3.3.15}$$

where

$$J_{cn}^* := J_n + \lceil [(n/d_c) Q \pi^{-2\alpha-2} (\alpha + 3)(2\alpha + 3)/\alpha]^{1/(2\alpha+3)} \rceil. \tag{3.3.16}$$

Theorem 3.3 *Let Assumption 3.1 hold and $S^T(1) = 0$. Suppose that the anchor $f_0 \in S(\alpha', Q')$, $\alpha' > \alpha$, $Q' < \infty$. Consider d_u and d_c defined in (3.3.5). If d_u is finite, then the density oracle-estimator (3.3.10), based on uncensored observations, attains the lower bound (3.3.6). If d_c is finite, then the oracle-estimator (3.3.15), based on censored observations, attains the lower bound (3.3.7). Further, the proposed Fourier coefficient estimators are unbiased and satisfy*

$$\mathbb{V}\{\tilde{\theta}_{uj}^*\} = \mathbb{E}\{(\tilde{\theta}_{uj}^* - \theta_j)^2\} = n^{-1}\sigma_{uj}^2, \quad \sigma_{uj}^2 = d_u(1 + o_j(1)), \tag{3.3.17}$$

and

$$\mathbb{V}\{\tilde{\theta}_{cj}^*\} = \mathbb{E}\{(\tilde{\theta}_{cj}^* - \theta_j)^2\} = n^{-1}(\pi j)^2 \sigma_{cj}^2, \quad \sigma_{cj}^2 = d_c(1 + o_j(1)). \tag{3.3.18}$$

Expressions (3.3.17) and (3.3.18) shed extra light on the ill-posedness of censored observations. We observe that the variance of $\tilde{\theta}_{cj}^*$ increases proportionally to j^2 while for the uncensored counterpart it is bounded from above. This is what defines the dominance of uncensored observations and ill-posedness of censored observations. This also explains complexity of their aggregation.

The nice feature of the series oracle-estimators (3.3.10) and (3.3.15) is that for small samples they justify the use of E-estimator of Sect. 1.2.3. Let us recall its algorithm for the cosine basis $\{\varphi_j, j = 0, 1, \ldots\}$ on $[0, 1]$.

Algorithm of E-Estimation: Let $f(x), x \in [0, 1]$ be a square integrable function of interest. There are three steps in the algorithm.

Step 1. The function can be written as $f(x) = \sum_{j=0}^{\infty} \theta_j \varphi_j(x)$. Here $\theta_j := \int_0^1 f(x)\varphi_j(x)dx$ are Fourier coefficients of f. Suggest a sample mean estimator $\hat{\theta}_j$ of the Fourier coefficients θ_j. Then calculate a corresponding sample variance estimator $\hat{\sigma}_{jn}^2$ of the variance $\mathbb{V}(\hat{\theta}_j)$.

Step 2. The E-estimator is defined as $\hat{f}(x) := \sum_{j=0}^{\hat{J}} \hat{\theta}_j I(\hat{\theta}_j^2 > c_{TH}\hat{\sigma}_{jn}^2)\varphi_j(x)$. Here the empirical cutoff is $\hat{J} := \text{argmin}_{0 \le J \le c_{J0}+c_{J1}\ln(n)}\{\sum_{j=0}^{J}[2\hat{\sigma}_{jn}^2 - \hat{\theta}_j^2]\}$, and c_{J0}, c_{J1} and c_{TH} are parameters (nonnegative constants).

Step 3. If there are bona fide restrictions on f (for instance, the probability density is nonnegative and integrated to one, or it is known that the function is monotonic) then a projection of $\hat{f}$ on the bona fide function class is performed.

Note that Steps 2 and 3 in construction of the E-estimator are the same for all non-parametric statistical problems. As a result, as soon as a sample mean estimator of Fourier coefficients is proposed, this Fourier coefficient estimator yields the corresponding nonpara-metric E-estimator. Further, the book's R software allows the user to change default values of the parameters c_{J0}, c_{J1} and c_{TH}.

The blockwise shrinkage methodology of adaptation to unknown smoothness of f^T is explained in Sects. 1.2 and 2.3 where the shrinking coefficients are defined via sums of squared Fourier coefficient estimators and U-statistics. Here we will learn a bit different technique that also yields efficient estimation. For some settings this technique may be more convenient for statistical analysis. Suppose that the oracle recommends to use an unbiased Fourier coefficient estimator $\bar{\theta}_j$ of θ_j satisfying $\mathbb{E}\{(\bar{\theta}_j - \theta_j)^2\} = dn^{-1}(1 + o_n(1) + o_j(1))$. Similarly to Sects. 1.2 and 2.3, set $b_1 := J_n + 1, b_{k+1} := b_k + \lceil(1 + 1/\ln(\ln(n + 20)))^k\rceil$, $k = 1, 2, \ldots, B_k := \{j : b_k \le j < b_{k+1}\}, L_k := b_{k+1} - b_k, k_n$ is the smallest integer such that $b_{k_n} \ge n^{1/3}\ln(\ln(n + 20))$,

$$\bar{\Theta}_k' := L_k^{-1} \sum_{j \in B_k} \bar{\theta}_j^2. \tag{3.3.19}$$

Also denote by $\bar{d}$ an estimate of d such that $\mathbb{E}\{(\bar{d} - d)^2\} = o_n(1)$. For instance, we can set $\bar{d} := n^{-1}\sum_{l=1}^{n} \Delta_l[S^C(V_l)]^{-2}$ for uncensored observations, and $\bar{d} := n^{-1}\sum_{l=1}^{n} \Delta_l[f^C(V_l)]^{-2}$ for censored observations. Denote by $\bar{\sigma}_{jn}^2$ the empirical variance of $\bar{\theta}_j$. Then the blockwise shrinkage estimator, that adapts to parameters (α, Q) and matches the MISE of a corresponding oracle-estimator, is

$$\bar{f}^*(t) := \sum_{j=0}^{J_n} \bar{\theta}_j I(\bar{\theta}_j^2 > c_{TH}\bar{\sigma}_{jn}^2)\varphi_j(t)$$

$$+ \sum_{k=1}^{k_n} \sum_{j \in B_k} \frac{\bar{\Theta}_k' - \bar{d}n^{-1}}{\bar{\Theta}_k'} I(\bar{\Theta}_k' \ge n^{-1}(\bar{d} + 1/\ln(k + 3)))\varphi_j(t). \tag{3.3.20}$$

The new here is that $\bar{\Theta}'_k - \bar{d}n^{-1}$ replaces an U-statistic used in Sects. 1.2 and 2.3. The U-statistic is unbiased estimator of the Sobolev functional $\Theta_k := L_k^{-1} \sum_{j \in B_k} \theta_j^2$, while $\bar{\Theta}'_k - \bar{d}n^{-1}$ is asymptotically unbiased estimator.

3.3.3 Aggregation of Uncensored and Censored Observations

The properties (3.3.17), (3.3.18) and Lemma 2.1 in the Sect. 2.2.1.2 yield the following unbiased aggregation of the two Fourier coefficient oracle-estimators proposed in the previous subsection for uncensored and censored subsamples,

$$\tilde{\theta}^*_{aj} := \tilde{\theta}^*_{uj} \frac{(\pi j)^2 \sigma_{cj}^2}{(\pi j)^2 \sigma_{cj}^2 + \sigma_{uj}^2} + \tilde{\theta}^*_{cj} \frac{\sigma_{uj}^2}{(\pi j)^2 \sigma_{cj}^2 + \sigma_{uj}^2}, \qquad (3.3.21)$$

with the variance satisfying

$$\mathbb{V}\{\tilde{\theta}^*_{aj}\} = \mathbb{E}\{(\tilde{\theta}^*_{aj} - \theta_j)^2\} = \sigma_{uj}^2 \frac{(\pi j)^2 \sigma_{cj}^2}{(\pi j)^2 \sigma_{cj}^2 + \sigma_{uj}^2}$$

$$=: \sigma_{uj}^2 (1 - \nu_j), \quad 0 < \nu_j < (\pi j)^{-2} [d_u/d_c] (1 + o_j(1)). \qquad (3.3.22)$$

Formula (3.3.18) implies that for small j we see only the onset of ill-posedness. Accordingly, the frequency-domain aggregation (3.3.21) is feasible, and formula (3.3.22) explains its benefits for low frequencies. Furthermore, the E-estimator based on the aggregated Fourier coefficient estimators can be used for small samples.

So far we have considered the case when S^C and f^C are known. If the latter is not the case, the method of moments estimator of S^C defined in Sect. 3.1 and the density estimator of f^C, based on censored observations, may be used.

We will test the estimators and discuss the aggregation in the next subsection.

3.3.4 Numerical Study

Three rows of diagrams in Fig. 3.2 present in their left diagrams results of different simulations of right-censored (RC) observations that are explained in the caption, and the right diagrams present density E-estimates (because only E-estimates are considered, in what follows we may refer to them as estimates). The four density estimates are the oracle-estimate (the short-dashed line) based on underlying direct observations of T, the uncensored-data estimate (the dotted line), the censored-data estimate (the dot-dashed line), the aggregated estimate (the long-dashed line). The estimate of f^C is shown by the circles at observed V_l, $l = 1, 2, \ldots, n$. Note that all estimates are data-driven and use neither information about distribution of C nor about smoothness of an estimated density. This figure can be repeated with different distributions and parameters as explained in the caption. The shown simulations are chosen to highlight several interesting aspects of the estimation problem. The top row presents the case when 26% of observations are censored (the theoretical $\mathbb{P}(\Delta = 0) = 0.25$), the underlying density f^T is the Bathtub (recall that this is the second corner function, see Fig. 1.8), and the censoring distribution is Uniform(0, 1.5). The sample size $n = 100$ and the number of censored observations $n - N = 26$. The left-top diagram shows the underlying survival function (the dashed line) and the Kaplan-Meier estimate. Despite the small sample size and relatively large rate of censoring, this particular simulation describes the underlying S^T very well except of the right tail with $t \geq 0.8$ due to just few available observations. Now let us look at the density estimates in the right-top diagram and compare them with the underlying bathtub density shown by the solid line. The short-dashed line is the oracle-estimate based on the underlying (hidden) observations of T, this data-driven estimate is used as a benchmark. It is very good for the available sample size and its integrated squared error ISEO = 0.00403. Note that in the title of the diagram only the last letter O is used for the ISEO to make the title shorter. The density oracle-estimate and the Kaplan-Meier estimate indicate that the particular simulation is reasonable. The circles show E-estimates of $f^C(V_l), l = 1, \ldots, n$ used by the censored-data density estimator of f^T. We conclude that the estimate correctly outlines the underlying Uniform(0, 1.5) density f^C using just 26 censored observations. Repeated simulations show that this outcome is a success.

Now let us look at the three density E-estimates based on the right-censored data. The dotted line is the uncensored-data estimate of f^T, it is based on uncensored observations shown by the circles in the top-left diagram with $\Delta = 1$ and on the estimated survival function of C. Its $ISEU = 0.0187$ is very respectful and the estimate is very good. Visualization of the uncensored observations supports the estimate. Now let us look at the 26 censored observations shown in the left-top diagram by the circles with $\Delta = 0$. Visual analysis does not help us to see the underlying density because censored observations are ill-posed. Indeed, according to (3.1.2), censored observations allow us to evaluate S^T, and then its derivative yields the density. This is why it is difficult to visualize the underlying density in censored observations, and also recall that censored observations are similar to the CSCNO discussed in Chap. 2. Despite the extremely small sample size for nonparametric estimation and the ill-posedness, the dot-dashed line, which is the proposed density estimate based on the censored observations, is not bad and fairly well shows the bathtub shape. Note the corresponding ISEC = 0.039 which is just the twofold ISEU. It will be explained shortly why such an outcome is possible for the censored-data E-estimator. The long-dashed line is the proposed aggregated estimate, it is better than the uncensored-data estimate, and its ISEA = 0.0129 is smaller than the ISEU = 0.0187.

The middle row of diagrams shows a similar experiment with the same underlying density and the larger rate of censoring, here the theoretical $\mathbb{P}(\Delta = 0) = 0.38$. The sample size $n = 200$ is larger than in the first experiment. The left diagram shows that the larger sample size dramatically improves right tail of the Kaplan-Meier survival estimate. At the same time, note that there are significantly more large censored observations than large

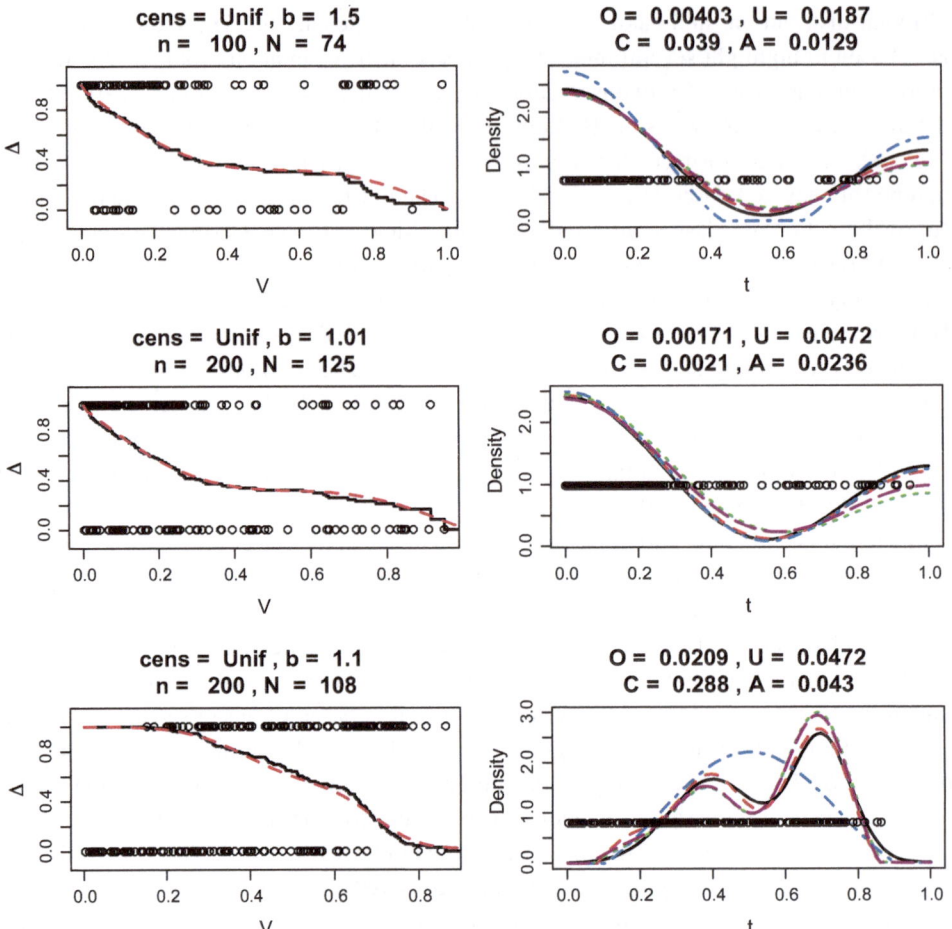

Fig. 3.2 Density estimation for three simulated examples shown in the corresponding rows. In each row the left diagram shows pairs (V_l, Δ_l), $l = 1, \ldots, n$ overlaid by the underlying survival function (the dashed line) and the Kaplan-Meier estimate (the solid stepwise line). The title also shows distribution of the censoring C, the sample size n and the number of uncensored observations $N := \sum_{l=1}^{n} \Delta_l$. In these three experiments the censoring distributions are Uniform(0,b). A right diagram shows estimates of $f^C(V_l)$ by the circles. An underlying density f^T is shown by the solid line, it is the Bathtub in the two top diagrams and the Bimodal in the bottom diagram. The short-dashed, dotted, dot-dashed and long-dashed lines are the oracle, uncensored-data, censored-data and aggregated E-estimates, respectively. A diagram also shows integrated squared errors of these estimates using letters O, U, C, and A, respectively. {The argument corn allows to choose underlying distributions of T, set cens $="$Unif$"$ to choose Uniform(0, b) and cens $="$Expon$"$ for Exponential(λ) distribution of C, here λ is the mean of C controlled by the argument lambda. Arguments cJ0C and cJ1C are used by the E-estimator for censored-data density estimator. Argument nsim is the number of repeated simulations to get mean and median ISEs for the four density estimates} [n = c(100, 200, 200), corn = c(2, 2, 4), cens $="$Unif$"$, b = c(1.5, 1, 01, 1.1), lambda = c(0.5, 1, 1.5), cJ0 = 3, cJ1 = 0.8, cTH = 4, cJ0C = 1, cJ1C = 0.2, nsim = 1]

uncensored ones. Let us check how the made changes in the experiment and the specific right-censored sample affect the density estimates. The oracle-estimate is much improved (compare the corresponding ISEOs and recall that only the last letter O is shown in the titles). Surprisingly, the uncensored-data estimate and especially its right tail are worse than what we see in the right-top diagram. This is also reflected by the corresponding ISEUs. The good news is the censored data-estimate which is comparable with the oracle-estimate. The aggregated estimate is also very good but worse than the censored-data one. This is because the aggregated estimate always more closely follows the uncensored-data estimate due to ill-posedness of censored observations. Is the censored-data estimate better due to the special right tail of the data shown in the left-middle diagram? We will explore this question shortly.

The bathtub density is the low-frequency function, and accordingly we see only the onset of ill-posedness of censored observations. Let is check estimation of the high-frequency Bimodal corner density presented in the bottom row of diagrams. Here $\mathbb{P}(\Delta = 0) = 0.5$, and for the particular simulation the rate of censoring is 46%. The Kaplan-Meier estimate is very good. The right-bottom diagram shows us the very good oracle-estimate and the surprisingly good uncensored-data estimate of the density. The censored-data estimate is poor, but it does its best in approximating the Bimodal using low frequencies. There are simply not enough censored observations to estimate high frequencies of the Bimodal density that can show its two modes. The aggregated estimator performs well under the circumstances and even outperforms a bit the uncensored-data estimator.

Now let us look one more time at the three aggregated density estimates. Note that they are not a linear combination of the uncensored and censored estimates. Namely, a class of linear aggregations $\{\bar{f}_a : \bar{f}_a = \lambda \tilde{f}_u + (1 - \lambda) \tilde{f}_c, \lambda \in [0, 1]\}$ is well known in the statistical literature, see the Notes. This aggregation is not applicable here because for right-censored data the theory requires individual aggregation of Fourier coefficients of $\tilde{f}_u$ and $\tilde{f}_c$, and this is what we see in the three aggregated density estimates.

Now let us continue the discussion, initiated by the second experiment in Fig. 3.2, about effect of the right tail of observations on the estimates. Are there situations when for small samples the censored-data estimator may outperform the uncensored-data estimator? Fig. 3.3 presents such examples. The two top rows of diagrams exhibit particular outcomes for the experiment identical to the two top in Fig. 3.2 only here the uniform C is replaced by C with density $f^C(t) = [1 + 0.9 \cos(\pi t)]I(t \in [0, 1])$. The main difference between the shown in Fig. 3.3 simulated samples and those in Fig. 3.2 is the right tail of uncensored observations. In the top (let us refer to it as the first) experiment there are no uncensored observations larger than 0.5. Accordingly, the Kaplan-Meier survival estimate provides no information about S^T for a half of the support of T. The situation is not much better in the second simulation. Now let us look at the density estimates. For the first experiment we get a poor estimate of the monotone density f^C, see the circles. This is always a possibility for right-censored data, and it is of interest to check robustness of the proposed density E-estimator. As it could be expected, the uncensored-data estimate is poor, and its right tail is too low. At the same

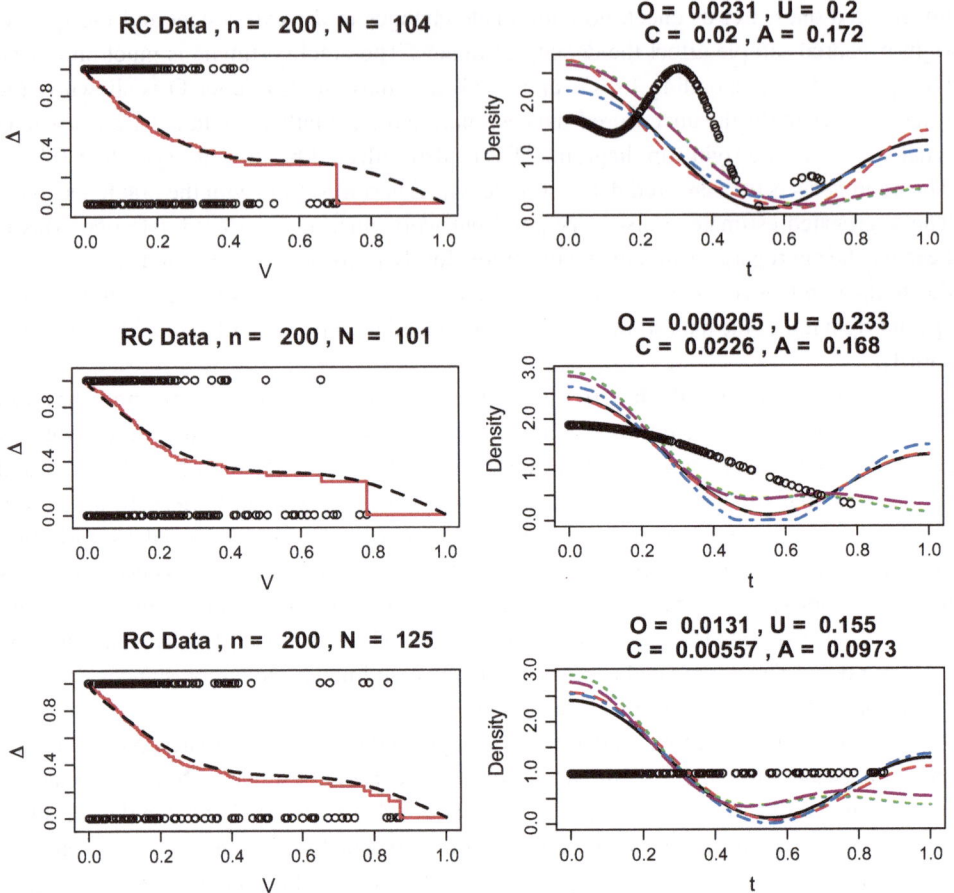

Fig. 3.3 Density estimation for three simulated examples with high censoring of right tail. The structure of diagrams is the same as in Fig. 3.2. Experiments for the two top rows of diagrams use f^C defined by the argument denC, and in the bottom experiment C is Uniform(0,b). [n = c(100, 200, 200), corn = c(2, 2, 2), denC =''1+ 0.9*cos(pi*x)'', cens =''Unif'', b = 0.9, lambda = 0.5, cJ0 = 3, cJ1 = 0.8, cTH = 4, cJ0C = 1, cJ1C = 0.2, nsim = 1]

time, the censored-data estimate is very good and almost perfectly indicates right tail of the Bathtub density. This is a very positive outcome. The aggregated estimate is better than the uncensored-data estimate, but worse than the censored one. As we already know, this is due to the aggregation algorithm which gives more weight to the uncensored observations. Let us make another comment that sheds extra light on the estimates. A direct calculation shows that for the problem at hand the coefficients of difficulty (3.3.5) for the uncensored and censored observations are $d_u = 7.9$ and $d_c = 0.4$, respectively. At first glance this creates a huge advantage for the censored-data estimator, but the core issue is that this estimator is ill-posed and formulas (3.3.17) and (3.3.18) allow us to appreciate that. Indeed, while the constant in variance of the uncensored-data Fourier coefficient estimator $\tilde{\theta}^*_{uj}$ is $d_u = 7.9$, for censored-data Fourier coefficient estimator $\tilde{\theta}^*_{cj}$ it is $(\pi j)^2 d_c$ which yields values 3.9, 15.8 and 35.5 for $j = 1, 2$ and 3, respectively. This is why censored observations are called ill-posed and overall the aggregated estimate more closely follows the uncensored-data estimate.

For the second simulation in Fig. 3.3 shown in the left-middle diagram, the density oracle-estimate is almost perfect, and this tells us about fair realizations of the underlying lifetime of interest T. The estimate of f^C is also good. Further, due to a larger number of censored observations in the right half of the support (compare with the first simulation), the censored-data estimate is good and its right tail is almost perfect. The right tail of the uncensored-data estimate is poor and reflects the same poor performance of the Kaplan-Meier estimate of S^T. The aggregated estimate is again better than the uncensored-data estimate.

The bottom diagrams in Fig. 3.3 present results of the experiment where C is Uniform(0, 0.9). Note that the support of C is smaller than the support of T, and no consistent estimation of the distribution of T is possible. Nonetheless, we still can use the four E-estimators and check their performance. In the left-bottom diagram we observe just 5 uncensored observations larger than 0.5. This is reflected by the poor Kaplan-Meier survival estimate. At the same time, we observe a relatively large number of censored observations with values larger than 0.5. Now let us look at the density estimates in the right-bottom diagram. The oracle-estimator is not perfect but relatively fair. This implies that the underlying sample from T is fair. The uncensored-data estimate has a poor right tail, and we already know why. At the same time, the censored-data estimate is very good, and look at its monotonically increasing right tail. The aggregated estimate again outperforms the uncensored-data estimate.

Based on these simulations we may conclude that the censored-data estimator may be very good for right-censored samples with sparse right tail uncensored observations. Moreover, its visual analysis may shed light on right tail of the underlying distribution.

So far we only observed particular simulations. What will be if they are repeated a large number of times and the mean and median empirical integrated squared errors are analyzed? Figs. 3.2 and 3.3 allow us to do exactly this using the argument nsim. Set nsim=k and the simulations of the figure will be repeated k times, the first simulation will be shown as a corresponding figure, and then two tables with 3 rows and 5 columns will appear on the monitor. The rows correspond to the three rows in the figure, the first four columns present statistics for the oracle, uncensored-data, censored-data, and aggregated estimators, the last column presents the theoretical rate of censoring $\mathbb{E}\{(1 - \Delta)\}$. Among the two tables on the monitor, the top one shows sample means of ISEs while the bottom one shows sample medians. This feature of the figures allows the user to compare different experiments, choose optimal parameters of the estimators, and learn more about the right-censoring.

For instance, using calls > ch3(Fig=2,nsim=5000) and > ch3(Fig=3,nsim=5000), we conduct two intensive numerical studies, based on 5000 repeated simulations, for the six default experiments shown in these two figures. Let us enumerate the experiments from 1 to 6, so experiment 3 is the bottom one in Fig. 3.2 and experiment 5 is the middle one in Fig. 3.3. Results are shown in Table 3.1. The general conclusion is that right-censoring significantly complicates density estimation, but there are intrinsic phenomenons in that effect. First, please note the difference between the first cohort of three experiments and the second cohort. For the first cohort the MISE of uncensored-data estimator is about several times larger than the MISE of oracle-estimate, while for the second cohort it is about ten-fold. Now let us look at the medians and note that they are less affected by the censoring. Next we compare uncensored-data and censored-data estimators. We see the confirmation of our preliminary conclusions for particular simulations shown in the figures, namely the relative performance of censored-data estimator improves dramatically for the second cohort of experiments with sparse right tails of uncensored observations. This is why in Fig. 3.3 we can often observe simulations for which the censored-data estimator outperforms the uncensored-data estimator. Finally, we see that the aggregated estimator performs as it could be expected from the theory, and for the second cohort it reliably outperforms the two other estimators.

Based on the numerical studies, we can make the following conclusions for density estimation: (i) Under a mild assumption, asymptotically uncensored observations dominate censored ones due to ill-posedness of censored observations; (ii) For small samples and high rates of censoring the aggregation of uncensored and censored observations may be beneficial; (iii) Aggregation may not benefit estimation of high-frequency densities but it definitely does not hurt the estimation; (iv) The censored-data estimate may help in analysis of censored data with sparse right-tail observations; (v) It is an important open problem of how to improve right tail of an aggregated estimator, and explanation of a possible approach for its solution is postponed until the end of this subsection.

Now let us consider density estimation for the practical example of a small cell lung cancer clinical study data. The data contains right-censored survival lifetimes in days and

Table 3.1 Numerical analysis of the six experiments in Figs. 3.2 and 3.2 based on nsim = 5000 repeated simulations. Each entry in columns 2–5 is written as a ratio where: The numerator is the sample mean integrated squared error of the estimator shown in the top row; The denominator is the sample median of integrated squarer errors. The last column shows theoretical rates of censoring

Experiment	Oracle estimator	Uncensored-data estimator	Censored-data estimator	Aggregated estimator	$\mathbb{E}\{1 - \Delta\}$
1	0.036/0.015	0.06/0.021	0.42/0.53	0.06/0.022	0.25
2	0.015/0.007	0.07/0.033	0.24/0.23	0.07/0.023	0.38
3	0.057/0.052	0.13/0.12	0.52/0.63	0.18/0.12	0.50
4	0.015/0.08	0.17/0.17	0.28/0.36	0.15/0.14	0.52
5	0.015/0.08	0.17/0.17	0.28/0.36	0.15/0.14	0.52
6	0.015/0.08	0.16/0.17	0.25/0.36	0.13/0.12	0.42

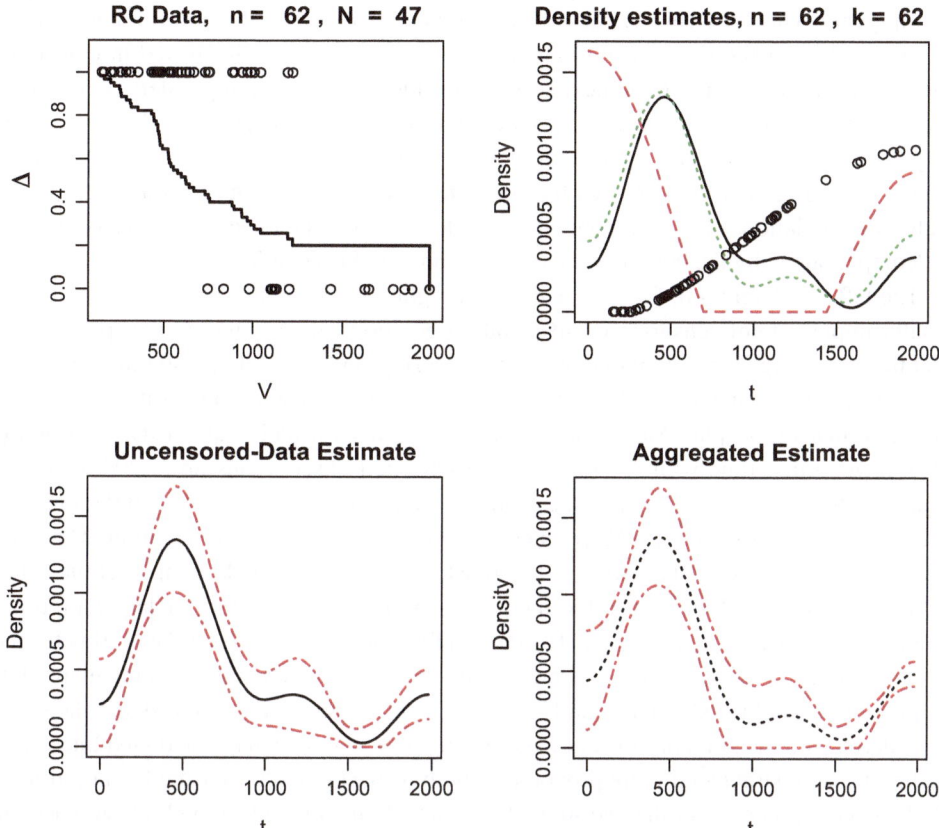

Fig. 3.4 Analysis of small cell lung cancer data. The data and the Kaplan-Meier estimate are shown in the left-top diagram. In the right-top diagram the solid, dashed and dotted lines are the uncensored-data, censored-data and aggregated estimates of f^T, respectively, and the circles show estimates of $f^C(V_l)$. The left-bottom diagram shows the uncensored-data estimate (the solid line) together with its 95% confidence band (the dot-dashed lines). The right-bottom diagram shows the aggregated estimate (the dotted line) together with its 95% confidence band (the dot-dashed lines). {Argument DATA allows to analyze any R data written as a matrix with two columns of V and Δ. Argument k allows to use only k first ordered (smallest) values of V by the E-estimator as explained in Sect. 3.3.5. Argument alpha allows to change the confidence coefficient.} [DATA="DataCancerA", $k = 62$, alpha $= 0.05$, cJ0 $= 3$, cJ1 $= 0.8$, cTH $= 4$, cJ0C $= 1$, cJ1C $= 0.2$]

age in years. The censoring is caused by administrative end of the clinical study, and it is independent of the survival lifetime and the age. Here we are interested in density of the survival lifetime, and in the next section consider the regression problem. Let us note that about 15% of all lung cancer cases are the small cell lung cancer and this is the most aggressive type of lung cancer with extremely short survival times after the cancer diagnosis. Figure 3.4 shows the right-censored lifetimes and estimates. The data resembles the sparse right-tail simulations of Fig. 3.3, only now we do not know the underlying density and this is "compensated" by the confidence bands explained in Sect. 1.2.3.

The left-top diagram in Fig. 3.4 exhibits the right-censored data. We are dealing with a small sample, rate of censoring is 24%, and observations in the right tail are sparse. Note that the seven largest observations of V are censoring times, and they point upon a subset of relatively large underlying lifetimes T. The largest uncensored observation is 1221 d, and note that the Kaplan-Meier estimate provides no information about the distribution of T beyond that time while there are still 7 censored observations up to 2000 d. In the right-top diagram the circles show the estimates of $f^C(V_l), l = 1, \ldots, n$, and the lines are estimates of $f^T(t)$ for $t \in [0, V_{(n)}]$ where $V_{(n)}$ denotes the largest observation. The solid line is the uncensored-data estimate. It indicates two strata of underlying lifetimes. The stratum of smaller lifetimes has the mode near 500, and the stratum of larger lifetimes begins with values exceeding 1600. The dashed line shows the censored-data estimate, and it very articulately points upon two strata of lifetimes. At the same time, keeping in mind its ill-posedness and the extremely small size of the censored-data, its overall shape should be considered with the grain of salt. The aggregated estimate is shown by the dotted line which makes the right strata a bit more pronounced with respect to the uncensored-data estimate. A plausible explanation of the two strata, that can be found in publication Maksymiuk et al. (1994) in the *Journal of Clinical Oncology*, is that the survival of participants in the study was primarily defined by the binary stage of cancer, limited or extensive. This is what the density estimates tell us about. The bottom diagrams allow us to look at the confidence bands for the two estimates. The interesting outcome is that the aggregated estimate has the smaller margin of error for the right tail.

Finally, as it was promised, let us return to the open problem of better estimation of the right tail for samples like the ones shown in Fig. 3.3. As we already know from the presented examples and numerical studies, aggregation in the frequency domain of Fourier coefficient estimators, based on uncensored and censored subsamples, improves estimation for right-censored data. At the same time, Fig. 3.3 indicates that the aggregation in frequency domain is not completely satisfactory for addressing the right tail issue. To shed light on the reason, let us look one more time at formula (3.3.21). Here the "global" Fourier coefficient estimators $\tilde{\theta}_{uj}^*$ and $\tilde{\theta}_{cj}^*$, that is Fourier coefficients for the density on its support and the corresponding orthonormal basis on the support, are aggregated according to their "global" variances. Asymptotically he variances are reciprocals of the corresponding "global" Fisher informations $\mathcal{J}_{uj} = [\int_0^1 (f^T(t)/S^C(t))dt]^{-1}$ and $\mathcal{J}_{cj} = [(\pi j)^2 \int_0^1 (S^T(t)/f^C(t))dt]^{-1}$. Accordingly, these Fisher informations are not tailored toward the right tail and represent the support of T as a whole. A feasible remedy for the right tail issue is to look at "pointwise" Fisher informations and the corresponding frequency-spatial aggregation. Namely, consider a fixed time t_0 and decreasing intervals $(t_0 - s_n, t_0 + s_n)$ around t_0 with $s_n = o_n(1)$. Then introduce a sequence of bases φ_{nj} on the local shrinking intervals $(t_0 - s_{nj}, t_0 + s_{nj})$, consider Fourier coefficients of f^T with respect to that local bases around point t_0, and calculate the ratio $\mathcal{J}_u(j, f^T, t_0)/\mathcal{J}_c(j, f^T, t_0)$ of Fisher informations for Fourier coefficient with index j based on uncensored and censored observations. Under a mild assumption we have the following formula for that ratio, $\mathcal{J}_u(j, f^T, t_0)/\mathcal{J}_c(j, f^T, t_0) = [(\pi j)^2 S^C(t_0) S^T(t_0)]/[f^T(t_0) f^C(t_0)]$. The formula tells us that the relative spatial Fisher information contained in censored subsample with respect to uncensored subsample may dramatically increase in the right tail because $S^T(t_0) S^C(t_0) \to 0$ as t_0 increases. The latter is the underlying idea of the frequency-spatial aggregation that may improve estimation of the right tail. Of course, special bases are needed to utilize the idea. Wavelets, and specifically multiwavelets that are created to work with a function and its derivative, may be a good choice. See references in the Notes.

3.3.5 Density Estimation Over Unknown Support

So far it was assumed that support of the lifetime of interest T is known. This subsection relaxes this assumption. Namely, it is assumed that $S^T(b) = 0$ for some finite b, but the value of b is unknown. The problem is to estimate f^T under the MISE (mean integrated square error) criterion over the interval $[0, b]$. Because the nuisance parameter b is unknown, the stated problem is equivalent to estimating f^T over the half-line $[0, \infty)$, that is using the MISE criterion $\mathbb{E}\{\int_0^\infty (\check{f}(t) - f^T(t))^2 dt\}$ for a density estimator $\check{f}$.

To shed light on complexity of the stated problem, recall that the survival function of the observed V is $S^V(t) = S^T(t) S^C(t)$, and the mixed joint density $f^{V,\Delta}(t, \delta)$ of the pair (V, Δ) is

$$f^{V,\Delta}(t, \delta) = [f^T(t) S^C(t)]^\delta [f^C(t) S^T(t)]^{1-\delta}, \quad (t, \delta) \in [0, \infty) \times \{0, 1\}. \qquad (3.3.23)$$

The formula is symmetric with respect to T and C because C right-censors T while T right-censors C. This symmetry is at the root of the unknown support issue. Indeed, first of all note that if $S^C(b - \epsilon) = 0$ and $S^T(b - \epsilon) > 0$ for some $\epsilon > 0$, then consistent estimation of f^T is impossible. Second, it is clear from (3.3.23) that to estimate f^T one needs to estimate S^C, and then $S^T(b) = 0$ creates the right-tail issue for estimation of S^C even if it is known that $S^C(b) > 0$. In other words, if $S^C(b) > 0$ and accordingly consistent estimation of f^T is possible, then the lifetime of interest T is "responsible" for the right-tail issue with estimation of S^C. Another interesting conclusion from this discussion is that, using the symmetry, we will be able to consider the case when $S^C(b) = 0$, that is the case of the same supports for T and C.

Assumption 3.2 The lifetime of interest T and the censoring lifetime C are independent continuous random variables. For some finite and in general unknown positive constant b relations $S^T(b) = 0$ and $S^C(b) > 0$ hold. The density $f^T(t)$ has a bounded derivative on $[0, b]$.

Given Assumption 3.2, we introduce in turn two nonparametric density estimators based on a sample of size n from (V, Δ). Recall that ζ denotes a generic positive constant. The following deterministic sequence γ_n will be used in definition of the estimators,

$$\gamma := \gamma_n \in (\zeta[\ln^2(n)/n]^{1/3}, 1/4). \tag{3.3.24}$$

The sequence γ_n (to simplify formulas we may skip the subscript n and write γ) is used to define the following deterministic and stochastic sequences,

$$k_\gamma := \lfloor (n+1)(1-\gamma) \rfloor, \quad \tilde{b} := \tilde{b}(n, \gamma) := V_{(k_\gamma)}. \tag{3.3.25}$$

Here $\lfloor x \rfloor$ is the largest integer not larger than x, and $V_{(l)}$ is the lth order statistic, that is $V_{(1)} \leq \dots \leq V_{(n)}$. Further, to simplify formulas, it is assumed that pairs $(V_l, \Delta_l), l = 1, \dots, n$ are arranged in ascending order of V_l, that is $V_l = V_{(l)}$.

The underlying idea is to use a series density estimator with a basis on the stochastic interval $[0, \tilde{b}]$. In other words, the density estimator is constructed over the interval defined by the order statistic $V_{(k_\gamma)}$. To shed light on the deterministic sequences γ_n and k_γ, let us introduce the empirical survival function

$$\hat{S}^V(v) := n^{-1} \sum_{l=1}^{n} I(V_l \geq v), \tag{3.3.26}$$

and present two relations for γ_n and k_γ,

$$S^V(\tilde{b}) = \frac{n - k_\gamma + 1}{n+1} \in [\gamma_n, \gamma_n + \frac{1}{n+1}], \tag{3.3.27}$$

$$\hat{S}^V(\tilde{b}) = \frac{n - k_\gamma + 1}{n} = S^V(\tilde{b})(1 + \frac{1}{n}). \tag{3.3.28}$$

We see that $S^V(\tilde{b}) = \gamma_n(1 + o_n(1))$ and $\hat{S}^V(\tilde{b}) = \gamma_n(1 + o_n(1))$ almost sure. These relations are the motivation behind the introduced γ_n and k_γ.

Now we define the density estimator,

$$\tilde{f}(t) := \tilde{f}(t, \tilde{b}, \gamma_n) := \sum_{j=0}^{J(n,\gamma_n)} \tilde{\theta}_j \eta_{\tilde{b}j}(t) I(t \in [0, \tilde{b}]). \tag{3.3.29}$$

Here $J(n, \gamma_n) := 1 + \lceil \gamma_n n^{1/3} \rceil$, η_{rj} are elements of the cosine basis on $[0, r]$,

$$\tilde{\theta}_j := n^{-1} \sum_{l=1}^{k_\gamma-1} \frac{\Delta_l \eta_{\tilde{b}j}(V_l)}{\max(\hat{S}^C(V_l), 1/[c_S \ln(n)])}, \tag{3.3.30}$$

$$\hat{S}^C(t) := e^{-n^{-1} \sum_{l=1}^{n}(1-\Delta_l)I(V_l \le t)/\hat{S}^V(V_l)}, \tag{3.3.31}$$

$\hat{S}^V$ is defined in (3.3.26), and in (3.3.30) the c_S is a positive constant. Note that $\hat{S}^C$ is well defined because $\hat{S}^V(V_l) \ge n^{-1}$.

Recall that ζ denotes generic positive constants that may be different even in the same line.

Theorem 3.4 *Suppose that Assumptions 3.2 and (3.3.24) hold. Then the density estimator (3.3.29) satisfies the following two relations,*

$$\mathbb{E}\{\int_0^{\hat{b}} (\tilde{f}(t) - f^T(t))^2 dt\} \le \zeta \gamma_n^{-2} n^{-2/3}, \tag{3.3.32}$$

and almost sure

$$\int_{\tilde{b}}^{\infty} (\tilde{f}(t) - f^T(t))^2 dt \le \zeta \gamma_n. \tag{3.3.33}$$

Further, if $\gamma_n = O_n(1)n^{-2/9}$, then

$$\mathbb{E}\{\int_0^{\infty} (\tilde{f}(t) - f^T(t))^2 dt\} \le \zeta n^{-2/9}. \tag{3.3.34}$$

The estimator (3.3.29) does not use the available additional information that $\tilde{b} < \hat{b} := V_{(n)} < b$ almost sure. This remark allows us to introduce a modified estimator $\hat{f}(t) := \tilde{f}(t, \hat{b}, \gamma_n)$ which estimates the density over the larger interval $[0, \hat{b}]$ and still, due to (3.3.30), uses only first $k_\gamma - 1$ observations of V_l. Then the inequality (3.3.34) of Theorem 3.4 is still valid.

Let us make several comments about the theory. First, for a direct sample of size n from T, the optimal rate of estimating a differentiable density f^T is $n^{-2/3}$, and if we restrict estimation to a fixed subinterval of $[0, b]$, then for censored data this rate is attainable by estimates based on the Kaplan-Meier estimator of S^C. By an appropriate choice of γ_n we can either "almost" match the above-mentioned optimal rate over increasing intervals $[0, V_{(k_\gamma)}]$, or by choosing a faster decreasing γ_n attain the very respectful rate $n^{-2/9}$ of the MISE convergence over an unknown support of T. Second, Assumption 3.2 has no restriction on tails of f^T and f^C that are often used in analysis of the effect of right tail on density estimation over its support, see the Notes. Third, numerical simulations show that $\hat{f}$ performs better for small samples due to utilizing the more accurate estimate of b.

The made assumption $S^C(b) > 0$ is traditional in the survival analysis literature devoted to consistent estimation, but still it is possible to relax it and consider a more complicated case $S^C(b) = 0$. Then, as the next theorem shows, rate of the MISE convergence slows down but it is still a power function in n.

Theorem 3.5 *Let Assumption 3.2 hold but in place of $S^C(b) > 0$ it is assumed that*

$$\int_0^b \frac{f^T(t)}{S^C(t)} dt < \infty. \qquad (3.3.35)$$

Consider a density estimator $\bar{f}$ defined as in lines (3.3.29)–(3.3.31) only with $J(n, \gamma_n) := 1 + \lfloor \gamma_n^{8/3} n^{1/3} \rfloor$ and in (3.3.30) the estimate $\hat{S}^C$ is truncated from below by γ_n in place of $1/\ln(n)$. Suppose that $\gamma_n \in (\zeta n^{-1/6}, 1/4)$. Then the following relations hold,

$$\mathbb{E}\{\int_0^{\tilde{b}} (\bar{f}(t) - f^T(t))^2 dt\} \leq \zeta n^{-2/3} \gamma_n^{-16/3}, \qquad (3.3.36)$$

$$\int_{\tilde{b}}^\infty (\bar{f}(t) - f^T(t))^2 dt \leq \zeta \gamma_n^{1/2} \qquad (3.3.37)$$

almost sure, and if $\gamma_n = O_n(1) n^{-4/35}$ then

$$\mathbb{E}\{\int_0^\infty (\bar{f}(t) - f(t))^2 dt\} \leq \zeta n^{-2/35}. \qquad (3.3.38)$$

To shed light on the assumption (3.3.35), note that if $S^C(b) = 0$, then we have the familiar property $\int_0^b \frac{f^C(t)}{S^C(t)} dt = \infty$ of the hazard rate function. Accordingly, (3.3.35) holds if the right tail of f^T is lighter than the right tail of f^C, say $f^T(t) < \zeta_1(b-t)^{\beta+\nu}$ and $f^C(t) > \zeta_2(b-t)^\beta$ for $t \in (t - \zeta_3, b)$ and $\beta > 0$, $\nu > 1$. Beta distributions are specific "boundary" examples.

Figure 3.5 illustrates the problem and the E-estimator that uses the Fourier coefficient estimator (3.3.30). The left-top diagram shows simulated right-censored data with the underlying f^T being the Normal (the third corner density) and f^C being Uniform(0, 1.1). As we see, 43% of observations are censored, and there are no observations of T, and even to the larger degree of C, in the right tail. This complicates the estimation of f^T. The right-top diagram shows us the underlying survival function S^C (the solid line) over the range of observed lifetimes V, and the circles show the estimates $\hat{S}^C(\Delta_l V_l)$. Note how the stepwise estimate approximates the underlying continuous S^C. The horizontal dashed line shows $1/[c_S \ln(n)]$ which is used to bound the estimate from below, the software allows the user to change c_S via the argument cS. The four density E-estimates (the dashed lines) for different values of γ are shown in the remaining four diagrams together with their confidence bands (the dotted lines). These diagrams also show the underlying density (the solid line) supported on [0, 1], parameters γ and $k := k_\gamma$, the integrated squared error ISE over [0, $V_{(n)}$], the BiasSq := $\int_{V_{(n)}}^1 [f^T(t)]^2 dt$, and the total ISET over [0, 1]. The right-bottom diagram with $\gamma = 0$ also shows the density E-estimate (the dot-dashed line) based on the underlying S^C, and its total ISETO. As we see, for this particular simulation the best γ is 0.01. It is highly recommended to repeat the simulation and explore other possible outcomes. Also, for each experiment, that is each triplet (f^T, f^C, n), it is possible to conduct a numerical study using parameter

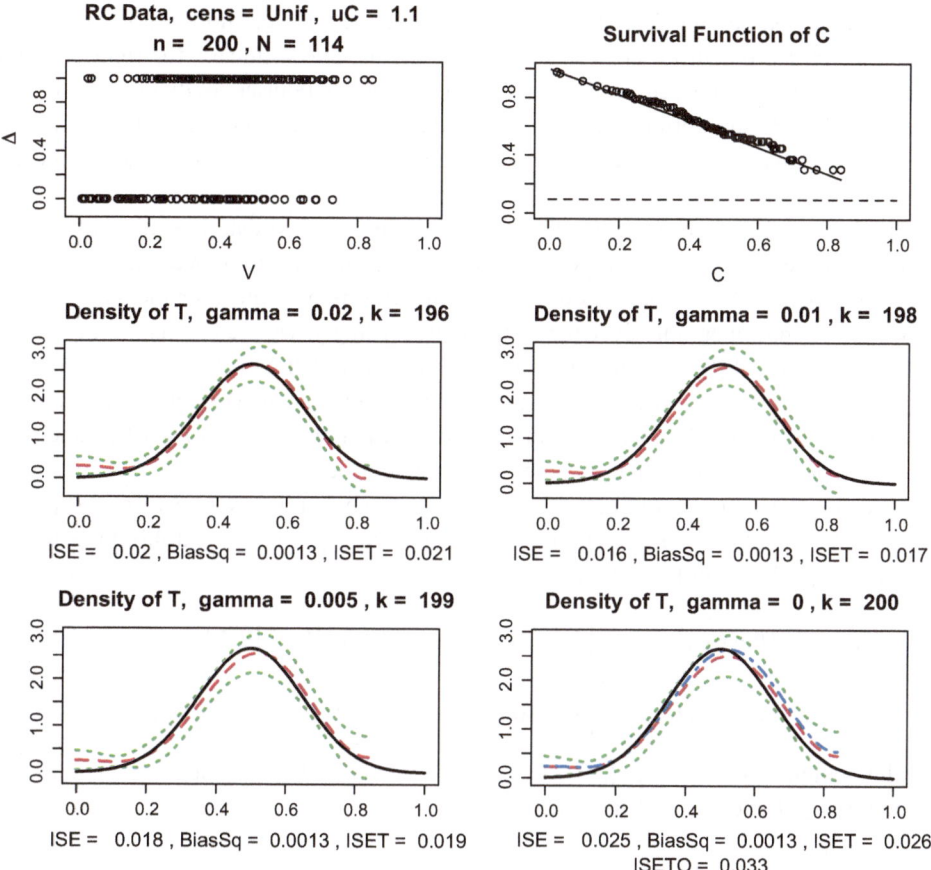

Fig. 3.5 Density estimation with unknown support. The underlying densities are shown by the solid line, density estimates by the dashed line, confidence bands by the dotted lines. In the right-bottom diagram the dot-dashed line is the estimate based on the underlying S^C. [n = 200, corn = 3, cens ="Unif", uC = 1.1, gamma = c(0.02, 0.01, 0.005, 0), cJ0 = 3, cJ1 = 0.8, cTH = 4, cS = 2, nsim = 1]

nsim and choose optimal parameter γ via analysis of sample mean and median values of (ISE, BiasSq, ISET) for the four values of γ.

3.4 Hazard Rate Estimation

Hazard rate is another classical characteristic of the studied lifetime of interest T. Recall that the hazard rate of a continuous lifetime of interest T is defined as $h^T(t) := f^T(t)/S^T(t)$ and also often referred to as the force of mortality. We already know from Sect. 1.2.4 how to estimate h^T based on direct observations of T. As we will see shortly, it is a relatively straightforward problem to suggest an efficient hazard rate estimator for right-censored data. Moreover, for right-censored data efficient hazard rate Fourier coefficient estimator is simpler and more straightforward than for the density. At the same time, efficient hazard rate estimation based on censored observations and efficient aggregation of uncensored and censored observations are open problems, see the Notes.

The theory of efficient hazard estimation for right-censored data is similar to the one presented in Sect. 3.3.1 for density estimation. The principal difference, that we already know from Sect. 1.2.4, is that no consistent estimation is possible over the support of T because h^T is not integrable over the support. Accordingly, we are estimating the hazard rate over an interval $[0, r]$ with a finite restriction r such that $S^T(r) > 0$. In the previous chapters we used the approach of converting estimation over an interval into estimation over the unit interval $[0, 1]$. Here let us directly explore estimating $h^T(t)$ for $t \in [0, r]$ under the MISE criterion. This will be a teachable problem to consider.

Following the approach of Sect. 3.3.1, introduce an anchor hazard rate $h_0(t), t \in [0, \infty)$ which is continuous and positive on $[0, r]$, and define the sequence of local shrinking, toward the anchor h_0, classes of hazard rates,

$$\mathcal{F}'_n := \mathcal{F}_n(h_0, \alpha, Q, r) := \Big\{ h : h(t) = h_0(t) + g(t)I(t \in [0, r]),$$

$$g \in \mathcal{S}_1(\alpha, Q, r), \quad \max_{t \in [0,r]} |g(t)| \le \min_{t \in [0,r]} h_0(t)/\ln(\ln(n+20)) \Big\}. \tag{3.4.1}$$

Recall notation $\eta_{rj}(t), t \in [0, r], j = 0, 1, \ldots$ for elements of the cosine basis on $[0, r]$, and define for $\alpha \ge 1$ and $k = 0, 1$ the function classes

$$\mathcal{S}_k(\alpha, Q, r) := \Big\{ g : g(t) = \sum_{j=k}^{\infty} \theta_j \eta_{rj}(t), \ t \in [0, r]$$

$$\sum_{j=k}^{\infty} (1 + (\pi j/r)^{2\alpha}) \theta_j^2 \le rQ \Big\}. \tag{3.4.2}$$

The class $\mathcal{S}_0(\alpha, Q, r)$ is called the global Sobolev class, and in (3.4.1) we use the sequence $\ln(\ln(n+20))$ because it is larger than 1 for all sample sizes n. As we will see shortly in Theorem 3.6, we need to use the local shrinking Sobolev class (3.4.1) because sharp constant of the MISE convergence depends on an underlying hazard rate. Let us also stress that the anchor h_0 is not necessarily the underlying hazard rate of interest, it simply anchors all underlying hazard rates h^T in its vanishing in L_∞-norm vicinity.

Theorem 3.6 *Consider a right-censored sample of size n from pair $(V, \Delta) := (\min(T, C), I(T \le C))$. The problem is to estimate hazard rate h^T of the lifetime of interest T over interval $[0, r]$ under the MISE criterion. Let Assumption 3.1 hold, $S^V(r) > 0$, and the oracle knows the right-censored data, density f^C and the function class $\mathcal{F}'_n$. Then*

$$\inf_{\tilde{h}^*} \sup_{h^T \in \mathcal{F}'_n} \mathbb{E}\{(n/d_r)^{2\alpha/(2\alpha+1)} \int_0^r (\tilde{h}^*(t) - h^T(t))^2 dt\} \geq P'(1 + o_n(1)), \tag{3.4.3}$$

where the infimum is over all possible oracle-estimators $\tilde{h}^*$,

$$P' := P'(\alpha, Q, r) := \frac{Q^{1/(2\alpha+1)} r\alpha^{2\alpha/(2\alpha+1)}(2\alpha+1)^{1/(2\alpha+1)}}{[\pi(\alpha+1)]^{2\alpha/(2\alpha+1)}}, \tag{3.4.4}$$

$$d_r := d_r(h^T, S^C) := r^{-1} \int_0^r h^T(t)[S^T(t)S^C(t)]^{-1} dt. \tag{3.4.5}$$

The lower bound (3.4.3) is sharp and attainable by an oracle-estimator based solely on uncensored observations. Furthermore, if additionally the anchor $h_0 \in S_0(\alpha', Q', r)$ with $\alpha' > \alpha$ and $Q' < \infty$, then the oracle can propose an estimator attaining the lower bound without knowledge of the anchor.

It immediately follows from Theorem 3.6 that the lower bound (3.4.3) also holds for oracle-estimators based on uncensored observations. This supports the Kaplan-Meier paradigm about dominance of uncensored observations. Further, the problem of hazard rate estimation based on censored observations is ill-posed similarly to the density estimation. Further, similarly to the density estimation, the optimal rate for hazard rate estimation is $n^{-2\alpha/(2\alpha+1)}$. The latter should not be a big surprise because hazard rate is the ratio between density and survival function, $h^T(t) = f^T(t)/S^T(t)$, and the survival function is estimated more accurately than the density. The main difference is in the coefficient of difficulty (3.4.5) for hazard rate estimation. Recall that for density the corresponding coefficient of difficulty is $d_r^* := r^{-1} \int_0^r f^T(t)[S^C(t)]^{-1} dt \leq d_r$. Further, the coefficients of difficulty show that if for density estimation there is a chance that $d_\infty^* < \infty$, for the hazard rate estimation $d_r \to \infty$ as $r \to \infty$. At the same time, efficient estimation of the hazard rate is simpler. Let us explain how the hazard rate can be estimated.

We begin with reviewing several useful probability formulas. Recall that we observe a sample from $(V, \Delta) := (\min(T, C), I(T \leq C))$ where the lifetime of interest T and the censoring lifetime C are independent continuous random variables. Then we get

$$S^V(t) := \mathbb{P}(\min(T, C) > t) = \mathbb{P}(T > t, C > t) = S^T(t)S^C(t), \tag{3.4.6}$$

and

$$f^{V,\Delta}(t, \delta) = [f^T(t)S^C(t)]^\delta [f^C(t)S^T(t)]^{1-\delta} I(t \geq 0, \delta \in \{0, 1\}). \tag{3.4.7}$$

Using these results we get the following useful formula for the hazard rate,

$$h^T(t) := \frac{f^T(t)}{S^T(t)} = \frac{f^{V,\Delta}(t, 1)}{S^C(t)S^T(t)} = \frac{f^{V,\Delta}(t, 1)}{S^V(t)}. \tag{3.4.8}$$

Note that the hazard rate of interest is expressed as the ratio of the density and the survival function for the observed variables. This is what makes series estimation of the hazard rate straightforward. Indeed, recall our notation η_{rj} for elements of the cosine basis on $[0, r]$. Then the Fourier coefficient of the hazard rate can be written as

$$\theta_j := \int_0^r h^T(t)\eta_{rj}(t)dt = \int_0^r \frac{f^{V,\Delta}(t,1)\eta_{rj}(t)}{S^V(t)}dt$$

$$= \mathbb{E}\left\{\frac{\Delta I(V \leq r)\eta_{rj}(V)}{S^V(V)}\right\}. \tag{3.4.9}$$

This yields the following plug-in sample mean Fourier coefficient estimator,

$$\hat{\theta}_j := n^{-1}\sum_{l=1}^n \frac{\Delta_l \eta_{rj}(V_l)I(V_l \leq r)}{\hat{S}^V(V_l)}, \tag{3.4.10}$$

where

$$\hat{S}^V(v) := n^{-1}\sum_{l=1}^n I(V_l \geq v) \tag{3.4.11}$$

is the empirical survival function of V. Note that $\hat{S}^V(V_l) \geq n^{-1}$ and hence can be used in denominators. Further, a direct calculation shows that

$$\lim_{n\to\infty}\lim_{j\to\infty} n\mathbb{V}\{\hat{\theta}_j\} = r^{-1}\int_0^r \frac{f^{V,\Delta=1}(t)}{[S^V(t)]^2}dt$$

$$= r^{-1}\int_0^r \frac{h^T(t)}{S^T(t)S^C(t)}dt = d_r. \tag{3.4.12}$$

We can conclude that the Fourier coefficient estimator (3.4.10) is efficient. The latter, together with Theorem 3.6 and the results of Sect. 3.3, allow us to state the following. First, the nonparametric series estimator (3.3.10), with $\tilde{\theta}_{uj}^*$ being replaced by the $\hat{\theta}_j$ defined in (3.4.10) and d_u being replaced by the d_r defined in (3.4.5), is efficient and attains the lower bound of Theorem 3.6. Second, the corresponding blockwise shrinkage estimator (3.3.20) is efficient. Finally, the Fourier coefficient estimator (3.4.10) allows us to use the E-estimation algorithm for construction of the hazard rate E-estimator for small samples.

Let us explain how the interval $[0, r]$ of the estimation can be chosen. One of the approaches is to estimate the coefficient of difficulty d_r by the following method of moments estimator,

$$\hat{d}_r := n^{-1}r^{-1}\sum_{l=1}^n \frac{\Delta_l I(V_l \leq r)}{[\hat{S}^V(V_l)]^{-2}}. \tag{3.4.13}$$

Recall that for the case of direct observations of a lifetime T supported on $[0, 1]$, the coefficient of difficulty for density estimation is 1. This setting, with the unit coefficient of

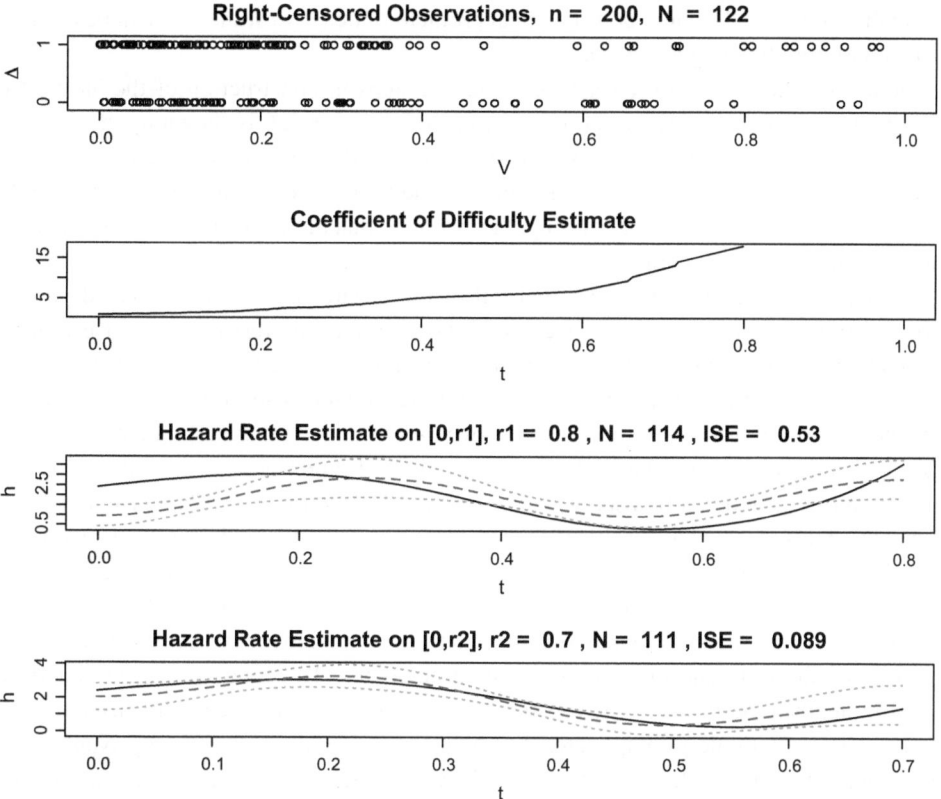

Fig. 3.6 Estimation of the hazard rate based on right-censored observations. Simulated sample of size $n = 200$ from (V, Δ) is shown in the top diagram, $N := \sum_{l=1}^{n} \Delta_l = 122$ is the number of uncensored observations. Estimate $\hat{d}_t, t \in [0, \max(r1, r2)]$ of the coefficient of difficulty is shown in the second from the top diagram. Two bottom diagrams show hazard rate estimates for $r = r1 = 0.8$ and $r = r2 = 0.7$, respectively. The curves are the underlying hazard rate (the solid line), the E-estimate (the dashed line), and the confidence band (the dotted lines). The $N := \sum_{l=1}^{n} \Delta_l I(V_l \le r)$ shows the number of uncensored observations within $[0, r]$. {Censoring distribution is either the default Uniform$(0, u_C)$ with $u_C = 1.5$, or Exponential(λ_C) with the default $\lambda_C = 1.5$ where λ_C is the mean. To choose an exponential censoring, set *cens* = $"Expon"$. Confidence level is controlled by the argument *alpha*. } [$n = 300$, corn $= 2$, cens $="Unif"$, $uC = 1.5$, *lambdaC* $= 1.5$, $r = c(0.8, 0.7)$, *alpha* $= 0.05$, *cJ0* $= 4$, *cJ1* $= 0.5$, *cTH* $= 4$]

difficulty for the well-understood statistical problem of density estimation, can be used as a benchmark for choosing a reasonable r.

Figure 3.6 sheds light on the above-explained steps in construction of the hazard rate E-estimator. A simulated right-censored sample of size $n = 200$ is shown by the circles in the top diagram. The sample size is relatively small, and note that there are only $N = 122$ uncensored observations. In this simulation T is the Bathtub, C is Uniform(0, 1.5), and that combination yields the severe censoring. Further, note that there are only a few observations in the right tail of the support.

As we know, estimation of the right tail of the hazard rate is a complicated problem even asymptotically, and our next step is to choose a feasible restriction r for estimating the hazard rate over interval $[0, r]$. The second from the top diagram allows us to evaluate complexity of the problem at hand and to choose a reasonable restriction r. Namely, here we see the estimate $\hat{d}_t$ for $t \in [0, \max(r1, r2)]$ where $r1$ and $r2$ are the arguments of Fig. 3.6. Recall that for estimating the Bathtub density f^T and direct observations the coefficient of difficulty is 1 and can be used as the benchmark. We know from Chap. 1 that $n = 200$ typically yields reasonable density estimates. However, the estimated hazard rate coefficient of difficulty is much larger if $r > 0.4$ and near 18 for $r = 0.8$. We may conclude that we are dealing with the extremely complicated hazard rate estimation problem.

The two bottom diagrams show E-estimates for $r = 0.8$ and $r = 0.7$, respectively. These particular values are chosen to show possible outcomes. The solid line is the underlying hazard rate, the dashed line is the E-estimate, the dotted lines are the confidence band. The case $r = 0.8$ is extremely complicated due to the large coefficient of difficulty. Nonetheless, note that the estimate shows us the "spirit" of the underlying hazard rate. At the same time, look at how large the confidence band is, it reflects the extremely large coefficient of difficulty. The outcome is better for $r = 0.7$. The reader is advised to repeat this figure with different arguments and get first-hand experience in dealing with hazard rate estimation for right-censored data.

3.5   Nonparametric Regression

We are considering nonparametric regression with right-censored (RC) responses and right-censored predictors in turn. In the former case the regression of interest is $m(x) = \mathbb{E}\{T|X = x\}$ and the sample is from $(X, V, \Delta) := (X, \min(T, C), I(T \le C))$. In the latter case the regression of interest is $m(t) = \mathbb{E}\{Y|T = t\}$ and the sample is from $(V, \Delta, Y) := (\min(T, C), I(T \le C), Y)$. These regressions are often of interest in survival analysis, and there is a vast literature devoted to the topic and presented in the Notes.

3.5.1 Right-Censored Response

The main message of this subsection is that, in contrary to the density and hazard rate estimation, censored responses are not ill-posed and should not be ignored for regression. This assertion contradicts the standard methodology of the dominance of uncensored observations. Indeed, the dominance of uncensored observations is at the core of the seminal Cox's regression methodology of partial likelihood as well as the popular Buckley-James imputation procedure when censored lifetimes are replaced by their conditional expectations calculated using uncensored observation. Accordingly, we begin with the theory of nonparametric regression based on censored responses, and the terminology is explained shortly.

The underlying regression model is as follows. There is an underlying pair of interest (X, T) where X is the predictor and lifetime T is the response, and the problem is to estimate nonparametric regression $m(x) = \mathbb{E}\{T|X = x\}$. The response T is not observed directly. Instead, we observe a sample of size n from (X, V, Δ) where $V := \min(T, C)$, $\Delta := I(T \leq C)$ and C is the censoring variable. The right-censoring partitions the sample into two complementary subsample with uncensored and censored responses when we observe realizations of $(X, T, \Delta = 1)$ and $(X, C, \Delta = 0)$, respectively. It is also possible to say that the uncensored observations are sampled from $(\Delta X, \Delta V, \Delta)$ and censored observations from $(\Delta'X, \Delta'V, \Delta)$, $\Delta' := 1 - \Delta$. This terminology for the two complementary subsamples and/or observations is used in this subsection. For instance, the term censored-data estimator means that the estimator is based on the censored subsample.

To formulate the main theoretical result, we are using the oracle's approach when the design density f^X and the density f^C of the censoring variable are known. We begin with assumptions. The first assumption is about the underlying distributions.

Assumption 3.3 The conditional density $f^{T|X}(t|x)$ is supported on $\mathcal{R}_b := [0, b] \times [0, 1]$. Censoring variable C is a continuous lifetime, its density f^C is positive and continuous on $[0, b)$, and C is independent of (X, T). Predictor X is a continuous variable with density f^X which is continuous, positive and supported on $[0, 1]$.

Our next assumption is about the oracle. Recall that Sobolev classes $\mathcal{S}_k(\alpha, Q), k = 0, 1$ are defined in (3.3.2).

Assumption 3.4 The oracle knows an anchor conditional survival function $S_0^{T|X}(t|x)$, $(t, x) \in \mathcal{R}_b$. Anchor $S_0^{T|X}(t|x)$ is continuous in (t, x) and differentiable in t on $\mathcal{R}_b$, and for any positive constant $a < b$

$$\min_{(t,x)\in\mathcal{R}_a} S_0^{T|X}(t|x) \geq u_1(a) > 0, \quad \max_{(t,x)\in\mathcal{R}_a} \partial S_0^{T|X}(t|x)/\partial t \leq -u_2(a) < 0. \tag{3.5.1}$$

Set $m_0(x) := \int_0^\infty S_0^{T|X}(t|x)dt$. For each sample size n the oracle knows that an underlying $S^{T|X}$ belongs to the local shrinking Sobolev class

$$\mathcal{F}_n := \{S^{T|X} : \int_0^\infty S^{T|X}(t|x)dt \in \mathcal{M}(\alpha, Q, m_0, n)\}, \qquad (3.5.2)$$

where

$$\mathcal{M}(\alpha, Q, m_0, n) := \Big\{m : m(x) = m_0(x) + g(x),$$

$$g(x) \in \mathcal{S}_1(\alpha, Q), \ |g(x)| \le 1/\ln(\ln(n+20)), \ x \in [0, 1]\Big\}. \qquad (3.5.3)$$

Let us explain the assumptions. The conditional survival function $S_0^{T|X}(t|x)$ anchors all possible underlying survival functions whose regression functions satisfy the additive perturbation (3.5.3). It also defines the anchor regression m_0. Because conditional survival function must be bona fide (nonnegative and nonincreasing in t), restriction (3.5.1) on the anchor is introduced. Also note that the second inequality in (3.5.1) implies that the anchor conditional density $f_0^{T|X}(t|x)$ is positive on $\mathcal{R}_a$. Let us note that a may depend on n. It is possible to relax the assumption about independence of C and (T, X) and replace it by conditional independence of T and C given X.

Now we are in a position to present a sharp lower bound for MISE of the oracle who uses only censored observations.

Theorem 3.7 *Consider a nonparametric regression problem of estimating* $m(x) = \mathbb{E}\{T|X = x\}$ *by the oracle who knows the nuisance functions* f^X, f^C *and the function class* $\mathcal{F}_n$ *defined in (3.5.2). The oracle uses only censored observations from a right-censored sample of size n from* $(X, V, \Delta) := (X, \min(T, C), I(T \le C))$. *Suppose that Assumptions 3.3 and 3.4 hold and*

$$d_c' := \int_0^1 \frac{\int_0^\infty \frac{S^{T|X}(t|x)}{f^C(t)}dt}{f^X(x)} dx < \infty. \qquad (3.5.4)$$

Then

$$\inf_{\tilde{m}^*} \sup_{S^{T|X} \in \mathcal{F}_n} [n/d_c']^{2\alpha/(2\alpha+1)} \mathbb{E}\{\int_0^1 (\tilde{m}^*(x) - m(x))^2 dx\} \ge P(1 + o_n(1)). \qquad (3.5.5)$$

Here the infimum is taken over all possible oracle-estimators and P is equal to the right side of (3.3.3). The lower bound is sharp and attainable by a censored-data oracle-estimator. Further, if the anchor $m_0 \in \mathcal{S}_0(\alpha', Q')$, $\alpha' > \alpha$, $Q' < \infty$ *then the lower bound is attainable by a censored-data oracle-estimator that does not use the anchor.*

Several comments are due. First, we know from Sect. 1.2.5 that $n^{-2\alpha/(2\alpha+1)}$ is the optimal rate of regression estimation for the case of a directly observed sample from (X, T). Theorem 3.7 asserts that using censored observations yields the same rate. Accordingly, for the regression problem censored observations are no longer ill-posed and the idea of aggregation is fertile. Second, the lower bound (3.5.5) is attainable by oracle-estimators even if the supremum is taken over the global Sobolev class $S_0(\alpha, Q)$ defined in (3.3.2). Note that this property matches the result known for the case of direct data. Third, for direct observations and a classical regression $Y = m(X) + v(X)\xi$ with standard Normal ξ, we would see in the lower bound (3.5.5) the functional $d := \int_0^1 v^2(x)[f^X(x)]^{-1}dx$ in place of d'_c. This allows us to conclude that using only censored observations is similar to the classical regression with normal regression errors and the scale function

$$v(x) = \left[\int_0^\infty \frac{S^{T|X}(t|x)}{f^C(t)} dt \right]^{1/2}. \tag{3.5.6}$$

This is an interesting outcome of the theory which sheds a new light on regression with right-censored responses. Fourth, the integral (3.5.4) can be finite even if T is supported on $[0, \infty)$. In that case the conditional survival $S^{T|X}(t|x)$ should decrease in t a "bit" faster than $f^C(t)$, for instance if for large t we have $S^{T|X}(t|x)/f^C(t) \leq \zeta t^{-1-\beta}$, $\beta > 0$. The corresponding example will be considered shortly.

Now let us verify that the lower bound is sharp and attainable by a series oracle-estimator. First of all, we need to understand how to construct an efficient Fourier coefficient estimator that uses censored observations and f^C. Recall that the likelihood (the mixed joint density) for a censored observation is

$$f^{X,V,\Delta}(x, t, 0) = f^X(x) f^C(t) S^{T|X}(t|x). \tag{3.5.7}$$

Accordingly, we need to understand how to express the Fourier coefficient

$$\theta_j := \int_0^1 m(x)\varphi_j(x)dx = \int_0^1 \mathbb{E}\{T|X = x\}\varphi_j(x)dx \tag{3.5.8}$$

via the conditional survival function $S^{T|X}$.

The hint how to do that is to recall the classical formula

$$\mathbb{E}\{T\} = \int_0^\infty S^T(t)dt, \tag{3.5.9}$$

and then repeat its proof that is based on integration by parts. Using this hint we continue (3.5.8),

$$\theta_j = \int_0^1 \left[\int_0^\infty t f^{T|X}(t|x)dt \right]\varphi_j(x)dx$$

$$= \int_0^1 \left[[-t S^{T|X}(t|x)]_{t=0}^{t=\infty} + \int_0^1 S^{T|X}(t|x)dt \right]\varphi_j(x)dx$$

$$= \int_0^1 \int_0^\infty S^{T|X}(t|x)\varphi_j(x)dtdx. \tag{3.5.10}$$

Note that we have established the analog $\mathbb{E}\{T|X = x\} = \int_0^\infty S^{T|X}(t|x)dt$ of (3.5.9).

Using (3.5.7) we can continue (3.5.10) and get

$$\theta_j = \int_0^1 \int_0^\infty \frac{f^{X,V,\Delta}(x,t,0)\varphi_j(x)}{f^X(x)f^C(t)}dtdx = \mathbb{E}\{\frac{\Delta'\varphi_j(X)}{f^X(X)f^C(V)}\}. \tag{3.5.11}$$

Formula (3.5.11) yields the sample mean Fourier coefficient oracle-estimator based on censored observations,

$$\hat{\theta}_{cj}^* := n^{-1}\sum_{l=1}^n \frac{\Delta_l'\varphi_j(X_l)}{f^X(X_l)f^C(V_l)}. \tag{3.5.12}$$

This oracle-estimator is unbiased and a direct calculation shows that

$$\mathbb{V}\{\hat{\theta}_{cj}^*\} = n^{-1}d_c'(1 + o_j(1)). \tag{3.5.13}$$

Let us present a series oracle-estimator (compare with the estimators in Sect. 3.3.2) which is based on censored observations and attains the lower bound (3.5.5). Set

$$\hat{m}^*(x) := \sum_{j=0}^{J_n} \hat{\theta}_{cj}^* I((\hat{\theta}_{cj}^*)^2 > c_{TH}\hat{\sigma}_{jn}^2)\varphi_j(x)$$

$$+ \sum_{j=J_n+1}^{J_{cn}'} (1 - (j/J_{cn}')^\alpha)\hat{\theta}_{cj}^*\varphi_j(x), \tag{3.5.14}$$

where J_n is the same as in (3.3.10) and J_{cn}' is equal to the right side of (3.3.11) with d_u being replaced by d_c'. If $m_0 \in S_0(\alpha', Q')$, $\alpha' > \alpha$, $Q' < \infty$, then $\hat{m}^*$ is efficient and attains the lower bound (3.5.5). Further, similarly to the previous sections, we can use the corresponding E-estimator for small samples and the blockwise shrinkage for sharp-minimax adaptive estimation. Further, in place of unknown densities f^X and f^C we can use the estimates of Sects. 1.2 and 3.3, respectively. Finally, let us stress that the product $f^X f^C$ is the natural nuisance function for a censored subsample.

Nonparametric regression based on uncensored observations is better known in the survival analysis literature. This regression is straightforward because the likelihood $f^{X,V,\Delta}(x,t,1) = f^X(x)S^C(t)f^{T|X}(t|x)$ is proportional to the conditional density $f^{T|X}$ defining the nonparametric regression. Then it is straightforwardly verified that the sample mean Fourier coefficient oracle-estimator

$$\hat{\theta}_{uj}^* := n^{-1}\sum_{l=1}^n \frac{\Delta_l V_l \varphi_j(X_l)}{f^X(X_l)S^C(V_l)} \tag{3.5.15}$$

is unbiased, and can be used by the plug-in E-estimator for small samples. The aggregation of Fourier coefficient estimators in frequency domain is the same as in Sect. 3.3.3.

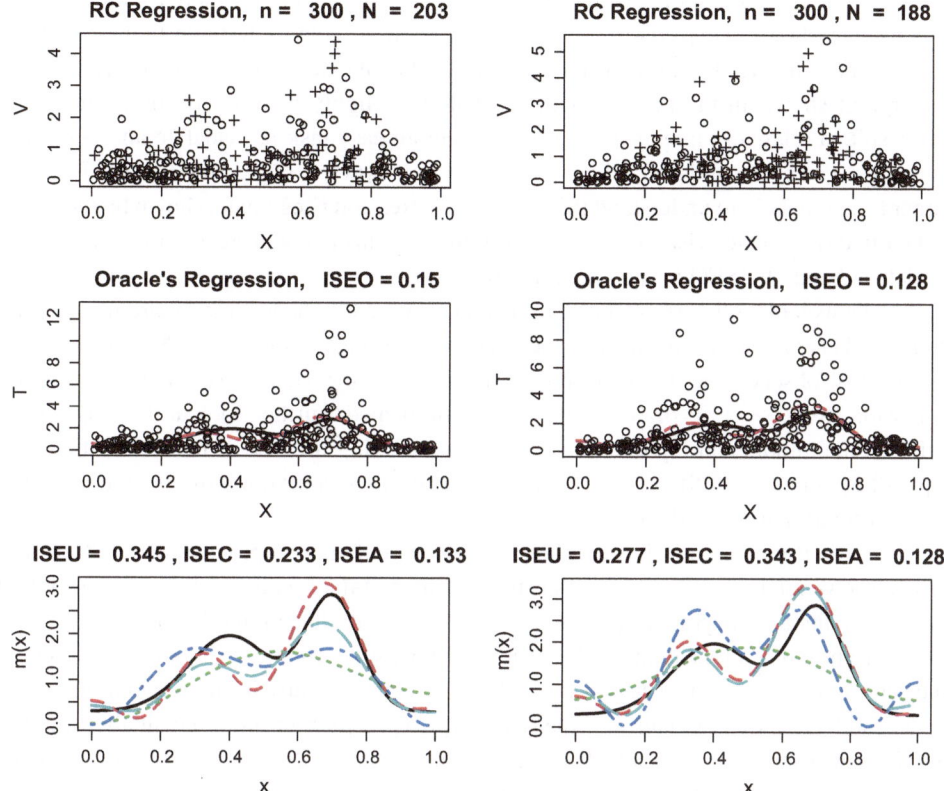

Fig. 3.7 Two simulated nonparametric regressions with predictor X and right-censored response T. The experiments use the same underlying model with the regression $m(x) = a + f_4(x)$ where f_4 is the Bimodal (number 4) corner function, T has exponential distribution with mean $m(X)$, X is Uniform(0,1), C is Exponential(λ) (another possibility is Uniform(0,b), use the argument cens="Unif"). The circles and crosses show uncensored and censored responses, respectively, and $N = \sum_{l=1}^{n} \Delta_l$. The solid line is the underlying regression, the short-dashed line is the oracle-estimate based on underlying observations of T, the dotted line is the uncensored-data estimate, the dot-dashed line is the censored-data estimate, the long-dashed line is the aggregated estimate. {While the black-white diagrams look overcrowded, the curves are colored on the monitor.} [n = 300, corn = 4, a = 0.3, cens ="Expon", lambda = 2, b = 4, cJ0 = 3, cJ1 = 0.8, cTH = 4, nsim = 1]

Let us verify feasibility of using censored observations for small samples. As we know from Sect. 3.3 and analysis of the bottom diagram in Fig. 3.2 with the estimated Bimodal density, censored observations are of a little help for estimating high-frequency densities like the Bimodal due to ill-posedness of censored observations for density estimation. The regression theory states that this is no longer the case for censored responses. Let us check this assertion via analysis of a simulated regression with the regression function having the Bimodal density shape.

Two particular simulations from the same underlying regression model with censored responses are shown in two columns of Fig. 3.7, its caption explains the underlying model and the diagrams. First, let us look at the left column, and begin with the underlying (not censored) scattergram shown in the middle diagram. This is a highly heteroscedastic scattergram, and despite a relatively large sample size it is not an easy task to visualize an underlying regression. Does the solid line, which is $m(x) := a + f_4(x)$, $a = 0.3$ and f_4 is the Bimodal corner function, help to understand the data? The regression E-estimate, shown by the short-dashed line (it is better visualized in the bottom diagram) is not perfect but reasonable for the data at hand, its ISEO is shown in the title.

Now let us look at the RC scattergram from (X, V, Δ) in the left-top diagram where the circles and crosses show uncensored and censored lifetimes of interest T. Note that almost one third of observations are censored. Here the task of visualization of the underlying regression $\mathbb{E}\{T|X = x\}$ is dramatically more complicated. Further, look at the ranges of T and V in the middle and top diagrams. For the underlying lifetime of interest T the range is $[0, 12.5]$ versus $[0, 4.5]$ for the observed V, and this is a severe issue for the available right-censored regression data.

The left-bottom diagram shows us the E-estimates and their empirical integrated squared errors. The worst among the estimates is the uncensored-data estimate shown by the dotted line and having the integrated squared error ISEU $= 0.345$. It is unimodal, but on the bright side its right tail is fatter than the left one, and thus it "correctly" smooths the Bimodal. The censored-data estimate (the dot-dashed) line is better because it shows two pronounced modes, and ISEC$=0.233$. This is a pleasant surprise keeping in mind the smaller number of censored responses. Also, recall from Sect. 3.3.3 that using censored observations for estimation of the Bimodal density we could not observe two modes. On the other hand, note that the two estimates are smaller in magnitude than the underlying regression. The aggregated estimate (the long-dashed line) is a pleasant surprise with the ISEA $= 0.133$ smaller than the oracle's ISEO $= 0.15$. We see the pronounced shape with two distinct modes resembling the Bimodal, their magnitudes are still smaller than the underlying regression but much improved with respect to other estimates.

Now let us look at the second simulation analyzed in the right column of Fig. 3.7. It is again difficult to visualize the underlying regression, hence let us look at the estimates. The oracle-estimate has two pronounced modes and the relatively small increasing tails, and its modes are larger than the underlying modes. The uncensored-data estimate is unimodal, and the censored-data estimate is bimodal with increasing tails mimicking the oracle. The pleasant surprise is the aggregated estimate which is dramatically better than the two others and it nicely mimics the oracle-estimate.

This figure can be repeated nsim times and yield information about mean and median of empirical integrated squared errors of the estimates. Using nsim=5000, we get for the oracle, uncensored-data, censored-data and aggregated estimates the sample mean integrated squared errors 0.14, 0.36, 0.28, 0.25, the sample median integrated squared errors 0.13, 0.33, 0.26, 0.24, and the mean rate of censoring 36%. The main and pleasant surprise is the good performance of the censored-data estimator which supports the theory.

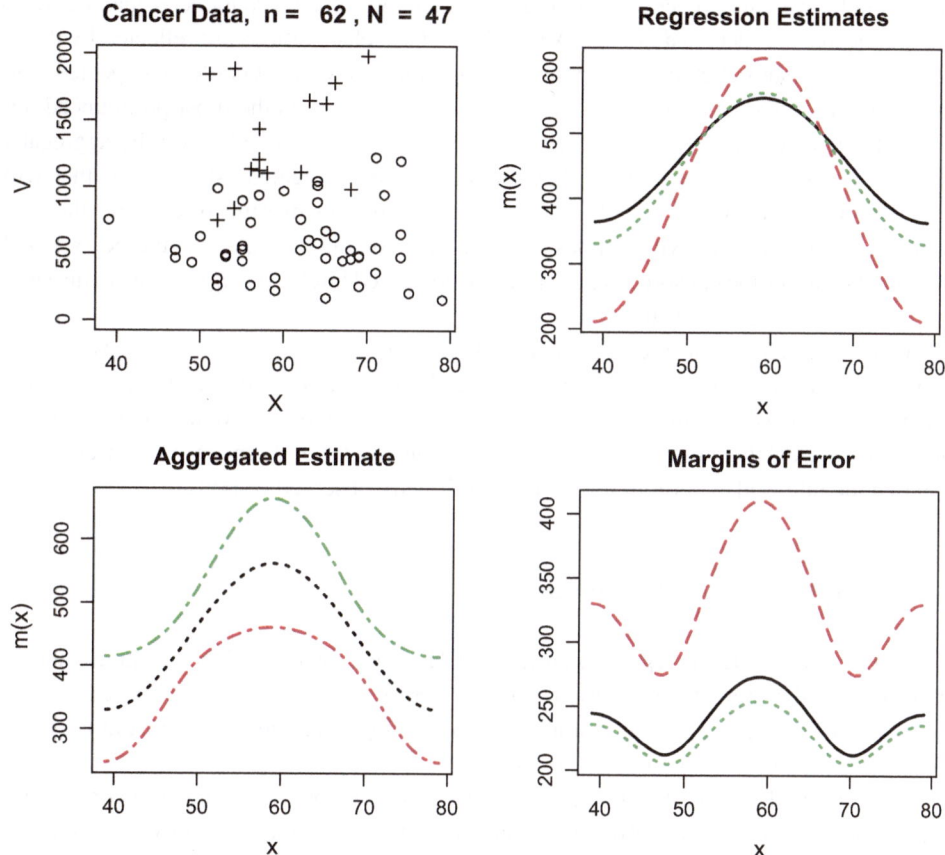

Fig. 3.8 Regression estimates for the lung cancer clinical study. Here T is the survival time in days from the cancer diagnosis and X is the age in years. The circles and the crosses show the uncensored and censored lifetimes, $N = \sum_{l=1}^{n} \Delta_l$. The solid, dashed and dotted lines correspond to the uncensored-data, censored-data and aggregated estimates, respectively. The dot-dashed lines show the 95% confidence band. {Argument DATA allows to analyze any R data written as a matrix with three columns of X, V and Δ.} [DATA $=$"DataCancerA", cJ0 $= 3$, cJ 1 $= 0.8$, cTH $= 4$, cJ0C $= 1$, cJ1C $= 0.2$, alpha $= 0.05$]

Now let us look at analysis of regression for the small cell lung cancer study. The small cell lung cancer is extremely aggressive, and not smoking is one of the main factors to prevent it. We already analyzed densities of survival and censoring lifetimes in Sect. 3.3. Here we are interested in regression of the survival time T on the age X. According to the literature, C and (T, X) are independent. The left-top diagram in Fig. 3.8 shows the scattergram of the right-censored data. Several interesting observations can be made from the data. First, note that the lifetimes are relatively small for the five youngest and four oldest participants. Second, the largest lifetimes are the censored lifetimes (the crosses) for the middle age participants. Accordingly, we can expect that the regression should have a pronounced maximum for the middle age and decreasing tails. And indeed, these observations are reflected by the all regression estimates shown in the right-top diagram. Further, as it could be expected from the scattergram, the censored-data estimate (the dashed line) is the most pronounced, but due to the smaller number of censored lifetimes (the crosses) its effect on the aggregated estimate is minor. The left-bottom diagram shows the aggregated estimate together with its 95% confidence band. The right-bottom diagram presents margins of error for the three estimates using the same type of lines as in the right-top diagram. As it could be expected, margin of error for the censored-data estimate (the dashed line) is the largest due to the small number of censored observations.

We may conclude that the proposed methodology of using censored observations for nonparametric regression with right-censored responses is feasible and can be recommended for analysis of real data. Further, let us stress the important theoretical outcome that censored observations yield the same rate of optimal regression estimation as uncensored observations, and that the rate is also optimal for the case of directly observed responses.

3.5.2   Right-Censored Predictor

The aim is to estimate the regression $m(t) := \mathbb{E}\{Y|T = t\}$ where Y is the response and the lifetime T is the predictor which is right-censored by the censoring lifetime C. Accordingly, we are estimating the regression function m using a sample of size n from the triplet $(V, \Delta, Y) := (\min(T, C), I(T \leq C), Y)$.

It is worthwhile to recall that regression with a modified predictor may be a very complicated problem. For instance, if the predictor is modified by a random additive measurement error ξ, that is we observe a sample from $(T + \xi, Y)$, then the regression becomes ill-posed and the rate of MISE convergence may be even logarithmic in n if ξ is Gaussian. Accordingly, it is of interest to understand what can be done for the case of right-censored predictor.

Assume that the pair (T, Y) and C are independent, the three variables are continuous and nonnegative, T has a continuous and positive density f^T on the support $[0, 1]$, and $S^C(1) > 0$. Write,

$$f^{V,Y,\Delta}(t, y, 1) = f^T(t) f^{Y|T}(y|t) S^C(t), \quad t \in [0, 1]. \tag{3.5.16}$$

Note that we are interested in the conditional density $f^{Y|T}$ which defines the regression of interest m. Namely,

$$m(t) = \mathbb{E}\{Y|T = t\} = \int_0^1 y f^{Y|T}(y|t)dy. \tag{3.5.17}$$

The formula (3.5.16) yields

$$f^{V,Y|\Delta}(t, y|1) = \frac{f^T(t) f^{Y|T}(y|t) S^C(t)}{\mathbb{P}(\Delta = 1)}$$

$$= \Big[\frac{f^T(t) S^C(t)}{\int_0^1 f^T(z) S^C(z) dz}\Big] f^{Y|T}(y|t) = f^{V|\Delta}(t|1) f^{Y|T}(y|t), \quad t \in [0, 1]. \tag{3.5.18}$$

In its turn, (3.5.18) yields

$$f^{Y|T}(y|t) = \frac{f^{V,Y|\Delta}(t, y|1)}{f^{V|\Delta}(t|1)} = f^{Y|V,\Delta}(y|t, 1), \quad t \in [0, 1]. \tag{3.5.19}$$

Relation (3.5.19) is the key for understanding the regression estimation based on uncensored predictors and the proposed below regression E-estimator. Indeed, whenever $f^{V|\Delta}(t|1)$ is positive, the conditional density of Y given $V = t$ in uncensored cases, that is given $\Delta = 1$, is the same as the underlying conditional density of Y given $T = t$. Accordingly, the regression E-estimator of Sect. 1.2.5, based on uncensored observations, is not only consistent but also rate optimal. Further, formula (3.5.19) shows how one can estimate the conditional density $f^{Y|T}$.

This is the good news. The bad news is that the problem of estimation based on censored predictors is a complicated problem resembling regression with measurement error in predictor, see the Notes.

Now let us check how the proposed uncensored-data regression E-estimator performs for small samples. Figure 3.9 shows a particular simulation explained in the caption. Note the high level of censoring with only $N = 48$ uncensored cases shown by the circles. Further, pay attention to the right tail of the scattergram because there are no observations with V exceeding 0.9. Further, note a very small number of uncensored observations (the circles) in the right half of the support of T. This is why the E-estimate is calculated for the interval $[0, V_{(n)}]$. The regression estimate (the dashed line) is good and correctly shows the bathtub profile of the underlying regression. It is recommended to repeat this figure with different underlying distributions and parameters, and get a better understanding of this challenging RC regression model.

3.6   Conditional Distribution of Right-Censored Lifetime

This section considers estimation of the conditional survival of a continuous lifetime of interest T given a continuous covariate of interest X supported on $[0, 1]$, and then estimation of the conditional hazard rate given a multivariate covariate with the emphasis on

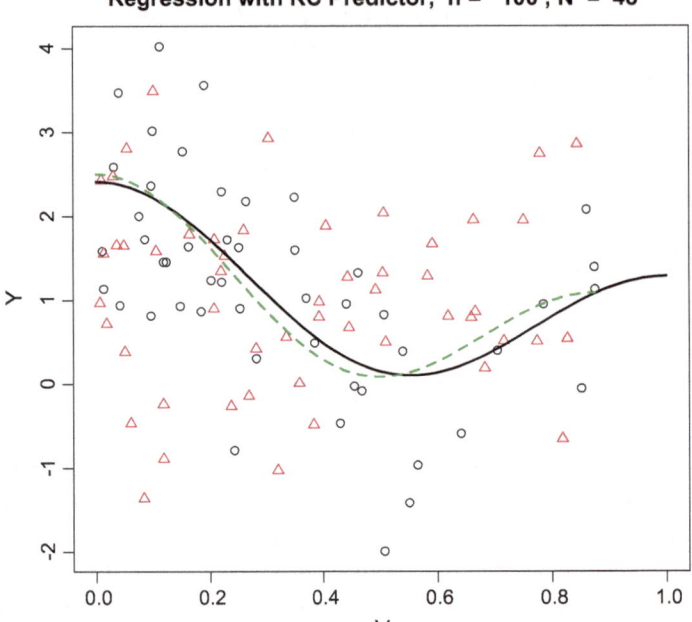

Fig. 3.9 Regression with right-censored (RC) predictors. The regression is $Y = m(T) + \sigma\xi$ where T is Uniform(0, 1), $m(t)$ is the Bathtub corner function, and ξ is standard normal. The predictor T is right-censored by the censoring variable C with Uniform(0, b) distribution. Accordingly, the available observations are from the triplet $(V, \Delta, Y) := (\min(T, C), I(T \leq C), Y)$. The circles and triangles show observations with uncensored predictor (when $\Delta = 1$) and censored predictor (when $\Delta = 0$), respectively. The solid line is the underlying regression function, the dashed line is the E-estimate calculated over interval $[0, V_{(n)}]$. The sample size n and the number of uncensored predictors $N := \sum_{l=1}^{n} \Delta_l$ are shown in the title. {The regression function is chosen by argument *corn*, parameter σ is chosen by argument *sigma*.} [$n = 100$, *corn* = 2, *sigma* = 1, b = 1, cJ0 = 4, cJ1 = 0.5, cTH = 4]

the possibility of dimension reduction. The lifetime of interest T is not observed directly because it is right-censored by a continuous censoring lifetime C.

For the case of a univariate covariate (predictor) X, the main assumption is that T and C are independent given X. Then, as we will see shortly, for efficient estimation the oracle uses right-censored data and the *natural nuisance function*

$$p(t, x) := f^X(x) S^{T|X}(t|x) S^{C|X}(t|x). \tag{3.6.1}$$

The latter is an important hint. Indeed, there are several underlying nuisance functions involved in the studied right-censored model including f^X, f^C, and $f^{C|X}$. The oracle tells us that only one function (3.6.1) is needed for efficient estimation of $S^{T|X}$. The natural nuisance function p includes as the factor the estimand $S^{T|X}$, and at first glance that looks strange. However, as we will see shortly, the presence of this factor streamlines estimation of the natural nuisance function.

For a data-driven estimation, mimicking the oracle, we need to understand how to estimate the natural nuisance function. Accordingly, we begin with the problem of its estimation. The proposed in the next subsection natural nuisance function estimator is a minor modification of the E-estimator which is of interest on its own, and it also helps the reader to get a better "feeling" of the series estimation. Then estimation of the conditional survival and conditional hazard function are explored in turn.

3.6.1   Natural Nuisance Function

We are already familiar with natural nuisance functions for a number of nonparametric problems. For instance, for direct observations and the regression problem it is the density f^X of the predictor X. The considered problem of estimating the nuisance function p defined in (3.6.1) is more complicated. First, we do not have direct observations of T and instead have a sample of size n from the triplet $(X, V, \Delta) := (X, \min(T, C), I(T \leq C))$. Second, the natural nuisance function (3.6.1) is a bivariate function. Third, it is the product of three functions with the estimand $S^{T|X}$ being one of them.

Nonetheless, there is a relatively simple methodology of estimating the natural nuisance function p. Namely, due to the assumed conditional independence between T and C given X, we get

$$S^{T|X}(t|x)S^{C|X}(t|x) = S^{V|X}(t|x). \tag{3.6.2}$$

This allows us to write down the natural nuisance function as

$$p(t, x) = f^X(x)S^{V|X}(t|x). \tag{3.6.3}$$

This formula indicates that the natural nuisance function is the characteristic of the directly observed pair (X, V), and this simplifies its estimation. At the same time, p is still a bivariate function. Does that yield the curse of dimensionality? Fortunately, the answer is "no". To see why, note that we can continue (3.6.3) and write

$$p(t, x) = f^X(x)\mathbb{E}\{I(V \geq t)|X = x\} = \mathbb{E}\{f^X(X)I(V \geq t)|X = x\}. \tag{3.6.4}$$

Accordingly, for each t the natural nuisance function is a nonparametric regression of the response $f^X(X)I(V \geq t)$ on the predictor X. As we know from Chap. 1, this is a univariate problem with the rate depending on the smoothness of f^X and $S^{V|X}(t|x)$ in x. This is another good news, but still we are dealing with the bivariate function.

Let us consider a series approach for estimating p. Using (1.1.2), (3.6.4) and the support $[0, 1]$ of X, we can write for the Fourier coefficient of the natural nuisance function $p(t, x)$

with a fixed t,

$$\kappa_j(t) := \int_0^1 p(t, x)\varphi_j(x)dx = \mathbb{E}\{I(V \geq t)\varphi_j(X)\}. \tag{3.6.5}$$

The pair (X, V) is directly observed. This allows us to propose the sample mean Fourier coefficient estimator of $\kappa_j(t)$,

$$\tilde{\kappa}_j(t) := n^{-1} \sum_{l=1}^n I(V_l \geq t)\varphi_j(X_l). \tag{3.6.6}$$

It is straightforward to get a simple upper bound for variance of this unbiased sample mean estimator, $\mathbb{V}\{\tilde{\kappa}_j(t)\} \leq 2n^{-1}$. This classical parametric rate n^{-1}, together with the fact that the upper bound for the variance does not depend on t and j, explain why we can avoid the curse of dimensionality in estimation of the bivariate natural nuisance function $p(t, x)$. This property is reminiscent of the possibility to estimate the univariate survival function $S^V(t)$ with the parametric rate n^{-1} by the sample mean estimator $\check{S}^V(t) := n^{-1} \sum_{l=1}^n I(V_l > t)$.

This is the good news. The remaining issue to resolve is how to estimate the bivariate natural nuisance function $p(t, x)$. Let us propose two solutions. The first one is to estimate it by a traditional projection estimator

$$\tilde{p}(t, x) = \sum_{j=0}^{J_p} \tilde{\kappa}_j(t)\varphi_j(x). \tag{3.6.7}$$

This is a nice approach for theoretical analysis, and J_p is chosen using a minimal assumption about smoothness of p. It also justifies using the E-estimator of Sect. 1.2.3 for small samples. The second approach is a compromise between the projection estimator (3.6.7) and the E-estimator. It uses the so-called *universal hard thresholding* for the Fourier coefficient estimators,

$$\hat{p}(t, x) = \sum_{j=0}^{J_p} \tilde{\kappa}_j(t)I(\tilde{\kappa}_j^2(t) > c_{THP}2\ln(n)\tilde{\sigma}_j^2(t))\varphi_j(x). \tag{3.6.8}$$

Here

$$\tilde{\sigma}_j^2(t) := (n-1)^{-1} \sum_{l=1}^n [I(V_l \geq t)\varphi_j(X_l) - \tilde{\kappa}_j(t)]^2 \tag{3.6.9}$$

is the sample variance of $\tilde{\kappa}_j(t)$, and the default value for the constant c_{THP} is 1.

The underlying idea of the universal thresholding is to remove all not significant underlying Fourier coefficients, and due to the factor $2\ln(n)$ the cutoff J_p may be very large and even proportional to n. The latter is the theoretical advantage of the universal thresholding. At the same time, we know that for small samples and the cosine basis there is no need for such large cutoffs and J_p proportional to $\ln(n)$ is a good choice. Let us stress that the main difference between the E-estimator of Sect. 1.2.3 and the (3.6.8) is that the E-estimator estimates J_p for each t while in (3.6.8) J_p is the parameter. Accordingly, the estimator (3.6.8) is faster to calculate.

Finally, let us note that both (3.6.7) and (3.6.8) are stepwise functions in t with jumps at observed values of V, and hence resemble the empirical survival function.

Figure 3.10 allows us to compare these estimators and gain a new experience in dealing with series estimation. The underlying simulation of the triplet (X, V, Δ) and the diagrams are explained in the caption. As we will see shortly, the natural nuisance function p should be known at points (V_l, X_l), $l = 1, 2, \ldots, n$, and this is why p and its estimates are shown at the observed values of V. The top diagram shows us values $p(V_l, X_l)$ by the circles, $\tilde{p}(V_l, X_l)$ by the crosses, and $\hat{p}(V_l, X_l)$ by the triangles. As we see, the projection estimate $\tilde{p}$ undersmooths the underlying p and overall exhibits larger deviations from the underlying natural nuisance function. At the same time, in some cases it almost perfectly matches the underlying natural nuisance function. The hard-thresholding estimate better fits the underlying natural nuisance function, but overall it oversmooths it. The middle and bottom diagrams show $p(V_l, x)$ and its estimates for $x = x_1 = 0.1$ and $x = x_2 = 0.7$, respectively. As we see, the hard-thresholding dramatically improves the estimation. What we observe is the classical nonparametric curve estimation issue of a correct "smoothing".

It is recommended to repeat Fig. 3.10 with different arguments J_p and c_{THP} for better understanding of how the series estimation performs.

In the next subsections we are using the estimator (3.6.8) of the natural nuisance function.

3.6.2   Conditional Survival Function

Consider the problem of estimating the conditional survival function $S^{T|X}(t|x) := \mathbb{P}(T > t|X = x)$ given n right-censored observations of the triplet $(X, V, \Delta) := (X, \min(T, C), I(T \leq C))$. Here X is the continuous predictor with density f^X which is differentiable, positive and supported on $[0, 1]$, T is the continuous lifetime of interest, and C is the continuous censoring lifetime. It is assumed that T and C are conditionally independent given X. Because $S^{T|X}(t|x) = \mathbb{E}\{I(T > t)|X = x\}$, we are dealing with a special nonparametric regression problem when the response is $I(T > t)$ and X is the predictor.

Let us show how the method of moments approach of Sect. 3.2, proposed for estimation of the univariate S^T, may be extended to estimation of the bivariate $S^{T|X}$. We begin with analysis of the *conditional cumulative hazard* $H^{T|X}$. Write,

$$H^{T|X}(t|x) := \int_0^t h^{T|X}(u|x)du = \int_0^t \frac{f^{T|X}(u|x)}{S^{T|X}(u|x)}du$$

$$= \int_0^t \frac{f^{T|X}(u|x)S^{C|X}(u|x)}{S^{V|X}(u|x)}du. \tag{3.6.10}$$

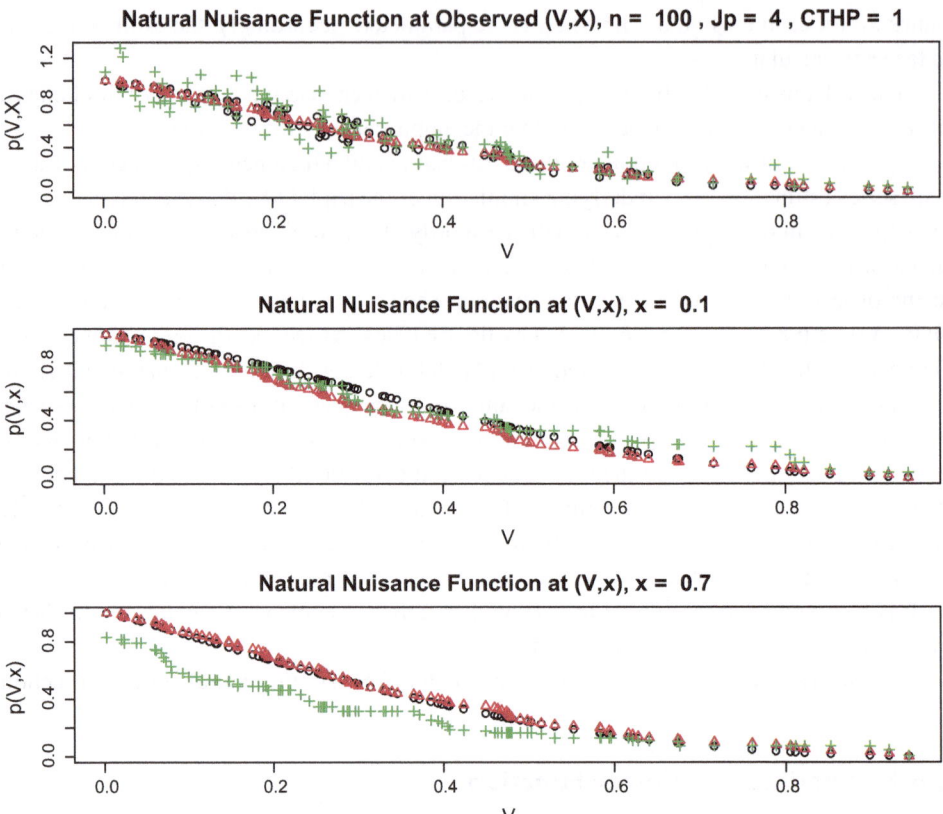

Fig. 3.10 Estimation of the natural nuisance function $p(t, x) = f^X(x)S^{T|X}(t|x)S^{C|X}(t|x)$. The sample of size n is from the triplet $(X, V, \Delta) := (X, \min(T, C), I(T \leq C))$. Predictor X has the Uniform(0,1) distribution, the lifetime of interest T is generated according to the Beta$(2, 2 + aX)$ distribution, and C is Uniform$(0, b + b_1 X)$. The circles show the underlying natural nuisance function, the crosses show the projection estimate (3.6.7), and the triangles show the hard-thresholding estimate (3.6.8) at points (V_l, X_l) in the top diagram and at points (V_l, x) with $x = x_1$ and $x = x_2$ in the middle and bottom diagrams, respectively. {Parameters $b, b_1, a, J_p, c_{THP}, x_1, x_2$ are controlled by the arguments $b, b1, a, Jp, CTHP, x1, x2$, respectively.} [$n = 100, b = 1.5, b1 = 0.5, a = -1, Jp = 4, CTHP = 1, x1 = 0.1, x2 = 0.7$]

For a fixed t we can express the conditional cumulative hazard as a Fourier series in x using the cosine basis on [0, 1]. Recalling formula $f^{X,V,\Delta}(x, v, 1) = f^X(x) f^{T|X}(v|x)S^{C|X}(v|x)$, and using (3.6.10) we can write for the corresponding Fourier coefficient,

$$\theta_j(t) := \int_0^1 H^{T|X}(t|x)\varphi_j(x)dx = \mathbb{E}\left\{\frac{\Delta I(V \leq t)\varphi_j(X)}{f^X(X)S^{V|X}(V|X)}\right\}$$

$$= \mathbb{E}\left\{\frac{\Delta I(V \le t)\varphi_j(X)}{p(V, X)}\right\}. \tag{3.6.11}$$

Here p is the natural nuisance function (3.6.1). Using the right-censored data and the natural nuisance function, the sample mean Fourier coefficient oracle-estimator is

$$\hat{\theta}_j^*(t) := n^{-1}\sum_{l=1}^{n}\frac{\Delta_l I(V_l \le t)\varphi_j(X_l)}{p(V_l, X_l)}. \tag{3.6.12}$$

Because in general the natural nuisance function is unknown, its estimator $\check{p} := \max(\hat{p}, n^{-1})$ is used where $\hat{p}$ is defined in (3.6.8). We have truncated $\hat{p}$ from below so it can be used in denominators. This yields the Fourier coefficient estimator

$$\hat{\theta}_j(t) := n^{-1}\sum_{l=1}^{n}\frac{\Delta_l I(V_l \le t)\varphi_j(X_l)}{\check{p}(V_l, X_l)}. \tag{3.6.13}$$

Then the universal hard thresholding series estimator of the cumulative hazard is

$$\hat{H}^{T|X}(t|x) = \sum_{j=0}^{J_H}\hat{\theta}_j(t)I(\tilde{\theta}_j^2(t) > c_{THH}2\ln(n)\hat{\sigma}_j^2(t))\varphi_j(x). \tag{3.6.14}$$

Here

$$\hat{\sigma}_j^2(t) := (n-1)^{-1}\sum_{l=1}^{n}\left[\frac{\Delta_l I(V_l \le t)\varphi_j(X_l)}{\check{p}(V_l, X_l)} - \hat{\theta}_j(t)\right]^2 \tag{3.6.15}$$

is the sample variance of the Fourier coefficient estimator $\hat{\theta}_j(t)$.

As soon as the cumulative hazard is estimated, we get the plug-in estimator of the conditional survival function,

$$\hat{S}^{T|X}(t|x) := e^{-\hat{H}^{T|X}(t|x)}. \tag{3.6.16}$$

Figure 3.11 helps us to understand how the estimator performs for small samples. The top diagram exhibits the available observations $(X_l, V_l, \Delta_l) := (X, \min(T, C), I(T \le C))$, $l = 1, 2, \ldots, n$ with the right-censored lifetime of interest T. Cases with $\Delta_l = 1$ are shown by the circles and with $\Delta = 0$ by the triangles. The number of uncensored observations is $N = 80$, and we are dealing with the moderately high rate of censoring. The second from the top diagram looks familiar because its structure is the same as in the top diagram of Fig. 3.10. This diagram underscores how heavily observations of V are skewed to the right. Note that there are just few observations to the right of 0.65. Nonetheless, the hard-thresholding estimate of the nuisance function p is fairly good under the circumstances. The two bottom diagrams show by the solid line the underlying $S^{T|X}(t|x)$ as a function in t given a specific value of x shown in the title, and by the dashed step-wise function the estimate $\hat{S}^{T|X}$. The estimates are fair for this complicated problem.

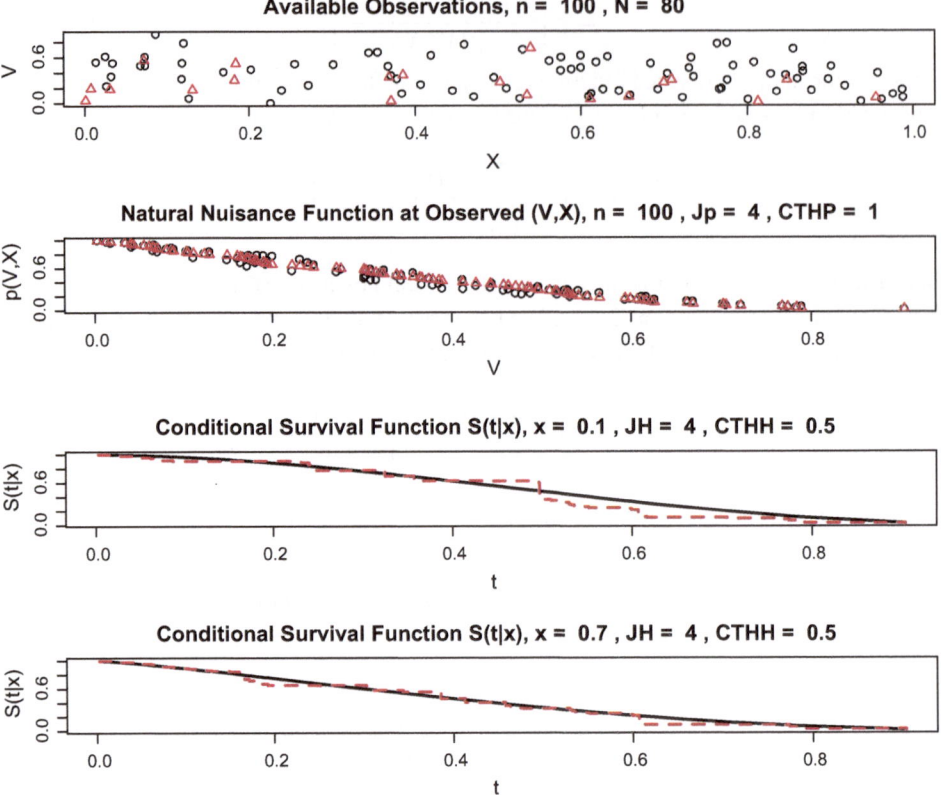

Fig. 3.11 Estimation of the conditional survival function $S^{T|X}$ based on a right-censored sample of size n from $(X, V, \Delta) := (X, \min(T, C), I(T \le C))$. The underlying experiment is the same as in Fig. 3.10. The top diagram shows simulated observations with the circles and triangle corresponding to uncensored and censored cases, respectively, and $N := \sum_{l=1}^{n} \Delta_l$. The second diagram's structure is identical to the top one in Fig. 3.10. The two bottom diagrams exhibit $S^{T|X}(t|x)$ by the solid line and its estimate $\hat{S}^{T|X}(t|x)$ by the stepwise dashed line for t over the range of V and for two values of x controlled by the arguments $x1$ and $x2$. {Arguments JH and $CTHH$ control parameters J_H and c_{THH} of the estimator (3.6.14).} [$n = 100$, $b=1.5$, $b1=0.5$, $a=-1$, $Jp=4$, $JH=4$, $CTHP=1$, $CTHH=0.5$, $x1=0.1$, $x2=0.7$]

It is useful to repeat Fig. 3.11 with different arguments Jp, JH, $CTHP$ and $CTHH$. Overall, decreasing Jp and increasing $CTHP$ smooths an estimate of the nuisance function, and similarly decreasing JH and increasing $CTHH$ smooths an estimate of the conditional survival function.

3.6.3   Conditional Hazard Rate

Conditional hazard rate (or we may say conditional hazard rate function) is another important characteristic of a relationship between a covariate (predictor) and a lifetime of interest. In this subsection we are considering a more general case of a multivariate covariate (predictor) that may include a categorical covariate. The main emphasis is on efficient estimation and the possibility of dimension reduction. For instance, if the lifetime of interest and the covariate are independent, then the problem becomes univariate and the estimator should recognize that. The subsection begins with the simpler case of a univariate continuous covariate which sheds light on the underlying theory and methodology, and then a multivariate covariate is considered.

3.6.3.1   Univariate Predictor

We begin with a familiar problem of statistical analysis of a relationship between an explanatory continuous variable (predictor) X and a continuous lifetime of interest (response) T based on a sample of size n from (X, T). Classical examples are age X at the time of a cancer surgery and how it affects time T until cancer relapse, or how credit score X affects time T until mortgage default. A traditional approach, used in the previous section, is to study regression $\mathbb{E}\{T|X = x\} := \int_0^\infty t f^{T|X}(t|x)dt$, where $f^{T|X}(t|x) := f^{X,T}(x, t)/f^X(x)$ is the conditional density of T given X, $f^{X,T}$ is the joint density of (X, T), and $f^X(x) := \int_0^\infty f^{X,T}(x, t)dt$ is the marginal (design) density of the predictor. In this subsection we use a more complicated and simultaneously more appealing approach for analysis of the relationship via the *conditional hazard rate function* (also referred to as the conditional failure rate in reliability theory, conditional force of mortality in actuarial science and sociology, or conditional age-specific rate in different fields of engineering and medical statistics)

$$h^{T|X}(t|x) := \frac{f^{T|X}(t|x)}{S^{T|X}(t|x)}, \quad S^{T|X}(t|x) := \int_t^\infty f^{T|X}(u|x)du. \tag{3.6.17}$$

In what follows, whenever no confusion can occur, we may refer to the conditional hazard rate function $h^{T|X}$ as conditional hazard rate or simply conditional hazard, while $H^{T|X}(t|x) := \int_0^t h^{T|X}(u|x)du$ is called the *conditional cumulative hazard*.

There is one important issue that distinguishes nonparametric estimation of the conditional hazard rate from other classical nonparametric problems like regression or density estimation. In classical problems an estimated function is assumed to be integrable, while a conditional hazard rate is not integrable, that is $\int_0^\infty h^{T|X}(t|x)dt = \infty$. Moreover, if T is supported on a finite interval $[0, b]$, then $\lim_{t \to b} h^{T|X}(t|x) = \infty$. This creates the curse of right tail. Another traditional complication in survival analysis is that the lifetime of interest T is right-censored. This is the setting that will be considered, and the problem is to estimate the conditional hazard rate under the mean integrated squared error (MISE) criterion.

We begin with the theory of conditional hazard rate estimation. There is a hidden sample $(X_1, T_1), \ldots, (X_n, T_n)$ of size n from a pair (X, T). Here T is the lifetime of interest and X is a univariate predictor (covariate) supported on $[0, 1]$. X and T are continuous random variables with a joint density $f^{X,T}(x, t)$ supported on $[0, 1] \times [0, \infty)$. Set $f^X(x) := \int_0^\infty f^{X,T}(x, t)dt$ for the marginal (design) density of X. It is assumed that f^X is continuous and positive on $[0, 1]$. We do not observe realizations $T_1, \ldots, T_n$ directly because they are right-censored by independent realizations $C_1, \ldots, C_n$ of a censoring lifetime C. Instead we observe a sample $(X_1, V_1, \Delta_1), \ldots, (X_n, V_n, \Delta_n)$ from the triplet (X, V, Δ) where $V := \min(T, C)$ and $\Delta := I(T \leq C)$ is the indicator of censoring.

For a given restriction r, which is a positive and finite constant, the aim is to estimate conditional hazard rate $h^{T|X}(t|x)$, defined in (3.6.17), over the rectangle $\mathcal{R} := \mathcal{R}(r) := [0, r] \times [0, 1]$. The used criterion for an estimator $\check{h}(t|x)$ of $h^{T|X}(t|x)$ is the mean integrated squared error (MISE) $\mathbb{E}\{\int_{\mathcal{R}}(\check{h}(t|x) - h^{T|X}(t|x))^2 dx dt\}$. Recall our notations η_{rj} and φ_j for elements of cosine bases on $[0, r]$ and $[0, 1]$, respectively. Then $\varphi_{ji}(t, x) := \eta_{rj}(t)\varphi_i(x)$, $(j, i) \in \{0, 1, \ldots\}^2$ are elements of the tensor-product basis on $\mathcal{R}$.

Now we can proceed to assumptions. The first one is necessary for consistent estimation.

Assumption 3.5 Given predictor X, lifetime of interest T and censoring lifetime C are independent.

Our next assumption is about smoothness of an estimated conditional hazard rate. Here some explanation is warranted. Conditional hazard rate $h^{T|X}(t|x)$ is a bivariate function. It is a tradition in nonparametric literature to assume that an estimated bivariate function is isotropic meaning that it is as smooth (has the same number of derivatives) in t as in x. For some settings this assumption is reasonable, but for a conditional hazard $h^{T|X}(t|x)$ there may be a difference between smoothness in t and x. For instance, consider a location model $T = m(X) + \xi$ where the random variable ξ is independent of X and its hazard rate is $h^\xi(t)$. Then $h^{T|X}(t|x) = h^\xi(t - m(x))$. Now note that smoothness of $h^{T|X}(t|x)$ in t is defined solely by smoothness of $h^\xi(t)$, while its smoothness in x depends on smoothness of $h^\xi(t)$ and smoothness of $m(x)$. Accordingly, it is prudent to assume that $h^{T|X}(t|x)$ may be an anisotropic bivariate function with different smoothness in t and x.

As we know from Chap. 1, for direct data and classical density or regression estimation problems the asymptotic MISE convergence does not depend on the estimand. It will be established shortly that an underlying conditional hazard affects sharp constant of the MISE convergence. Accordingly, in a lower bound for the MISE we are considering a *shrinking local anisotropic* function class of $h^{T|X}$ with an anchor $h_0(t|x)$ which may depend on n. This approach will allow us to understand how an underlying $h^{T|X}$ affects its estimation.

After these comments, let us introduce several functional classes. In what follows α_0 and α_1 are positive integer numbers that define the number of derivatives of the conditional hazard $h^{T|X}(t|x)$ in t and x, respectively. Introduce two anisotropic Sobolev classes of bivariate functions $g(t, x)$ on $\mathcal{R} := [0, r] \times [0, 1]$,

$$S_k := S_k(\alpha_0, \alpha_1, Q, r) := \left\{ g : \ g(t, x) := \sum_{j=k}^{\infty} \sum_{i=0}^{\infty} \theta_{ji} \varphi_{ji}(t, x), \ (t, x) \in \mathcal{R}, \right.$$

$$\left. \sum_{j=k}^{\infty} \sum_{i=0}^{\infty} [1 + (\pi j/r)^{2\alpha_0} + (\pi i)^{2\alpha_1}] \theta_{ji}^2 \le rQ < \infty \right\}, \ k \in \{0, 1\}. \tag{3.6.18}$$

The case $k = 0$ implies a classical anisotropic Sobolev class S_0, and if $g \in S_1$ then $\int_0^r g(t, x)dt = 0$. Introduce a continuous conditional hazard $h_0(t|x)$, $(t, x) \in [0, \infty) \times [0, 1]$ that will be referred to as the anchor. Now we can define the following shrinking local function class $\mathcal{F}_n$ of conditional hazards,

$$\mathcal{F}_n := \mathcal{F}_n(\alpha_0, \alpha_1, Q, r, h_0) := \left\{ h^{T|X} : \ h^{T|X}(t|x) = h_0(t|x) + g(t, x)I((t, x) \in \mathcal{R}); \right.$$

$$g \in S_1(\alpha_0, \alpha_1, Q, r); \ |g(t, x)| < 1/\ln(\ln(n + 20));$$

$$\min_{(t,x) \in \mathcal{R}} h_0(t|x) > 2/\ln(\ln(n + 20)), \quad \max_{x \in [0,1]} \int_0^r h_0(u|x)du < \infty;$$

$$\left. \sum_{j,i=0}^{\infty} (1 + j^{2\alpha_0 + \beta} + i^{2\alpha_1 + \beta}) \left[\int_{\mathcal{R}} h_0(t|x) \varphi_{ji}(t, x)dxdt \right]^2 < \infty, \ \beta > 0 \right\}. \tag{3.6.19}$$

In what follows we denote by $\mathcal{F}_n^* := \mathcal{F}_n(\alpha_0, \alpha_1, Q, r, h_0)$ a class (3.6.19) without the bottom line assuming that the anchor is a Sobolev function. This class will be used to establish the lower bound. To get an upper bound, the oracle may or may not use knowledge of the anchor. The latter case is of the main interest, and then the bottom line in assumption (3.6.19) is used. Let us stress that the anchor is not necessarily an underlying conditional hazard, its the only role is to define the local function class. Final remark is about the used notation. Parameters α_0 and α_1 describe smoothness of a function in t and x, and we use these numeric indexes because they are naturally extended to a multivariate predictor $\mathbf{x} := (x_1, \ldots, x_\tau)$ where we will use parameters $\alpha_0, \alpha_1, \ldots, \alpha_\tau$.

For the introduced function classes set

$$P := P(\alpha_0, \alpha_1, Q, r) = r\pi^{-\frac{4\alpha}{2\alpha+1}} [Q/P_1]^{\frac{1}{2\alpha+1}} P_2, \quad \alpha := [\alpha_0^{-1} + \alpha_1^{-1}]^{-1}, \tag{3.6.20}$$

$$P_1 := \int_{\{(u,v):u^{2\alpha_0}+v^{2\alpha_1} \le 1; u, v \ge 0\}} ([u^{2\alpha_0} + v^{2\alpha_1}]^{1/2} - [u^{2\alpha_0} + v^{2\alpha_1}])dvdu, \tag{3.6.21}$$

$$P_2 := \int_{\{(u,v):u^{2\alpha_0}+v^{2\alpha_1} \le 1; u, v \ge 0\}} (1 - [u^{2\alpha_0} + v^{2\alpha_1}]^{1/2})dvdu. \tag{3.6.22}$$

The α in (3.6.20) is called the effective smoothness.

Our second assumption is about the natural nuisance function which, together with right-censored data, is used by the oracle for efficient estimation.

Assumption 3.6 The natural nuisance function

$$p(t, x) := f^X(x) S^{C|X}(t|x) S^{T|X}(t|x) \qquad (3.6.23)$$

is continuous and bounded below from zero on $\mathcal{R}$, and its partial derivative in x is bounded and integrable on $\mathcal{R}$.

Assumption 3.6 is mild, involves only first-order differentiability in x, and not tied to smoothness of the estimated conditional hazard $h^{T|X}$.

Now let us explain several notations that will be used by the proposed oracle-estimator. First of all, the oracle-estimator must take into account the possibility that T and X are independent, because then the bivariate $h^{T|X}(t|x)$ becomes the univariate hazard $h^T(t)$, and correspondingly it can be estimated with faster rate of the MISE convergence. To achieve this aim without testing the hypothesis, we introduce two tensor-product arrays of indexes (frequencies) (j, i). The first one is $A_1 := \{0, 1, 2, \ldots\} \times \{0\}$, the second one is $A_2 := \{0, 1, 2, \ldots\} \times \{1, 2, \ldots\}$. We have $\varphi_{j0}(t, x) = \eta_{rj}(t)$, and then for $(t, x) \in \mathcal{R}$ we may write

$$h^{T|X}(t|x) = \sum_{(j,i)\in A_1} \theta_{ji}\varphi_{ji}(t, x) + \sum_{(j,i)\in A_2} \theta_{ji}\varphi_{ji}(t, x) =: h_1(t) + h_2(t, x). \qquad (3.6.24)$$

Note that if T and X are independent, then $h_1(t) = h^T(t)$ and $h_2(t, x) = 0$. Expansion (3.6.24) explains the underlying idea of dimension reduction from a bivariate conditional hazard to a univariate hazard, and it also sheds light on the used arrays of indexes. Recall that a similar approach was used in Sect. 2.4.

To solve the problem of adaptation to unknown smoothness of conditional hazard $h^{T|X}(t|x)$ in t and x, the indexes (or equivalently we may say frequencies) are grouped into special blocks, and then Fourier estimators from a block are shrunk by the same coefficient. Recall that this approach is called the blockwise shrinking (or simply blockwise) adaptation. We are using the familiar notation $J_n := c_{J0} + c_{J1} \ln(n)$, and in the next several formulas it will be convenient to use $s_n := 3 + \ln(\ln(n + 20))$. Introduce an increasing sequence of integers $b_1 = 0$, $b_2 = b_1 + 1, \ldots, b_{J_n+1} = b_{J_n} + 1$, and $b_{J_n+m} = b_{J_n+m-1} + \lceil(1 + s_n^{-1})^m\rceil$ for $m = 2, 3, \ldots$ Introduce blocks of integers $B_k := \{b_k, \ldots, b_{k+1} - 1\}$, $k = 1, 2, \ldots$, denote their cardinality as L_k, and define K_{0n} and K_{1n} as the largest integers such that $b_{K_{0n}+1} \leq n^{1/3} s_n + J_n + 1$ and $b_{K_{1n}+1} \leq n^{1/4} s_n + J_n + 1$. Now set $\mathbf{k} := (k_1, k_2)$, recall that arrays A_ν, $\nu = 1, 2$ were introduced above line (3.6.24), and for each array introduce its own array of blocks $B_{\nu\mathbf{k}} := \{B_{k_1} \times B_{k_2}\} \cap A_\nu$. Note that $B_{1(k_1,1)} = B_{k_1} \times \{0\}$, $B_{1(k_1,k_2)}$ is the empty set whenever $k_2 \geq 2$, $B_{2(k_1,1)}$ is always empty, and $B_{2(k_1,k_2)} = B_{k_1} \times B_{k_2}$ whenever $k_2 \geq 2$. We use this particular definition and notation for $B_{\nu\mathbf{k}}$ because they also will be used for a vector-predictor. Using this notation we can approximate (3.6.24) by a partial sum

$$h_n^{T|X}(t|x) := \sum_{\nu=1}^{2} \sum_{k_1=1}^{K_{0n}} \sum_{k_2=1}^{K_{1n}} \sum_{(j,i)\in B_{\nu\mathbf{k}}} \theta_{ji}\varphi_{ji}(t, x). \qquad (3.6.25)$$

This series approximation will be used shortly. Further, denote by $L_{v\mathbf{k}}$ the cardinality of $B_{v\mathbf{k}}$, and set $\rho_{\mathbf{k}} := 1/\ln(3 + k_1 k_2)$.

The underlying idea of the proposed conditional hazard estimator is to mimic the oracle who knows both data and the underlying model. This approach allows us to develop: (i) A lower bound for the MISE; (ii) An oracle-estimator that attains the lower bound; (iii) A data-driven estimator that mimics the oracle-estimator and matches its properties. The corresponding results are presented in turn.

We begin with the oracle's lower bound for the MISE.

Theorem 3.8 *Let Assumptions 3.5 and 3.6 hold, the anchor $h_0(t|x)$ is continuous on $\mathcal{R}$, and a sample of size n from $(X, V, \Delta) := (X, \min(T, C), I(T \leq C))$ is available. The oracle knows data, the underlying function class $\mathcal{F}_n^*$, the anchor h_0, the design density f^X, and the conditional survival $S^{C|X}$. Then*

$$\inf_{\check{h}^*} \sup_{h^{T|X} \in \mathcal{F}_n^*} (n/d)^{\frac{2\alpha}{2\alpha+1}} \mathbb{E}\{\int_{\mathcal{R}} (\check{h}^*(t|x) - h^{T|X}(t|x))^2 dx dt\} \geq P(1 + o_n(1)), \qquad (3.6.26)$$

where the infimum is taken over all possible oracle-estimators $\check{h}^*$,

$$d := d(h^{T|X}, S^{C|X}, f^X, r) := r^{-1} \int_{\mathcal{R}} \frac{h^{T|X}(t|x)}{f^X(x) S^{C|X}(t|x) S^{T|X}(t|x)} dx dt \qquad (3.6.27)$$

is the coefficient of difficulty, $\alpha := [\alpha_0^{-1} + \alpha_1^{-1}]^{-1}$ is the effective smoothness of the class $\mathcal{F}_n^*$, and P is defined in (3.6.20). The lower minimax bound is sharp and attainable by an oracle-estimator. Further, if the inequality in the bottom line of (3.6.19) holds, then the lower bound (3.6.26) is attainable for $h^{T|X} \in \{\mathcal{F}_n \cup S_0\}$ by oracle-estimators that do not use the anchor, and in particular by the following sharp-minimax oracle-estimator,

$$\tilde{h}^*(t|x) := \sum_{v=1}^{2} \sum_{k_1=1}^{K_{0n}} \sum_{k_2=1}^{K_{1n}} \sum_{(j,i) \in B_{v\mathbf{k}}} \left[I((\hat{\theta}_{ji}^*)^2 \geq c_{TH} \mathbb{V}\{\hat{\theta}_{ji}^*\}) I(\max(j, i) \leq J_n) \right.$$

$$\left. + \frac{\Theta_{v\mathbf{k}} I(\Theta_{v\mathbf{k}} \geq \rho_{\mathbf{k}} n^{-1}) I(\max(j, i) > J_n)}{\Theta_{v\mathbf{k}} + L_{v\mathbf{k}}^{-1} \sum_{(j,i) \in B_{v\mathbf{k}}} \mathbb{E}\{(\hat{\theta}_{ji}^* - \theta_{ji})^2\}} \right] \hat{\theta}_{ji}^* \varphi_{ji}(t, x). \qquad (3.6.28)$$

Here

$$\hat{\theta}_{ji}^* := n^{-1} \sum_{l=1}^{n} \frac{\Delta_l I(V_l \in [0, r]) \varphi_{ji}(V_l, X_l)}{p(V_l, X_l)} \qquad (3.6.29)$$

is the oracle-estimator of Fourier coefficients $\theta_{ji} := \int_{\mathcal{R}} h^{T|X}(t|x) \varphi_{ji}(t, x) dx dt$ and

$$\Theta_{v\mathbf{k}} := L_{v\mathbf{k}}^{-1} \sum_{(j,i) \in B_{v\mathbf{k}}} \theta_{ji}^2 \qquad (3.6.30)$$

is the Sobolev functional.

Let us comment on results of Theorem 3.8. The lower bound (3.6.26) asserts that the MISE decreases with rate $n^{-2\alpha/(2\alpha+1)}$ which is the same as for the classical bivariate regression of T on a pair of predictors (X_1, X_2) and the case of directly observed data. This outcome is encouraging because it shows that, in a more complicated problem of conditional hazard estimation based on right-censored data, the rate does not slow down. On the other hand, the coefficient of difficulty (3.6.27) sheds light on complexity of the problem. Namely, formula (3.6.27) explains how the quartet $(h^{T|X}, S^{C|X}, S^{T|X}, f^X)$ affects the MISE convergence. Further, note that each of the first three functions in the quartet makes the coefficient of difficulty (3.6.27) larger as r increases because $\int_0^\infty h^{T|X}(t|x)dt = \infty$ and $\max(S^{C|X}(t|x), S^{T|X}(t|x)) \to 0$ as $t \to \infty$. Formula (3.6.27) also quantifies complexity of the right tail estimation.

Now let us explain how to construct an efficient data-driven estimator that mimics the oracle-estimator (3.6.28). We begin with introducing the following estimator $\hat{p}$ of the natural nuisance function,

$$\hat{p}(t, x) := \max(\tilde{p}(t, x), 1/\ln(n)),$$

$$\tilde{p}(t, x) := n^{-1} \sum_{l=1}^{n} \sum_{i=0}^{\lceil n^{1/3} s_n \rceil} I(V_l \geq t)\varphi_i(X_l)\varphi_i(x). \tag{3.6.31}$$

Because $\hat{p}$ is separated from zero, it can be used in a denominator. Accordingly, we plug $\hat{p}$ in (3.6.29) and get the Fourier coefficient estimator

$$\hat{\theta}_{ji} := n^{-1} \sum_{l=1}^{n} \frac{\Delta_l I(V_l \leq r)\varphi_{ji}(V_l, X_l)}{\hat{p}(V_l, X_l)}. \tag{3.6.32}$$

Similarly we get the estimator of the coefficient of difficulty,

$$\hat{d} := \hat{d}(r) := n^{-1} \sum_{l=1}^{n} \frac{\Delta_l I(V_l \leq r)}{r[\hat{p}(V_l, X_l)]^2}. \tag{3.6.33}$$

To mimic the shrinking weights in (3.6.28) we calculate two statistics,

$$\hat{\Theta}_{\nu k} := \frac{2}{L_{\nu k} n(n-1)}$$

$$\times \sum_{(j,i)\in B_{\nu k}} \sum_{1 \leq l < m \leq n} \frac{\Delta_l \Delta_m \varphi_{ji}(V_l, X_l)\varphi_{ji}(V_m, X_m)I(\max(V_l, V_m) \leq r)}{\hat{p}(V_l, X_l)\hat{p}(V_m, X_m)}, \tag{3.6.34}$$

and

$$\tilde{\Theta}_{\nu k} := L_{\nu k}^{-1} \sum_{(j,i)\in B_{\nu k}} \hat{\theta}_{ji}^2. \tag{3.6.35}$$

The proposed conditional hazard estimator, which mimics (3.6.28), is

$$
\hat{h}(t|x) := \sum_{v=1}^{2} \sum_{k_1=1}^{K_{0n}} \sum_{k_2=1}^{K_{1n}} \sum_{(j,i)\in B_{vk}} \Big[I(\hat{\theta}_{ji}^2 \geq c_{TH}\hat{\sigma}_{ji}^2) I(\max(j,i) \leq J_n)
$$

$$
+ \min(1, \hat{\Theta}_{vk}/\tilde{\Theta}_{vk}) I(\hat{\Theta}_{vk} \geq \rho_k n^{-1}) I(\max(j,i) > J_n) \Big] \hat{\theta}_{ji} \varphi_{ji}(t,x). \qquad (3.6.36)
$$

Here $\hat{\sigma}_{ji}^2$ is the sample mean estimator of the variance $\mathbb{V}\{\hat{\theta}_{ji}\}$. If necessary, a projection on the class of nonnegative functions can be performed.

The next theorem describes properties of the proposed estimator. Recall that ζ denotes generic positive constants.

Theorem 3.9 *Let Assumptions 3.5 and 3.6 hold. Then the data-driven estimator (3.6.36) is efficient and*

$$
\sup_{h^{T|X}\in\{\mathcal{F}_n\cup\mathcal{S}_0\}} [n/d]^{2\alpha/(2\alpha+1)} \; \mathbb{E}\{\int_{\mathcal{R}} (\hat{h}^{T|X}(t|x) - h^{T|X}(t|x))^2 dxdt\}
$$

$$
\leq P(1 + o_n(1)), \qquad (3.6.37)
$$

that is the estimator not only matches performance of the oracle that knows the shrinking local class $\mathcal{F}_n$ defined in (3.6.19), but it also attains the oracle's lower bound over the global Sobolev class $\mathcal{S}_0 := \mathcal{S}_0(\alpha_0, \alpha_1, Q, r)$ defined in (3.6.18). If additionally the lifetime of interest T and the predictor X are independent and accordingly $h^{T|X} = h^T$, then the estimator attains the optimal univariate rate and

$$
\sup_{h^{T|X}\in\{\mathcal{F}_n\cup\mathcal{S}_0, \, h^{T|X}=h^T\}} \mathbb{E}\left\{ \int_{\mathcal{R}} (\hat{h}^{T|X}(t|x) - h^T(t))^2 dxdt \right\} \leq \zeta n^{-2\alpha_0/(2\alpha_0+1)}. \qquad (3.6.38)
$$

We conclude that the proposed data-driven estimator of the conditional hazard rate function: (i) Adapts to an underlying smoothness of the conditional hazard; (ii) Matches performance of the oracle under minimal assumptions on smoothness of nuisance functions, and those assumptions are not tied to smoothness of the conditional hazard; (iii) Takes into account a possible independence between the lifetime of interest T and the predictor X when the bivariate $h^{T|X}$ is equal to the univariate hazard h^T. In that case the estimator attains the optimal univariate rate of the MISE convergence; (iv) All these desired properties are achieved without any complementary procedures like a hypothesis testing or an intensive numerical calculation.

The developed theory of series estimation allows us to use the E-estimator for small samples. Further, the E-estimator can use the already tested estimator of the natural nuisance function of Sect. 3.6.1 in place of the projection estimator (3.6.31) which is more convenient for theoretical analysis. Further, we can also use the universal hard-thresholding estimator

with parameters J_h and c_{THh} for the conditional hazard estimation, and it will be of interest
to check how it performs.

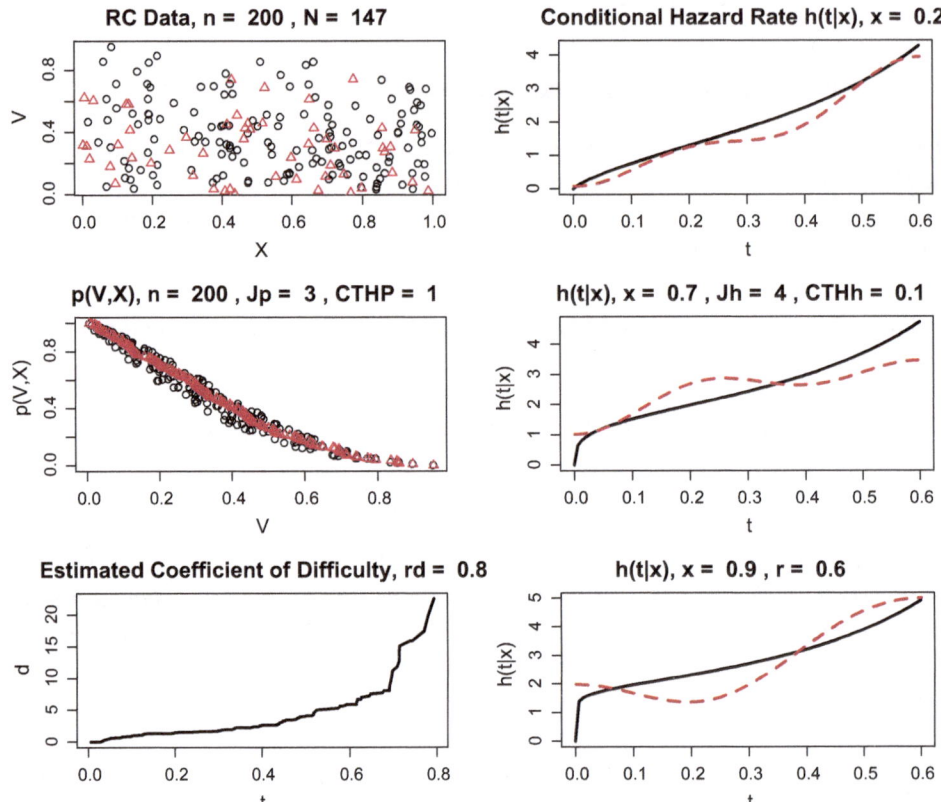

Fig. 3.12 Estimation of the conditional hazard rate $h^{T|X}$ based on a right-censored (RC) sample of
size n from $(X, V, \Delta) := (X, \min(T, C), I(T \le C))$. The underlying experiment is the same as in
Fig. 3.10. Structure of the left-top and left-middle diagrams is the same as in Fig. 3.11. The left-bottom
diagram shows the estimated coefficient of difficulty over interval $[0, r_d]$. The right diagrams show
slices of the underlying conditional hazard rate $h^{T|X}(t|x)$ (the solid line) and of its estimate (the
dashed line) for $t \in [0, r]$ and the indicated values of x controlled by the argument xx. [$n = 200$, $b =$
1.5, $b1 = 0.5$, $a = -1$, $rd = 0.8$, $r = 0.6$, $Jp = 3$, $Jh = 4$, $CTHP = 1$, $CTHh = 0.1$, $xx = c(0.2, 0.7,$
$0.9)$]

Figure 3.12 illustrates performance of the conditional hazard universal hard-thresholding
estimator for a simulated right-censored data. The underlying simulation experiment is the
same as in Fig. 3.10. It is difficult to get any opinion about the conditional hazard rate from
the data shown in the left-top diagram. Further, note that more than 25% of observations
are censored. The estimated and underlying natural nuisance functions are shown in the

left-middle diagram by the triangles and the circles, respectively. The left-bottom diagram shows the estimated coefficient of difficulty as a function in the restriction r over the interval $[0, r_d]$. The figure allows to change the range by using the argument r_d. Note how fast the coefficient of difficulty increases in r. Let us stress that this is the coefficient of difficulty for a bivariate nonparametric estimation problem. For comparison, estimation of a bivariate density supported on $[0, 1]^2$ has the coefficient of difficulty equal to 1. This is the benchmark to compare with, and it is worthwhile to return to Fig. 1.7 and recall a typical accuracy for the benchmark. The sample size $n = 200$ is considered small for bivariate problems, and the estimated coefficient of difficulty emphasizes the dramatical challenge of estimating the underlying conditional hazard rate.

Let us look at performance of the estimator over interval $[0, r]$ with the chosen restriction $r = 0.6$. Note that the estimated coefficient of difficulty is about 6. The right diagrams show slices of the underlying conditional hazard rate $h^{T|X}(t|x)$ (the solid line) and its estimate (the dashed line) for $t \in [0, r]$ and values of x shown in the titles. First of all, note the complicated shape of the underlying conditional hazards. Also note the dramatic effect of x on the conditional hazard for the smallest times t. Overall, these shapes are very difficult for nonparametric estimation. Now let is look at the estimates (the dashed lines). The estimate for $x = 0.2$ is surprisingly good keeping in mind the all complications. It correctly shows the underlying conditional hazard near $t = 0$ as well as the increasing nature in t of the conditional hazard. Now let us look at the right-middle diagram with $x = 0.7$. Note that here the underlying conditional hazard rate increases very rapidly for the smallest values of t. In other words, increase in x rapidly improves the chance of occurring the event of interest during the initial period of time. The estimate also indicates this phenomenon. The right-bottom diagram shows us even more dramatic initial hazard rate, and the estimate correctly stresses this. Overall, despite the complicated bivariate nature of the underlying conditional hazard rate and the very large coefficient of difficulty, the particular conditional hazard estimate is relatively good.

In Fig. 3.12 the universal hard-thresholding estimator is used. Intensive numerical simulations show that the E-estimator is better due its step of choosing the data-driven cutoff. At the same time, the universal hard-thresholding gives the reader a hands-on approach to choosing the cutoff and it tailors the experience in nonparametric series estimation. This is why it is highly recommended to repeat this figure with different arguments.

Let us finish this section by the following remark. The conditional hazard rate sheds an interesting light on the underlying distribution that is difficult to gain from analysis of the conditional survival function. Of course, it is simpler to estimate the latter, but visualization of the conditional hazard rate may tell us much more about the underlying lifetime of interest.

3.6.4   Mixed Multivariate Predictor

It is often the case that there are several covariates of interest, both continuous and categorical, that may affect distribution of the lifetime of interest. In this subsection we are considering this case, that is we consider a vector-predictor which may contain mixed (continuous and ordinal/nominal categorical) covariates. Namely, the predictor is $\mathbf{Z} := (\mathbf{X}, \mathbf{U})$ where $\mathbf{X} := (X_1, \ldots, X_\tau) \in [0, 1]^\tau$ is the vector of τ continuous covariates and $\mathbf{U} := (U_1, \ldots, U_m) \in \mathcal{M}_m := \prod_{k=1}^m \{0, 1, \ldots, M_k - 1\}$ is the vector of m categorical covariates. The corresponding mixed design density $f^{\mathbf{Z}}(\mathbf{z}) = f^{\mathbf{Z}}(\mathbf{x}, \mathbf{u}) := f^{\mathbf{X}|\mathbf{U}}(\mathbf{x}|\mathbf{u}) f^{\mathbf{U}}(\mathbf{u})$ is supported on $R_{\tau m} := [0, 1]^\tau \times \mathcal{M}_m$. Also set $M := \prod_{k=1}^m M_k$ and introduce two hyper-rectangles $\mathcal{R} := [0, r] \times [0, 1]^\tau$ and $R_{1\tau m} := [0, r] \times R_{\tau m}$.

Our main aim is to suggest an estimator whose MISE converges with the rate corresponding to the underlying smoothness and dimensionality of $h^{T|\mathbf{Z}}$ when the lifetime of interest T may depend on a subset of continuous covariates. Similarly to the case of a univariate predictor, we will use a series estimator based on a tensor-product basis defined on the set $R_{1\tau m}$ with the inner product

$$\langle g_1, g_2 \rangle := M^{-1} \sum_{\mathbf{u} \in \mathcal{M}_m} \int_{\mathcal{R}} g_1(t, (\mathbf{x}, \mathbf{u})) g_2(t, (\mathbf{x}, \mathbf{u})) d\mathbf{x} dt. \qquad (3.6.39)$$

The corresponding squared norm is denoted as $\|g\|^2 := \langle g, g \rangle$. A convenient basis is defined as follows. For vector $\mathbf{X}$ of continuous predictors on $[0, 1]^\tau$ we use the cosine tensor-product basis $\boldsymbol{\varphi}_{\mathbf{i}}(\mathbf{x}) := \prod_{k=1}^\tau \varphi_{i_k}(x_k)$ where $\mathbf{i} := (i_1, \ldots, i_\tau) \in \{0, 1, \ldots\}^\tau$. For a kth categorical covariate $u_k \in \{0, 1, \ldots, M_k - 1\}$ we use a discrete trigonometric basis

$$v_0(u_k, M_k) = 1, \quad v_\mu(u_k, M_k) := 2^{1/2} \cos(\pi(2u_k + 1)\mu/(2M_k)), \quad \mu = 1, \ldots, M_k - 1,$$

which allows us to define the tensor-product basis $\{\boldsymbol{v}_\mu(\mathbf{u}) := \prod_{k=1}^m v_{\mu_k}(u_k, M_k), \ \boldsymbol{\mu} \in \mathcal{M}_m, \mathbf{u} \in \mathcal{M}_m\}$. Finally, for the inner product (3.6.39) we use the tensor-product basis $\{\psi_{j\mathbf{i}\mu}(t, (\mathbf{x}, \mathbf{u})) := \eta_{rj}(t)\varphi_{\mathbf{i}}(\mathbf{x})\boldsymbol{v}_\mu(\mathbf{u}) =: \eta_{rj}(t)\varphi_{\mathbf{i}\mu}(\mathbf{x}, \mathbf{u}), (j, \mathbf{i}, \boldsymbol{\mu}) \in \mathcal{M}_{1+\tau, m} := \{0, 1, \ldots\}^{1+\tau} \times \mathcal{M}_m\}$. Using this basis and the Parseval theorem, a conditional hazard with a finite norm may be written as a Fourier series,

$$h^{T|\mathbf{Z}}(t|\mathbf{z}) = \sum_{(j, \mathbf{i}, \boldsymbol{\mu}) \in \mathcal{M}_{1+\tau, m}} \theta_{j\mathbf{i}\mu} \psi_{j\mathbf{i}\mu}(t, \mathbf{z}), \quad \text{where}$$

$$\theta_{j\mathbf{i}\mu} := \langle h^{T|\mathbf{Z}}, \psi_{j\mathbf{i}\mu} \rangle, \quad (t, \mathbf{z}) \in R_{1\tau m}. \qquad (3.6.40)$$

Now let us describe the approach used to solve the problems of dimension reduction and adaptation to smoothness of $h^{T|\mathbf{Z}}(t|\mathbf{x}, \mathbf{u})$ in continuous variables $(t, \mathbf{x})$. The approach is to use special blocks of indexes $(j, \mathbf{i}, \boldsymbol{\mu})$. Set $A'_0 := \{0, 1, \ldots\}$, and introduce tensor-product arrays of indexes $A_1 := A'_0 \times \{0\}^\tau \times \mathcal{M}_m$, $A_2 := \{A'_0 \times A'_0 \times \{0\}^{\tau-1} \times \mathcal{M}_m\} \setminus A_1$, $A_3 := \{A'_0 \times \{0\} \times A'_0 \times \{0\}^{\tau-2} \times \mathcal{M}_m\} \setminus \cup_{k=1}^2 A_k, \ldots, A_{1+\tau} := \{A'_0 \times \{0\}^{\tau-1} \times A'_0 \times \mathcal{M}_m\} \setminus$

$\cup_{k=1}^{\tau} A_k, A_{1+\tau+1} := \{\{A_0'\}^3 \times \{0\}^{\tau-2} \times \mathcal{M}_m\} \setminus \cup_{k=1}^{1+\tau} A_k, A_{1+\tau+2} := \{\{A_0'\}^2 \times \{0\} \times A_0' \times \{0\}^{\tau-3} \times \mathcal{M}_m\} \setminus \cup_{k=1}^{1+\tau+1} A_k, \ldots, A_{2^\tau} := \{\{A_0'\}^{1+\tau} \times \mathcal{M}_m\}\} \setminus \cup_{k=1}^{2^\tau-1} A_k$. The arrays are arranged in such a way that the number c_ν, $c_\nu \in \{1, 2, \ldots, 1+\tau\}$ of considered continuous variables in array A_ν is a stepwise function increasing in $\nu = 1, 2, \ldots, 2^\tau$, and note that continuous variables in A_ν include variable t together with $c_\nu - 1$ continuous covariates. Then similarly to (3.6.24) we can rewrite (3.6.40) as

$$h^{T|Z}(t|\mathbf{z}) = \sum_{\nu=1}^{2^\tau} h_\nu(t, \mathbf{z}), \quad \text{where} \quad h_\nu(t, \mathbf{z}) = \sum_{(j,\mathbf{i},\mu) \in A_\nu} \theta_{j\mathbf{i}\mu} \psi_{j\mathbf{i}\mu}(t, \mathbf{z}). \tag{3.6.41}$$

Using blocks B_k introduced in the previous subsection, for each A_ν we define its own array of tensor-product blocks $B_{\nu\mathbf{k}} := A_\nu \cap \{\{B_{k_1}, B_{k_2}, \ldots, B_{k_{1+\tau}}\}^{1+\tau} \times \mathcal{M}_m\}$, $\mathbf{k} := (k_1, \ldots, k_{1+\tau})$, and then denote by $L_{\nu\mathbf{k}}$ cardinality of $B_{\nu\mathbf{k}}$. Recall notation K_{0n}, introduced in the paragraph above line (3.6.25), and set $K_{\tau n}$ to be the largest integer such that $b_{1+K_{\tau n}} \leq n^{1/(2\tau+1)}$. Note that if $\tau = 1$ then K_{1n} is the same as for the univariate predictor. Also set $\rho_{\mathbf{k}} := 1/\ln(3 + \prod_{\nu=1}^{1+\tau} k_\nu)$. Then following (3.6.25) we can approximate conditional hazard (3.6.41) by

$$h_n^{T|Z}(t|\mathbf{z}) := \sum_{\nu=1}^{2^\tau} \sum_{k_1=1}^{K_{0n}} \sum_{k_2,\ldots,k_{1+\tau}=1}^{K_{\tau n}} \sum_{(j,\mathbf{i},\mu) \in B_{\nu\mathbf{k}}} \theta_{j\mathbf{i}\mu} \psi_{j\mathbf{i}\mu}(t, \mathbf{z}). \tag{3.6.42}$$

This is the partial sum whose Fourier coefficients we intend to estimate.

Now let us introduce a considered function class. For a vector $\boldsymbol{\alpha} := (\alpha_0, \alpha_1, \ldots, \alpha_\tau)$ of positive integers, whose elements represent number of derivatives of a function in its $1 + \tau$ continuous variables, we define an anisotropic Sobolev function class

$$\mathcal{S}(\boldsymbol{\alpha}, Q) := \Big\{ g : g(t, \mathbf{z}) \geq 0 \text{ for } (t, \mathbf{z}) \in [0, \infty) \times R_{\tau m},$$

$$g(t, \mathbf{z}) = \sum_{(j,\mathbf{i},\mu) \in \mathcal{M}_{1+\tau,m}} \theta_{j\mathbf{i}\mu} \psi_{j\mathbf{i}\mu}(t, \mathbf{z}) \text{ for } (t, \mathbf{z}) \in R_{1\tau m},$$

$$\sum_{(j,\mathbf{i},\mu) \in \mathcal{M}_{1+\tau,m}} [1 + (\pi j/r)^{2\alpha_0} + \sum_{k=1}^{\tau} (\pi i_k)^{2\alpha_k}] \theta_{j\mathbf{i}\mu}^2 \leq rQ < \infty \Big\}, \tag{3.6.43}$$

and its effective smoothness

$$\alpha := \frac{1}{\sum_{k=0}^{\tau} \alpha_k^{-1}}. \tag{3.6.44}$$

For this Sobolev class and direct data nonparametric regression with $(1 + \tau)$ continuous covariates, the optimal rate of the oracle's MISE convergence is $n^{-2\alpha/(2\alpha+1)}$. As we will see shortly, this result can be matched for conditional hazard with τ continuous covariates and m categorical ones.

Assumption 3.7 Given predictor $\mathbf{Z}$, lifetime of interest T and censoring lifetime C are independent. An underlying natural nuisance function $p(t, \mathbf{z}) := f^{\mathbf{Z}}(\mathbf{z}) S^{T|\mathbf{Z}}(t|\mathbf{z}) S^{C|\mathbf{Z}}(t|\mathbf{z})$ for all $t \in [0, r]$ belongs to the Sobolev class

$$\mathcal{S}_* := \left\{ g : g(t, \mathbf{z}) = \sum_{(\mathbf{i}, \mu) \in \mathcal{M}_{\tau, m}} \kappa_{\mathbf{i}\mu}(t) \varphi_{\mathbf{i}\mu}(\mathbf{z}) \text{ for } (t, \mathbf{z}) \in \mathcal{R}_{1\tau m}, \right.$$

$$\max_{t \in [0, r]} \sum_{(\mathbf{i}, \mu) \in \mathcal{M}_{\tau, m}} [1 + \sum_{k=1}^{\tau} (\pi i_k)^{2\tau}][\kappa_{\mathbf{i}\mu}(t)]^2 \le Q_* < \infty,$$

$$\left. \min_{(t, \mathbf{z}) \in \mathcal{R}_{1\tau m}} g(t, \mathbf{z}) > 0, \ g(t, \mathbf{z}) \ge 0 \text{ for } (t, \mathbf{z}) \in [0, \infty) \times \mathcal{R}_{\tau m} \right\}. \qquad (3.6.45)$$

The oracle knows the natural nuisance function p. Assumption 3.7 allows us to match the oracle by suggesting an appropriate estimate of p. To overcome the curse of dimensionality, the assumed Sobolev's smoothness in (3.6.45) depends on τ, but it is not tied to smoothness of $h^{T|\mathbf{Z}}$. Also note that the natural nuisance function is the product of three functions, and accordingly its smoothness in a continuous covariate x_i is defined by a coarsest in x_i function among the triplet $(f^{\mathbf{Z}}, S^{T|\mathbf{Z}}, S^{C|\mathbf{Z}})$. Further, because $S^{T|\mathbf{Z}}$ is a factor in the natural nuisance function, the estimand $h^{T|\mathbf{Z}}$ is at least as smooth in $\mathbf{x}$ as $p(t, \mathbf{z})$, and accordingly (3.6.45) implies that in (3.6.43) we have $\min(\alpha_1, \ldots, \alpha_\tau) \ge \tau$.

The introduced notions and notations for the case of a multivariate predictor are similar to the univariate case, and the same similarity will be observed between the proposed estimators. We begin with introducing a plug-in sample mean Fourier coefficient estimator

$$\hat{\theta}_{j\mathbf{i}\mu} := n^{-1} M^{-1} \sum_{l=1}^{n} \frac{\Delta_l I(V_l \le r) \psi_{j\mathbf{i}\mu}(V_l, \mathbf{Z}_l)}{\hat{p}(V_l, \mathbf{Z}_l)} \qquad (3.6.46)$$

and an estimator of the coefficient of difficulty,

$$\hat{d} := n^{-1} M^{-1} \sum_{l=1}^{n} \frac{\Delta_l I(V_l \le r)}{[\hat{p}(V_l, \mathbf{Z}_l)]^2}. \qquad (3.6.47)$$

Here $\hat{p}$ is an estimator of the natural nuisance function defined as

$$\hat{p}(t, \mathbf{z}) := \max(\tilde{p}(t, \mathbf{z}), 1/\ln(n)),$$

$$\tilde{p}(t, \mathbf{z}) := n^{-1} M^{-1} \sum_{l=1}^{n} \sum_{(\mathbf{i}, \mu) \in A'} I(V_l \ge t) \varphi_{\mathbf{i}\mu}(\mathbf{Z}_l) \varphi_{\mathbf{i}\mu}(\mathbf{z}), \qquad (3.6.48)$$

where $A' := \{0, 1, \ldots, n^{1/3\tau} \ln(\ln(n + 20))\}^{\tau} \times \mathcal{M}_m$. Next, similarly to (3.6.34) and (3.6.35) we introduce two statistics

$$\hat{\Theta}_{\nu\mathbf{k}} := \frac{2}{L_{\nu\mathbf{k}} n(n-1)}$$

$$\times \sum_{(j,\mathbf{i},\mu)\in B_{\nu\mathbf{k}}} \sum_{1\le l_1 < l_2 \le n} \prod_{s=1}^{2} \frac{\Delta_{l_s} \psi_{j\mathbf{i}\mu}(V_{l_s}, \mathbf{Z}_{l_s}) I(V_{l_s} \le r)}{\hat{p}(V_{l_s}, \mathbf{Z}_{l_s})} \tag{3.6.49}$$

and $\tilde{\Theta}_{\nu\mathbf{k}} := L_{\nu\mathbf{k}}^{-1} \sum_{(j,\mathbf{i},\mu)\in B_{\nu\mathbf{k}}} \hat{\theta}_{j\mathbf{i}\mu}^2$. The proposed conditional hazard estimator is

$$\hat{h}^{T|\mathbf{Z}}(t|\mathbf{z}) := \sum_{\nu=1}^{2^\tau} \sum_{k_1=1}^{K_{0n}} \sum_{k_2,\ldots,k_{1+\tau}=1}^{K_{\tau n}} \left[\sum_{(j,\mathbf{i},\mu)\in B_{\nu\mathbf{k}}} I(\hat{\theta}_{j\mathbf{i}\mu}^2 > 2\ln(n)\hat{d}n^{-1}) I(j \vee \mathbf{i} \le J_n) \right.$$

$$+ \min(1, \hat{\Theta}_{\nu\mathbf{k}}/\tilde{\Theta}_{\nu\mathbf{k}}) I(\hat{\Theta}_{\nu\mathbf{k}} > \rho_{\mathbf{k}} n^{-1}) I(j \vee \mathbf{i} > J_n) \Big] \hat{\theta}_{j\mathbf{i}\mu} \psi_{j\mathbf{i}\mu}(t, \mathbf{z}), \tag{3.6.50}$$

where $j \vee \mathbf{i} := \max(j, i_1, \ldots, i_\tau)$.

Theorem 3.10 *Let Assumption 3.7 hold. Then the conditional hazard estimator (3.6.50) is rate-optimal over an anisotropic Sobolev class (3.6.43) and*

$$\sup_{h^{T|\mathbf{Z}}\in S(\alpha,Q)} \mathbb{E}\{\|\hat{h}^{T|\mathbf{Z}} - h^{T|\mathbf{Z}}\|^2\} \le \zeta n^{-2\alpha/(2\alpha+1)}, \tag{3.6.51}$$

where α is the effective smoothness (3.6.44) of the Sobolev class. If the conditional hazard $h^{T|\mathbf{Z}}$ depends only on a subvector $(X_{k_1}, \ldots, X_{k_s})$, $s < \tau$ of continuous covariates, then the estimator's MISE satisfies (3.6.51) with α being replaced by a corresponding effective smoothness $\alpha^* := 1/[\alpha_0^{-1} + \sum_{i=1}^{s} \alpha_{k_i}^{-1}]$.

The rates outlined in Theorem 3.10 are optimal even for the oracle who knows an underlying dimensionality, and this implies that the estimator performs the desired dimension reduction and adaptation to smoothness of the underlying conditional hazard rate function. This is a nice bouquet of desired statistical properties.

3.7   Bivariate Survival Analysis

There are many examples in survival analysis when we are interested in a pair (T, X) of lifetimes. In biomedical research, study subjects from the same cluster (e.g. family or twins) share common genetic and/or environmental factors. Another example is the time to a deterioration level or the time to reaction of a treatment in pairs of lungs, kidneys, eyes or ears of humans, or time from initiation of treatment until first response in two successive courses of a treatment in the same patient. Two related diseases may happen in one patient and the times to healing are of interest. Two sequential recurrence times of a certain disease. Time from surgery to cancer relapse, and then time from the relapse to death. The amounts paid in different but related lines of business. The time from an insurable event (disability, fire) to informing the insurance company, and then the time that takes to settle the claim.

In all these examples right-censoring may occur for a number of reasons including withdraw from the study, a change of health status or contamination, or by death from a cause unrelated to the study. In actuarial science an important example is the joint life annuities issued to married couples who tend to be exposed to similar risks and likely to have the same living habits. A couple can cancel the annuity and this yields the right-censoring. A similar example is when a couple takes two loans or opens two credit cards and we are interested in times to default.

The main aim of this section is to explain how the joint survival function and the joint density of a pair of lifetimes of interest may be estimated. This is the notoriously complicated problem for right-censored lifetimes. For instance, it is well documented in the survival analysis literature that estimating the joint survival function of a pair of right-censored lifetimes presents challenges unparalleled to the univariate case where a product-limit Kaplan-Meyer's methodology yields optimal estimation. There is no straightforward extension of the univariate product-limit methodology to the bivariate case. Instead, a number of sophisticated and mathematically involved procedures, ranging from expectation-maximization (EM) algorithm, hazard gradient, partial differential equations and copula to nonparametric maximum likelihood estimation (MLE) and solving an inhomogeneous Volterra equation via Peano series, have been proposed, see the Notes. Just mentioning these methods makes it clear how complicated the problem is. Nonetheless, in this section it will be explained that the method of moments, proposed in Sect. 3.2. for right-censored univariate lifetime, may be extended to the pair of right-censored lifetimes.

Let us formulate the studied bivariate right-censored model. We observe a sample of size n from a quartet

$$(V, W, \Delta, \Gamma) := (\min(T, C), \min(X, D), I(T \leq C), I(X \leq D)). \qquad (3.7.1)$$

Here (T, X) is the pair of continuous lifetimes of interest, (C, D) is a pair of continuous censoring lifetimes, and it is assumed that the two pairs are independent. We are interested in estimating the joint survival function $S^{T,X}(t, x) := \mathbb{P}(T > t, X > x)$ and the joint density $f^{T,X}$ of the pair (T, X).

Please note that again we have a symmetry in the right-censoring when the pair (C, D) right-censors (T, X), and the pair (T, X) right-censors (C, D). Accordingly, if we know how to estimate the distribution of (T, X), we also know how to estimate the distribution of (C, D).

We are considering estimation of the joint survival function $S^{T,X}$ and the joint density $f^{T,X}$ in turn. Due to the familiar right-tail curse, these bivariate functions are estimated on a finite rectangle $\mathcal{R} := [0, r_T] \times [0, r_X]$ with some given positive finite restrictions r_T and r_X.

3.7.1 Estimation of Bivariate Survival Function

We begin with the basic assumption about underlying distributions.

Assumption 3.8 Pair of continuous lifetimes of interest (T, X) is independent of pair of continuous censoring lifetimes (C, D). The survival $S^{T,X}$ of the lifetimes of interest is estimated on a rectangle $\mathcal{R} = [0, r_T] \times [0, r_X]$. The joint survival function $S^{C,D}(t, x) = \mathbb{P}(C > t, D > x)$ of the censoring lifetimes is positive on $\mathcal{R}$, that is $S^{C,D}(r_T, r_X) > 0$. If $S^{C,D}$ is unknown, then it is additionally assumed that $S^{T,X}(r_T, r_X) > 0$.

Note that the made assumption about a positive joint survival function $S^{C,D}$ on $\mathcal{R}$ is traditional. Indeed, the rectangle $\mathcal{R}$ must be a subset of the support of (C, D) for consistent estimation of the distribution of (T, X) on $\mathcal{R}$. A similar conclusion holds for estimation of the distribution of censoring (C, D).

We begin with the methodology. Introduce a *bivariate cumulative hazard*

$$H^{T,X}(t, x) := -\ln(S^{T,X}(t, x)). \tag{3.7.2}$$

Using notation $f^{T>t,X}(z) := \partial\mathbb{P}(T > t, X \le z)/\partial z$ and a line integration we write,

$$H^{T,X}(t, x) = \int_0^t \left[\frac{\partial}{\partial x_1} H^{T,X}(x_1, 0)\right] dx_1 + \int_0^x \left[\frac{\partial}{\partial x_2} H^{T,X}(t, x_2)\right] dx_2$$

$$= \int_0^t \frac{f^T(x_1)}{S^T(x_1)} dx_1 + \int_0^x \frac{f^{T>t,X}(x_2)}{S^{T,X}(t, x_2)} dx_2 =: H_1(t) + H_2(t, x). \tag{3.7.3}$$

The underlying idea of the formula (3.7.3) is that the integrals H_1 and H_2 can be estimated by appropriate method moments estimators. We begin with estimation of $H_1(t)$. Using $S^V(t) = S^T(t)S^C(t)$ we write,

$$f^{V,\Delta}(t, 1) = f^T(t)S^C(t) = \frac{f^T(t)S^V(t)}{S^T(x)}. \tag{3.7.4}$$

This formula implies that the integrand in $H_1(t)$ may be written as

$$\frac{f^T(x_1)}{S^T(x_1)} = \frac{f^{V,\Delta}(x_1, 1)}{S^V(x_1)} \quad \text{whenever} \quad S^V(x_1) > 0. \tag{3.7.5}$$

The last relation allows us to write for $H_1(t)$,

$$H_1(t) = \int_0^t \frac{f^T(x_1)}{S^T(x_1)} dx_1 = \int_0^t \frac{f^{V,\Delta}(x_1, 1)}{S^V(x_1)} dx_1 = \mathbb{E}\left\{\frac{\Delta I(V \leq t)}{S^V(V)}\right\}. \tag{3.7.6}$$

To use this formula for sample mean estimation we need to know the univariate survival function $S^V(x)$ of the continuous lifetime V. This nuisance function is estimated straightforwardly because V is directly observed, and we use the empirical survival function

$$\hat{S}^V(t) := n^{-1} \sum_{l=1}^n I(V_l \geq t). \tag{3.7.7}$$

Note that $\hat{S}^V(V_l) \geq n^{-1}$ and hence can be used in denominators. Accordingly, we propose the following method of moments estimator of H_1 (compare with (3.2.6)),

$$\hat{H}_1(t) := n^{-1} \sum_{l=1}^n \frac{\Delta_l I(V_l \leq t)}{\hat{S}^V(V_l)} = \sum_{l=1}^n \frac{\Delta_l I(V_l \leq t)}{\sum_{k=1}^n I(V_k \geq V_l)}. \tag{3.7.8}$$

Estimation of the second integral $H_2(t, x)$ on the right side of (3.7.3) is more involved. Using $S^{V,W}(t, x) = S^{T,X}(t, x)S^{C,D}(x, y)$ we can write

$$\mathbb{P}(V > t, W \leq x, \Gamma = 1) = \int_0^x f^{V>t, W, \Gamma}(z, 1) dz$$

$$= \int_0^x S^{C,D}(t, z) f^{T>t, X}(z) dz = \int_0^x \frac{S^{V,W}(t, z) f^{T>t, X}(z)}{S^{T,X}(x, z)} dz. \tag{3.7.9}$$

Using (3.7.9) we conclude that

$$f^{V>t, W, \Gamma}(x, 1) = \frac{S^{V,W}(t, x) f^{T>t, X}(x)}{S^{T,X}(t, x)}. \tag{3.7.10}$$

The last relation yields

$$H_2(t, x) = \int_0^x \frac{f^{V>t, W, \Gamma}(z, 1)}{S^{V,W}(t, z)} dz = \mathbb{E}\left\{\frac{\Gamma I(V > t) I(W \leq x)}{S^{V,W}(t, W)}\right\}. \tag{3.7.11}$$

To use the last expectation for the method of moments estimation we need to propose an estimator of $S^{V,W}$. The pair (V, W) is observed directly, and hence we can use the empirical bivariate survival function

$$\hat{S}^{V,W}(t, x) := n^{-1}\left[1 + \sum_{k=1}^n I(V_k \geq t) I(W_k \geq x)\right]. \tag{3.7.12}$$

Note that $\hat{S}^{V,W}(V_1, W_2) \geq n^{-1}$, and hence it may be used in denominators. This allows us to propose the following method of moments estimator of H_2 for $(t, x) \in \mathcal{R}$,

$$\hat{H}_2(t, x) := n^{-1} \sum_{l=1}^{n} \frac{\Gamma_l I(V_l > t) I(W_l \leq x)}{\hat{S}^{V,W}(t, W_l)}$$

$$= \sum_{l=1}^{n} \frac{\Gamma_l I(V_l > t) I(W_l \leq x)}{1 + \sum_{k=1}^{n} I(V_k \geq t) I(W_k \geq W_l)}, \quad (t, x) \in \mathcal{R}. \qquad (3.7.13)$$

Further, we immediately get the wished method of moments estimator of the bivariate survival function $S^{T,X}(t, x)$ for $(x, t) \in \mathcal{R}$,

$$\hat{S}^{T,X}(t, x) = \exp\{-\hat{H}_1(t) - \hat{H}_2(t, x)\}$$

$$= \exp\left\{-\sum_{l=1}^{n}\left[\frac{\Delta_l I(V_l \leq t)}{\sum_{k=1}^{n} I(V_k \geq V_l)} + \frac{\Gamma_l I(V_l > t) I(W_l \leq x)}{1 + \sum_{k=1}^{n} I(V_k \geq t) I(W_k \geq W_l)}\right]\right\}. \qquad (3.7.14)$$

Due to the symmetry in right-censoring, we can also suggest the corresponding empirical bivariate survival function for (C, D). Because it will be later used in denominators, we add n^{-1} and set for $(t, x) \in \mathcal{R}$

$$\hat{S}^{C,D}(t, x) := n^{-1}$$

$$+ \exp\left\{-\sum_{l=1}^{n}\left[\frac{(1 - \Delta_l)I(V_l \leq t)}{\sum_{k=1}^{n} I(V_k \geq V_l)} + \frac{(1 - \Gamma_l)I(V_l > t) I(W_l \leq x)}{1 + \sum_{k=1}^{n} I(V_k \geq t) I(W_k \geq W_l)}\right]\right\}. \qquad (3.7.15)$$

Let us check performance of the bivariate survival function estimator (3.7.14). We begin with the case of independent T and X considered in Fig. 3.13. The caption explains the experiment, and note that here $S^{T,X} = S^T S^X$. At the same time, the two censoring variables C and D are dependent. The left-top diagram shows us the data. Here we observe four subsets with different values of (Δ, Γ) explained in the caption. Almost half of pairs (T, X) are censored, and this creates a complicated problem for the bivariate estimation. Further, the majority of observations are from the rectangle $[0, 0.5]^2$ and there is just one uncensored pair with both values larger 0.6. The left-middle diagram shows us values $S^{T,X}(V_l, W_l)$ for the observed pairs $(V_l, W_l), l = 1, 2, \ldots, n$ by the circles and the corresponding estimates by the triangles. Overall the estimates follow the joint survival function, and the corresponding errors are shown in the left-bottom diagram. The right column of diagrams sheds additional light on the estimate. The right-top diagram shows the joint survival function and its estimate at the vector of pairs $(z1, z2)$ with $z1_l = z2_l = l/(k-1), l = 0, 1, \ldots, k-1$, and the figure allows to choose any two vectors of $z1$ and $z2$. The quality of estimation is good. The right-middle diagram shows $S^{T,X}(t, z2)$ for the fixed $t = 0.1$ whose value may be changed. Finally, the right-bottom diagram shows the conditional survival $S^{X|T}(z|t) = \mathbb{P}(X > z|T > t)$ and its estimates for the three values of t shown in the title. For the considered simulation we have $S^{X|T} = S^X$ due to the independence between T and X. As we see, the three estimates

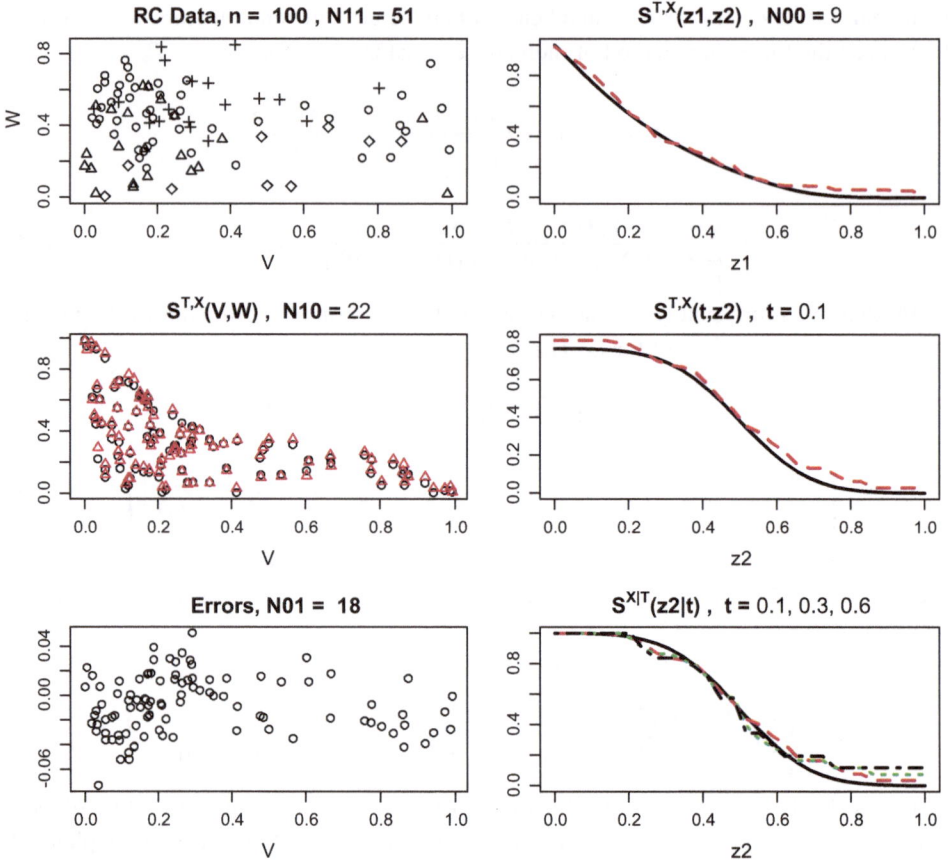

Fig. 3.13 Estimation of the bivariate survival function $S^{T,X}$. The case of independent T and X, and dependent censoring lifetimes C and D. The left-top diagram shows simulated data. Here T is the Bathtub, X is the Normal, C is Uniform$(0, b)$, and D is Uniform$(0, b + b_1 C)$, that is C and D are dependent. The left-top diagram shows by the circles, the triangles, the crosses and the rhombi pairs (V, W) corresponding to $\Delta \Gamma = 1, \Delta(1 - \Gamma) = 1, (1 - \Delta)\Gamma = 1, (1 - \Delta)(1 - \Gamma) = 1$, respectively. Further, the numbers of corresponding observations are denoted as $N11, N10, N01, N00$ and shown in the titles. The left-middle diagram shows $S^{T,X}(V_l, W_l)$ by the circles and its estimate $\hat{S}^{T,X}(V_l, W_l)$ by the triangles, $l = 1, \ldots, n$. The left-bottom diagram shows the errors $S^{T,X}(V_l, W_l) - \hat{S}^{T,X}(V_l, W_l)$. The right-top diagram shows $S^{T,X}(z1, z2)$ and $\hat{S}^{T,X}(z1_l, z2_l)$ at $z1_l = z2_l = l/(k - 1), l = 0, 1, \ldots, k - 1$ by the solid and dashed lines, respectively. The right-middle diagram shows $S^{T,X}(t, z2)$ (the solid line) and $\hat{S}^{T,X}(t, z2)$ (the dashed line) for $t = 0.1$. The right-bottom diagram shows by the solid line $S^X(z2)$ and $\hat{S}^{X|T}(z2|t)$ by the dashed, dotted and dot-dashed lines for $t = 0.1, 0.3$ and 0.6, respectively. {The corner densities of T and X are controlled by the argument *corn*. Vectors $z1$ and $z2$ must have the same length but otherwise may be different.} [$n = 100, corn = c(2, 3), b = 1.5, b1 = 0.5, k = 30, z1 = seq(0, 1, len = k), z2 = seq(0, 1, len = k), tset = c(0.1, 0.3, 0.6)$]

are close to each other, and accordingly this may allow us to conjecture about independence between T and X.

It is important to repeat this figure a number of times with different parameters to realize complexity of the problem and the relatively good performance of the estimator. Further, it is advisable to compare outcomes with results based on using Fig. 3.1 where estimation of univariate survival functions is analyzed.

Figure 3.14 allows us to check performance of the estimator for the case of dependent T and X. The left-top diagram of the data shows complexity of the data. First of all, it is difficult to believe that we see $n = 100$ observations. Second, there are just a few large observations with uncensored pairs (T, X). This creates the curse of right tail for right-censored observations. Finally, note that more than 40% of observations are censored. Now let us check how the estimator performs for the case of dependent lifetimes of interest. Estimation of $S^{T,X}$ at the available observations of (V, W) is respectively good, see the left-middle and left-bottom diagrams and compare with results in Sect. 1.2. Results shown in the right diagrams are also encouraging keeping in mind complexity of the bivariate estimation, the small sample size and the high rate of censoring. The reader is advised to repeat this figure with different parameters and get used to the problem.

As a final remark, recall that estimation of the joint survival function $S^{C,D}$ of the censoring variables is the dual problem. It is solved by the same estimator only with pair (Δ, Γ) being replaced by the pair $(\Delta', \Gamma') := (1 - \Delta, 1 - \Gamma)$.

3.7.2 Estimation of Bivariate Density

The aim of this subsection is to show that, with respect to directly observed data, right-censoring does not change the optimal rate of MISE convergence but affects the sharp constant and the coefficient of difficulty. Further, we will see that a series estimator is efficient, and it motivates the corresponding bivariate E-estimator.

First, let us recall the result of Sect. 1.2.6 about MISE convergence for the case of direct observations from lifetimes (T, X). If the joint density $f^{T,X}(t, x)$ is α_T-fold differentiable in t and α_X-fold differentiable in x, then based on a sample of size n it is possible to estimate the bivariate density with the MISE decreasing with the optimal rate $n^{-2\alpha/(2\alpha+1)}$ where

$$\alpha := \frac{\alpha_T \alpha_X}{\alpha_T + \alpha_X} \tag{3.7.16}$$

is the *effective smoothness*. For instance, if $\alpha_T = \alpha_X = \nu$, then $\alpha = \nu/2$ and we get the familiar rate $n^{-2\nu/(2\nu+2)}$.

Now we return to the case of right-censored observations when we observe a sample of size n from the quartet (V, W, Δ, Γ) defined in (3.7.1). We begin with several notations and formulas. Recall notation $\mathcal{R} := [0, r_T] \times [0, r_X]$ for the rectangle of interest. The mixed joint density of the quartet (V, W, Δ, Γ) is

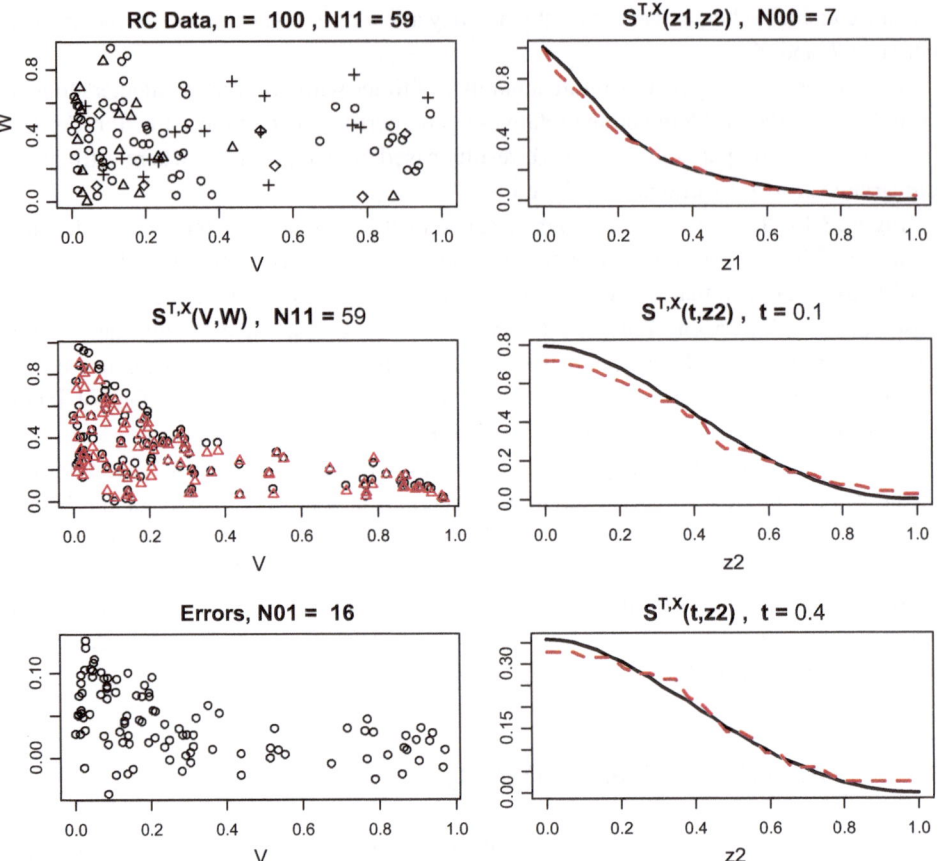

Fig. 3.14 Estimation of the bivariate survival function $S^{T,X}$. The case of dependent T and X. The left-top diagram shows simulated data. Here T is the Bathtub, X is the Beta($s_1, s_1 + f_c(T)$) with $s_1 = 2$ and $f_c(T)$ being the Normal (number 3) corner function, C is Uniform(0, b), and D is Uniform(0,$b + b_1 C$). The diagrams are similar to those in Fig. 3.13 only here the right-bottom diagram is similar to the right-middle diagram only with different value of t. {The corner densities of T and f_c in the beta distribution of X are controlled by the argument *corn*. Vectors $z1$ and $z2$ must have the same length but otherwise may be different.} [$n = 100, corn = c(2, 3), s1 = 2, b = 1.5, b1 = 0.5, k = 30, z1 = seq(0, 1, len = k), z2 = seq(0, 1, len = k), tset = c(0.1, 0.4)$]

$$f^{V,W,\Delta,\Gamma}(v, w, \delta, \gamma) := f^{V,W}(v, w)\mathbb{P}(\Delta = \delta, \Gamma = \gamma | V = v, W = w). \qquad (3.7.17)$$

Set

$$a_{ji} := a_{ji}(\alpha_T, \alpha_X, r_T, r_X) := 1 + (\pi j/r_T)^{2\alpha_T} + (\pi i/r_X)^{2\alpha_X}, \qquad (3.7.18)$$

$$P := \left[\frac{r_T r_X}{\pi^2}\right]^{2\alpha/(2\alpha+1)} \left[\frac{Q}{C_1(\alpha_T, \alpha_X)}\right]^{1/(2\alpha+1)} C_2(\alpha_T, \alpha_X), \qquad (3.7.19)$$

and define the coefficient of difficulty

$$d := d(f^{T,X}, S^{C,D}, r_T, r_X) := \frac{1}{r_T r_X} \int_{\mathcal{R}} \frac{f^{T,X}(t,x)}{S^{C,D}(t,x)} dt dx. \qquad (3.7.20)$$

Here for positive constants k and m,

$$C_1(k,m) := \int_{\{(u,v): u^{2k}+v^{2m} \le 1; u,v \ge 0\}} [(u^{2k} + v^{2m})^{1/2} - (u^{2k} + v^{2m})] du dv, \qquad (3.7.21)$$

$$C_2(k,m) := \int_{\{(u,v): u^{2k}+v^{2m} \le 1; u,v \ge 0\}} [1 - (u^{2k} + v^{2m})^{1/2}] du dv. \qquad (3.7.22)$$

For square-integrable functions on the rectangle $\mathcal{R}$ we will use the tensor-product cosine basis

$$\varphi_{j,i}(t,x) := \varphi_{ji}(t,x) := \eta_{r_T j}(t) \eta_{r_X i}(x), \quad (t,x) \in R, \ j,i = 0,1,\ldots, \qquad (3.7.23)$$

and recall that

$$\eta_{rj}(z) := \frac{1}{\sqrt{r}} \left[I(j = 0) + \sqrt{2} \cos\left(\frac{\pi jz}{r}\right) I(j \ge 1) \right].$$

Next, similarly to univariate problems, we are introducing the global Sobolev class

$$\mathcal{S}(\alpha_T, \alpha_X, Q, r_T, r_X)$$

$$:= \{g : g(x,y) = \sum_{j,i=0}^{\infty} \theta_{ji} \varphi_{ji}(t,x), \ \sum_{j,i=0}^{\infty} a_{ji} \theta_{ji}^2 \le Q\}, \qquad (3.7.24)$$

and the shrinking, toward the anchor bivariate density f_0, local Sobolev classes

$$\mathcal{F}_n := \mathcal{F}_n(f_0, \alpha_T, \alpha_X, Q, r_T, r_X)$$

$$:= \Big\{ f : \ f(t,x) = f_0(t,x) + g(t,x) I((t,x) \in \mathcal{R}_n), \ (t,x) \in [0,\infty)^2,$$

$$g \in \mathcal{S}(\alpha_T, \alpha_X, Q, r_T, r_X), \quad \max_{(t,x) \in \mathcal{R}_n} |g(t,x)| \le \min_{(t,x) \in \mathcal{R}_n} f_0(t,x) / \ln(\ln(n+20)),$$

$$\int_0^a g(t,x) dt = \int_0^b g(t,x) dx = 0 \Big\}, \qquad (3.7.25)$$

where $\mathcal{R}_n := [s_n, r_T(1 - s_n)] \times [s_n, r_X(1 - s_n)] \subset \mathcal{R}$, and $s_n := [\ln(\ln(n+20))]^{-2}$.
    The following theorem presents a sharp minimax oracle's lower bound for the MISE.

Theorem 3.11 *The problem is to estimate a joint density $f^{T,X}$ of lifetimes of interest (T, X) by an oracle-estimator $\tilde{f}_*$ on a rectangle $\mathcal{R} = [0, r_T] \times [0, r_X]$. Suppose that the joint survival function $S^{C,D}$ of censoring lifetimes (C, D) is known, Assumption 3.8 holds, and $f^{T,X}$ belongs to the shrinking local Sobolev class $\mathcal{F}_n$ defined in (3.7.25) where the anchor density*

f_0 *is continuous and positive on* $\mathcal{R}$. *The oracle knows right-censored sample* $(V_l, W_l, \Delta_l, \Gamma_l)$, $l = 1, 2, \ldots, n$ *from the quartet (3.7.1) and everything about the class* $\mathcal{F}_n$. *Then the following lower bound for minimax MISE holds,*

$$\inf_{\tilde{f}_*} \sup_{f^{T,X} \in \mathcal{F}_n} [n/d]^{2\alpha/(2\alpha+1)} \mathbb{E}_{f^{T,X}, S^{C,D}} \{ \int_{\mathcal{R}} (\tilde{f}_*(t, x) - f^{T,X}(t, x))^2 dt dx \}$$

$$\geq P(1 + o_n(1)). \tag{3.7.26}$$

Here the infimum is taken over all oracle-estimators $\tilde{f}_*$, *d is defined in (3.7.20), P is defined in (3.7.19), and* α *is the effective smoothness defined in (3.7.16). The lower bound is sharp and attainable by oracle-estimators and data-driven estimators. Moreover, the lower bound is attainable by an oracle-estimator based only on uncensored pairs of observations with* $\Delta_l \Gamma_l = 1$.

Note that the rate $n^{-2\alpha/(2\alpha+1)}$ is the same as for direct data, and the effect of right-censoring is described by the coefficient of difficulty (3.7.20). The coefficient of difficulty sheds light on complications in estimating tails of $f^{T,X}$ due to vanishing tails of $S^{C,D}$. Moreover, the coefficient of difficulty may increase to infinity as r_T or r_X increases. At the same time, if right tail of $f^{T,X}$ is lighter than that of $S^{C,D}$, then there is a chance to estimate the joint density over its support.

Now let us show how the oracle, who knows the underlying Sobolev class and the survival function $S^{C,D}$, suggests to estimate the density of interest $f^{T,X}$ over the rectangle $\mathcal{R}$. Further, to help us with finding a feasible data-driven estimator, the oracle will step-by-step remove facts known only to the oracle.

Using the assumed square integrability of $f^{T,X}$ on $\mathcal{R}$ and the above-introduced tensor-product cosine basis φ_{ji} on $\mathcal{R}$, the oracle writes down the bivariate density using the Fourier series

$$f^{T,X}(t, x) = \sum_{j,i=0}^{\infty} \theta_{ji} \varphi_{ji}(t, x), \quad (t, x) \in \mathcal{R},$$

$$\theta_{ji} := \int_{\mathcal{R}} f^{T,X}(t, x) \varphi_{ji}(t, x) dt dx, \tag{3.7.27}$$

Here θ_{ij} are Fourier coefficients of the bivariate density. To estimate them, the oracle suggests to use only uncensored pairs of observations with $\Delta \Gamma = 1$. Then the formula

$$f^{V,W,\Delta,\Gamma}(t, x, 1, 1) = f^{T,X}(t, x) S^{C,D}(t, x), \tag{3.7.28}$$

yields

$$\theta_{ji} = \int_{\mathcal{R}} \frac{f^{V,W,\Delta,\Gamma}(t, x, 1, 1) \varphi_{ji}(t, x)}{S^{C,D}(t, x)} dt dx$$

$$= \mathbb{E} \left\{ \frac{\Delta \Gamma \varphi_{ji}(V, W) I((V, W) \in \mathcal{R})}{S^{C,D}(V, W)} \right\}. \tag{3.7.29}$$

The oracle knows the joint survival function $S^{C,D}$, and using (3.7.29) proposes the following sample mean Fourier coefficient estimator based solely on uncensored pairs of observations,

$$\hat{\theta}^*_{ji} := n^{-1} \sum_{l=1}^{n} \frac{\Delta_l \Gamma_l \varphi_{ji}(V_l, W_l) I((V_l, W_l) \in \mathcal{R})}{S^{C,D}(V_l, W_l)}. \tag{3.7.30}$$

If $S^{C,D}$ is unknown, its estimate (3.7.15) can be used instead, and this yields the data-driven Fourier coefficient estimator based on both uncensored and censored observations,

$$\hat{\theta}_{ji} := n^{-1} \sum_{l=1}^{n} \frac{\Delta_l \Gamma_l \varphi_{ji}(V_l, W_l) I((V_l, W_l) \in \mathcal{R})}{\hat{S}^{C,D}(V_l, W_l)}. \tag{3.7.31}$$

Several more steps are needed to introduce a bivariate series density estimator. Set

$$a_n := \left[\frac{\pi^2 Q}{r_T r_X C_1(\alpha_T, \alpha_X) \, d} \, n \right]^{2\alpha/(2\alpha+1)}, \tag{3.7.32}$$

where $C_1(\alpha_T, \alpha_X)$ is defined in (3.7.21) and the coefficient of difficulty d in (3.7.20). Recall that the coefficient of difficulty depends on both $f^{T,X}$ and $S^{C,D}$, and it is used in the denominator of (3.7.32). The oracle proposes to evaluate the coefficient of difficulty by a truncated from below sample mean estimator

$$\hat{d} := \max\left(\frac{1}{\ln(n+20)}, \tilde{d} \right), \quad \tilde{d} := \frac{1}{n} \sum_{l=1}^{n} \frac{\Delta_l \Gamma_l I((V_l, W_l) \in \mathcal{R})}{r_T r_X [\hat{S}^{C,D}(V_l, W_l)]^2}. \tag{3.7.33}$$

Then the oracle plugs $\hat{d}$ in (3.7.32) and get the estimator

$$\hat{a}_n := \left[\frac{\pi^2 Q}{r_T r_X C_1(\alpha_T, \alpha_X) \, \hat{d}} \, n \right]^{2\alpha/(2\alpha+1)}. \tag{3.7.34}$$

Further, denote by $\hat{\sigma}^2_{ji}$ the sample mean estimator of the variance of $\hat{\theta}_{ji}$, and recall that a_{ji} are defined in (3.7.18).

Now we are ready to present an oracle-estimator of the bivariate density $f^{T,X}$ based on right-censored data and parameters (α_T, α_X, Q) of the underlying functional class. The oracle-estimator is

$$\tilde{f}^{T,X}_*(t, x) := \sum_{j,i \geq 0} \hat{\lambda}_{ji} \hat{\theta}_{ji} \varphi_{ji}(t, x), \tag{3.7.35}$$

where the shrinkage weights are

$$\hat{\lambda}_{ji} := I(\hat{\theta}^2_{ji} > c_{TH} \hat{\sigma}^2_{ji}) I(\max(j, i) \leq J_n)$$

$$+ [1 - (a_{ji}/\hat{a}_n)^{1/2}] I(\max(j, i) > J_n) I(a_{ji} \leq \hat{a}_n). \tag{3.7.36}$$

Note that, to help the statistician, the oracle does not use the anchor density f_0 which was used in establishing the lower bound. To offset this knowledge in establishing an upper

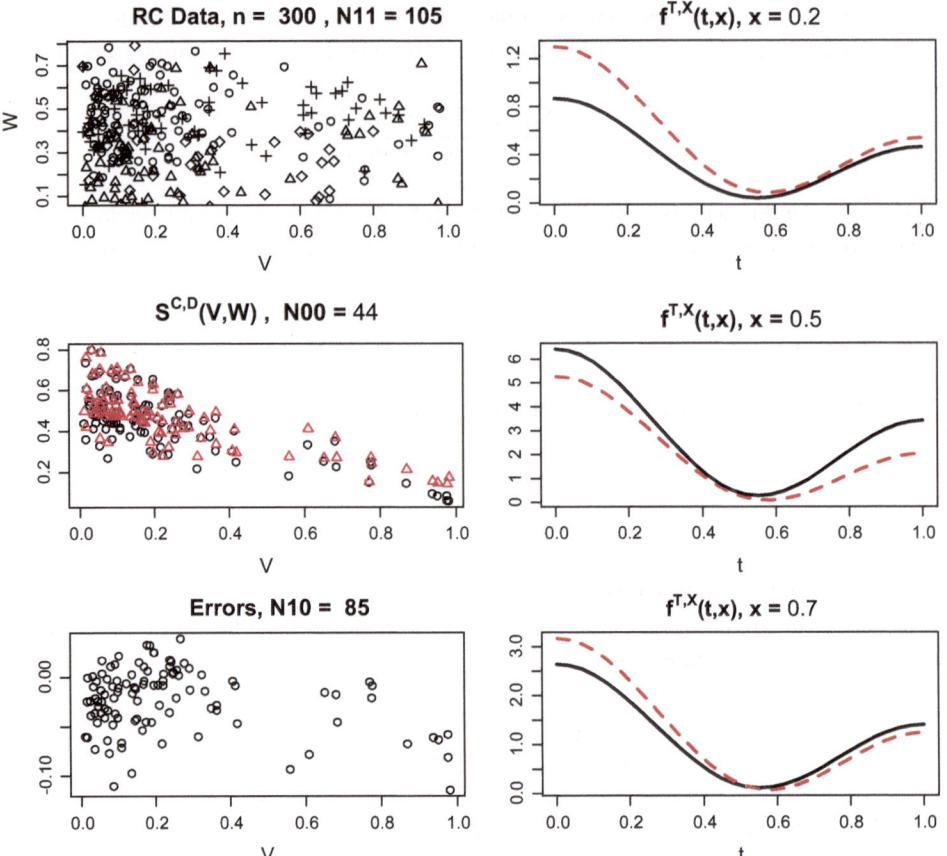

Fig. 3.15 Estimation of bivariate density $f^{T,X}$ based on right-censored (RC) data. Simulated sample of size n from $(V, W, \Delta, \Gamma) := (\min(T, C), \min(X, D), I(T \leq C), I(X \leq D))$ is shown in the left-top diagram. The mutually independent random variables T, X, C and D are from the Bathtub, the Normal, Uniform$(0, b)$ and Uniform$(0, b1)$ distributions, respectively. The left-top diagram shows by the circles, the triangles, the crosses and the rhombi pairs (V, W) corresponding to $\Delta\Gamma = 1$, $\Delta(1 - \Gamma) = 1$, $(1 - \Delta)\Gamma = 1$, $(1 - \Delta)(1 - \Gamma) = 1$, respectively. Further, the numbers of corresponding observations are denoted as $N11, N10, N01, N00$. The left-middle diagram shows the joint survival function $S^{C,D}(V_l, W_l)$ by the circles and its estimate $\hat{S}^{C,D}(V_L, W_L)$ by the triangles, $l = 1, \ldots, n$. The left-bottom diagram show the errors $S^{C,D}(V_l, W_l) - \hat{S}^{C,D}(V_l, W_l)$. Diagrams in the right column show slices of the joint density $f^{T,X}$ (the solid line) and of its E-estimate (the dashed line) for fixed values of x that are controlled by the argument $xset$. [$n = 300$, $corn = c(2, 3)$, $b = 1.1$, $b1 = 1.1$, $xset = c(0.2, 0.5, 0.7)$, $cJ0 = 3$, $cJ1 = 0.8$, $cTH = 4$]

bound, the oracle makes a traditional assumption that on the rectangle $\mathcal{R}$ the anchor is smoother than an underlying density.

Theorem 3.12 *Let the assumption of Theorem 3.11 hold and additionally the anchor $f_0 \in \mathcal{S}(\alpha_T + 1, \alpha_X + 1, Q', r_T, r_X), Q' < \infty$. Then the MISE of oracle-estimator (3.7.36) attains the lower bound (3.7.26) and*

$$\sup_{f \in \mathcal{F}^*} [n/d]^{2\beta/(2\beta+1)} \mathbb{E}_{f^{T,X},S^{C,D}} \left\{ \int_{\mathcal{R}} (\tilde{f}_*(x, y) - f^{T,X}(x, y))^2 dx dy \right\}$$

$$= P(1 + o_n(1)), \tag{3.7.37}$$

where $\mathcal{F}^*$ is union of the global and local shrinking Sobolev classes (3.7.24) and (3.7.25). Further, there exists a data-driven estimator whose MISE also satisfies (3.7.37).

This result proves that the oracle's lower bound (3.7.26) is sharp and attainable. Further, the series oracle-estimator (3.7.36) allows us to use the bivariate E-estimation methodology, defined in Sect. 1.2.6, for small samples.

Figure 3.15 illustrates the problem of bivariate density estimation by the E-estimator. Here we are considering a larger sample size than in Fig. 3.13. The main reason is that the E-estimator is based on the estimator of the joint survival function $S^{C,D}$, and we need a reasonably large number $N00$ of censored pairs. The survival function, its estimate and the errors are shown in the left-middle and left-bottom diagrams. Note that $N00 = 44$, and nonetheless we get a very reasonable estimate of the joint survival function of the censoring pair (C, D). The right column of diagrams shows us slices $f^{T,X}(t, x), t \in [0, 1]$ of the underlying joint density (the solid line) for the three particular values of x shown in the titles, and the corresponding slices of the bivariate density E-estimate (the dashed line). The three slices of the E-estimate correctly show the underlying bathtub shape. Now note the different scales in the diagrams, and that the slices well indicate the underlying magnitudes of the joint density corresponding to values of $f^X(x)$ at points $x = 0.2, x = 0.5$ and $x = 0.7$.

It is highly recommended to repeat this figure with different arguments and get first hand experience in estimation of bivariate density based on right-censored data.

3.8   Mean Residual Life

The hazard rate function tells us about probability of occurring the event of interest in the next instant given survival up to time t. Given survival up to time t, there are other characteristics that may be of interest. One of them is the mean residual (remaining) lifetime until the event of interest given survival up to time t. The corresponding characteristic is called the *mean residual life* (MRL) at time t, and it is defined as the conditional expectation

$$m^T(t) := \mathbb{E}\{(T - t)|T > t\}. \tag{3.8.1}$$

For times t such that $S^T(t) = 0$, by convention we set $m^T(t) := 0$. Note that $m^T(0) = \mathbb{E}\{T\}$ and it is the classical mean of T. MRL is often used in fields such as medical study, survival analysis, actuarial science, and reliability research to help investigators make decisions. Of course, all other familiar characteristics of a random variable, like moments, skewness or kurtosis, can be of interest under the assumed survival up to time t.

Let us briefly present several known theoretical facts about the MRL function. First, for a continuous T with density f^T we have the following relation between m^T and S^T for any t such that $S^T(t) > 0$,

$$m^T(t) = \frac{\int_t^\infty u f^T(u) du}{S^T(t)} - t = \frac{\int_t^\infty S^T(u) du}{S^T(t)}. \tag{3.8.2}$$

Accordingly, if we know the survival function S^T, we can calculate the MRL. Under a mild assumption, the inverse is also true, and the following equality holds,

$$S^T(t) = [m^T(0)/m^T(t)]e^{-\int_0^t [1/m^T(u)] du}. \tag{3.8.3}$$

If a sample $T_1, \ldots, T_n$ from T is available, then the empirical MRL is

$$\tilde{m}^T(t) := \frac{\int_t^\infty \tilde{S}^T(u) du}{\tilde{S}^T(t)} = \frac{\sum_{l=1}^n (T_l - t) I(T_l > t)}{\sum_{l=1}^n I(T_l > t)}, \quad \tilde{S}^T := n^{-1} \sum_{l=1}^n I(T_l > t). \tag{3.8.4}$$

Here $\tilde{S}^T$ is the empirical survival function. According to (3.8.2), $\tilde{m}^T$ is the plug-in estimator. Further, denote by $T_{(1)} \leq T_{(2)} \leq \ldots \leq T_{(n)}$ the ordered observations of T, and the (3.8.4) can be simplified,

$$\tilde{m}^T(t) = \frac{\sum_{l=k+1}^n (T_{(l)} - t)}{n - k} I(t \in [T_{(k)}, T_{(k+1)})). \tag{3.8.5}$$

Overall, the empirical MRL is well studied and it has excellent statistical properties, see the Notes.

The notion of MRL is straightforwardly extended to the vector of lifetimes. For instance, let (T, X) be a pair of continuous lifetimes. The *bivariate mean residual life* (BMRL) is defined as

$$\mathbf{m}^{T,X}(t, x) := (m^{T\|X}(t, x), m^{X\|T}(x, t))$$

$$:= (\mathbb{E}\{[T - t]|T > t, X > x\}, \mathbb{E}\{[X - x]|X > x, T > t\}). \tag{3.8.6}$$

Note the special notation $m^{T\|X}(t, x)$ and $m^{X\|T}(x, t)$ for the components of BMRL.

The BMRL is a useful characteristic of a pair of lifetimes which tells us about mean remaining lifetimes given that the both lifetimes survived until specific times. The classical examples are: (i) Husband and wife survived until age t and x, and we are interested in their mean remaining longevity; (ii) Mean residual times to a cataract surgery for each eye at a specific age; (iii) Mean residual times to relapses after treatments of cancer and heart failure at specific ages.

The two components of the BMRL are symmetric, and it is sufficient to understand how one of them may be estimated. Similarly to (3.8.2) we have

$$m^{X\|T}(x, t) = \frac{\int_x^\infty S^{T,X}(t, u)du}{S^{T,X}(t, x)}.$$ (3.8.7)

This formula yields the following empirical BMRL, compare with (3.8.4),

$$\tilde{m}^{X\|T}(x, t) := \frac{\int_x^\infty \tilde{S}^{T,X}(t, u)du}{\tilde{S}^{T,X}(t, x)},$$ (3.8.8)

$$\tilde{S}^{T,X}(t, x) := n^{-1}\sum_{l=1}^n I(T_l > t, X_l > x).$$ (3.8.9)

This ends our review of the case of direct observations of (T, X).

Now we are considering a typical in survival analysis case when the lifetimes are right-censored. In the univariate case this means that we observe a sample from pair $(V, \Delta) = (\min(T, C), I(T \le C))$ where C is the censoring variable (lifetime). In this case the analog of the empirical survival function $\tilde{S}^T$ is the Kaplan-Meier estimator $\tilde{S}^T_{KM}$ and it can be plugged in (3.8.2).

In a bivariate case, instead of a direct sample from (T, X), the available sample is from $(V, W, \Delta, \Gamma) := (\min(T, C), \min(X, D), I(T \le C), I(X \le D))$. Accordingly, to follow (3.8.8) we need to use a bivariate survival estimator.

We begin with an interesting case of the so-called *univariate right-censoring* when $\mathbb{P}(C = D) = 1$, that is T and X are censored by the same censoring variable C. Then, assuming that the pair (T, X) is independent of C, we get

$$S^{V,W}(t, x) = S^{T,X}(t, x)S^C(\max(t, x)).$$ (3.8.10)

Accordingly, we get the following estimator of $S^{T,X}$,

$$\bar{S}^{T,X}(t, x) := \frac{\tilde{S}^{V,W}(t, x)}{\tilde{S}^C_{KM}(\max(t, x))}.$$ (3.8.11)

Here $\tilde{S}^{V,W}$ is the empirical survival (3.8.9) based on the available direct observations of (V, W), and $\tilde{S}^C_{KM}$ is the Kaplan-Meier estimator (3.2.4) (the method of moments estimator of Sect. 3.2 may be used instead).

Now let us consider the general case of BMRL estimation for right-censored lifetimes. The proposed estimator does not follow (3.8.8). Instead it uses an estimator of the *conditional residual survival* (CRS)

$$S^{X\|T}(x,t) := \mathbb{P}(X > x | T > t) = \frac{S^{T,X}(t,x)}{S^T(t)}. \qquad (3.8.12)$$

Note the special notation $S^{X\|T}$ for the CRS on the left side of (3.8.12). Using formula

$$S^{T,X}(t,x) = S^T(t) S^{X\|T}(x,t), \qquad (3.8.13)$$

we may write down the component $m^{X\|T}$ of the BMRL as a functional of the CRS $S^{X\|T}$,

$$m^{X\|T}(x,t) = \frac{\int_x^\infty S^{T,X}(t,u)du}{S^{T,X}(t,x)} = \frac{\int_x^\infty S^{X\|T}(u,t)du}{S^{X\|T}(x,t)}. \qquad (3.8.14)$$

According to formula (3.7.13), the following estimator of the CRS function can be used,

$$\tilde{S}^{X\|T}(x,t) = \exp\left\{ -\sum_{l=1}^n \frac{\Gamma_l I(V_l > t) I(W_l \le x)}{1 + \sum_{s=1}^n I(V_s \ge t) I(W_s \ge W_l)} \right\}. \qquad (3.8.15)$$

This estimator of the CRS is of interest on its own. Further, for a bounded lifetime X the estimator (3.8.15) can be plugged in (3.8.14) to get the estimator $\tilde{m}^{X\|T}$ of BMRL. Otherwise, we can plug in $\hat{S}^{X\|T}(x,t) = \tilde{S}^{X\|T}(x,t) I(x < W_{(n)})$, or consider a restricted BMRL when in (3.8.7) the infinity in the integral is replaced by a finite restriction r.

Figure 3.16 sheds light on the problem and the proposed estimators $\tilde{S}^{X\|T}$ and $\tilde{m}^{X\|T}$. Here we see results for two different experiments explained in the caption. We begin with the left column where the lifetimes of interest T and X are independent while the censoring variables are dependent. The left-top diagram shows simulated data. The left-middle diagram shows the underlying conditional residual survival by the solid line and its estimate (3.8.15) by the dashed line. Note that for this experiment $S^{X\|T} = S^X$ due to the independence, but the estimator does not know that. We may visualize the rather complicated shape of the Bathtub survival function. The estimate is relatively good keeping in mind that this is the slice at $t = 0.5$ of the bivariate estimate $\tilde{S}^{X\|T}(x,t)$ based on just $N11 = 109$ uncensored pairs of the lifetimes of interest. The left-bottom diagram shows us the underlying component $m^{X\|T}$ of the BMRL by the solid line. Note that it is defined in (3.8.14) as the normed integral of the conditional residual survival function shown in the left-middle diagram by the solid line. It takes some experience to get used to that two functions and to the relationship between the two, and Fig. 3.16 can help to gain that experience. Now let us look at the estimated component of the BMRL shown by the dashed line. It is constructed identically to the underlying $m^{X\|T}$ only using the estimate $\tilde{S}^{X\|T}$, shown by the dashed line in the left-middle diagram, instead of $S^{X\|T}$. We can learn from the diagram how a deviation in the estimated conditional residual survival function affects the estimated BMRL. Namely, look at the left-middle diagram and note the pronounced deviation of the estimate from the underlying CRS in the vicinity of $x = 0.2$ This deviation produces the sharp peak seen in the left-bottom diagram. But overall we observe a nice example keeping in mind estimation of the bivariate functions. The right column presents a simulation with dependent lifetimes T and X and the

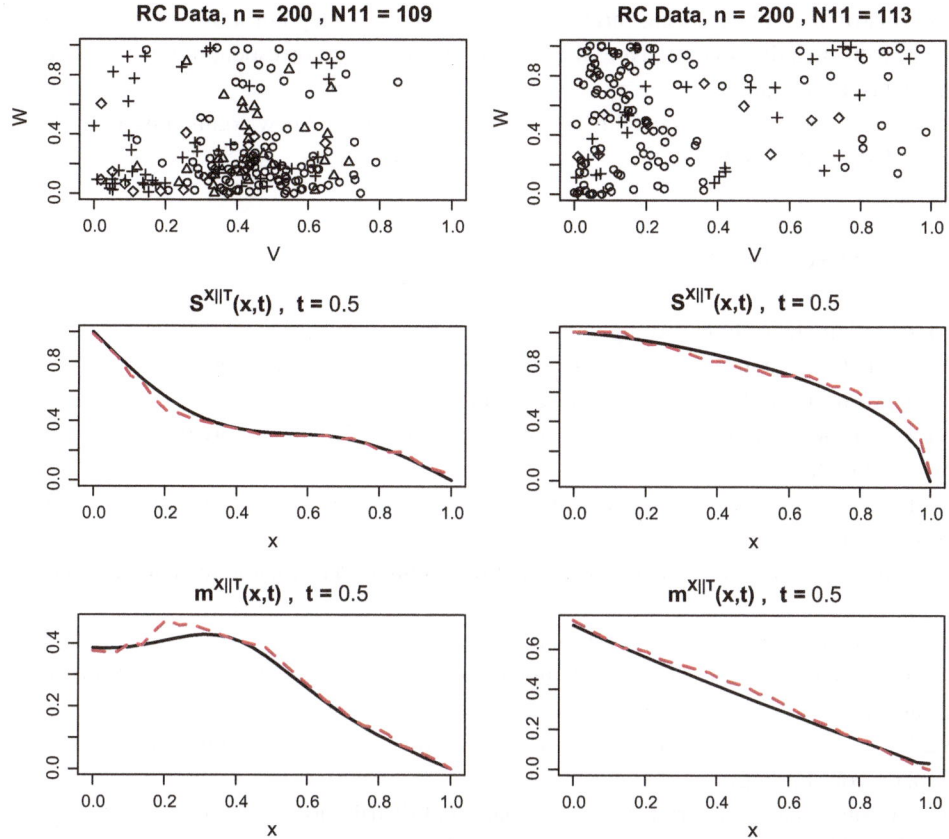

Fig. 3.16 Estimation of the conditional residual survival $S^{X\|T}$ and component $m^{X\|T}$ of the bivariate mean residual life for right-censored (RC) data. Two columns exhibit results for different simulations. In the left column T and X are independent according to the Normal and the Bathtub corner densities, respectively. Recall that they are the number 3 and the number 2 in the set of corner densities. The censoring variables C and D are dependent, they are Uniform(0,b) and Uniform(0, $b + b_1 C$), respectively. In the right column T and X are dependent, they are the Bathtub and Beta(0.5 + T, 0.5), respectively, while joint distribution of censoring variables is the same as in the left column. The top diagrams show simulated observations similarly to Fig. 3.15. The middle diagrams show by the solid line the conditional residual survival function $S^{X\|T}(x, t) := \mathbb{P}(X > x | T > t)$ for a fixed t shown in the title, and by the dashed line its estimate (3.8.15). The bottom diagrams show by the solid line the component $m^{X\|T}(x, t) := \int_x^\infty S^{X\|T}(u, t)du / S^{X\|T}(x, t)$ of the bivariate mean residual life, and by the dashed line its estimate. {The choice of two corner functions used in the left column and the corner density for f^T in the right column is controlled by the argument *corn*. The choice of shape 2 in the Beta distribution is controlled by the argument *shape2*. The choice of parameters t for the two columns is controlled by the argument *tset*}. [$n = 200$, $corn = c(3, 2, 2)$, $b = 1.1$, $b1 = 1.1$, $shape2 = 0.5$, $tset = c(0.5, 0.5)$]

same distribution of censoring variables. The right-middle diagram shows us the CRS $S^{X\|T}$ for X with Beta distribution. Note the specific shape of the corresponding $m^{X\|T}$ shown in the right-bottom diagram.

It is recommended to repeat this figure with the default and new arguments, and get used to the CRS and the BMRL.

3.9   Exercises

3.1.1 Explain the right-censoring model.

3.1.2 What are the uncensored observations and censored observations?

3.1.3* Verify formula (3.1.2). Hint: Use Assumption 3.1.

3.1.4 Consider a lifetime T right-censored by a lifetime C. Can we say that the C is right-censored by the T?

3.2.1 Definition of the survival function and its empirical estimate for the case of direct observations.

3.2.2 Show that the empirical survival function is unbiased estimator and calculate its variance.

3.2.3 Explain the Kaplan-Meier estimator (3.2.4).

3.2.4 Repeat Fig. 3.1 four times and compare cases of direct and right-censored observations. Write a report.

3.2.5* Using Fig. 3.1, find the better and worse density f^C for Kaplan-Meier estimator. Explain your findings.

3.2.6 Explain the symmetry between estimation of distributions of T and C.

3.2.7* Verify formula (3.2.5).

3.2.8 Explain the method of moments estimator (3.2.6) of the survival function S^T.

3.2.9 Explain the motivation behind the estimator (3.2.7) of S^V, and then find its bias and variance.

3.2.10 Does the estimator (3.2.7) satisfy the inequality $\hat{S}^V(V_l) \geq n^{-1}$? Is this property important?

3.2.11* What is the underlying idea of the estimator (3.2.8)?

3.2.12* Prove the assertion of Theorem 3.1.

3.2.13 What is the distribution of the number $N := \sum_{l=1}^n \Delta_l$ of uncensored observations?

3.3.1 Explain the function classes (3.3.1) and (3.3.2).

3.3.2* Why is the Sobolev class (3.3.2) so convenient for analysis of the MISE? Hint: Think about the integrated squared bias.

3.3.3 Are the anchor density and the underlying density of interest the same function? What is the role of the anchor in the theory of nonparametric density estimation?

3.3.4 Explain the assumptions of Theorem 3.2. Further, can we say that densities from the class $\mathcal{F}_n$ converge to the anchor in L_∞-norm?

3.3.5* What is the difference in quality of density estimation based on uncensored and censored observations? Hint: Use the results of Theorem 3.2.

3.3.6 Why is the density estimation, based on censored observations, called ill-posed?

3.3.7* Consider the two coefficients of difficulty presented in (3.3.5) for uncensored and censored observations. Is it possible to compare their values? Can they be finite? What will be if $S^C(1) = 0$?

3.3.8* Why is it alarming that for censored observations the Fisher information for θ_j vanishes with rate j^{-2}?

3.3.9 Why does Theorem 3.2 support the Kaplan-Meier paradigm about dominance of uncensored observations?

3.3.10 Rate of the density oracle-estimator, based on censored observations, is $n^{-2\alpha/(2\alpha+3)}$. The same rate is optimal for estimation of an α-fold differentiable in each variable multivariate density based on direct observations. What is the dimensionality of that density? Hint: The optimal rate for a k-variate density is $n^{-2\alpha/(2\alpha+k)}$.

3.3.11 Why does Theorem 3.2 create an opportunity for aggregating uncensored and censored observations?

3.3.12 Prove formula (3.3.8) for uncensored observations.

3.3.13* Verify that the oracle-estimator (3.3.9) of Fourier coefficients is unbiased. Then calculate its variance and compare with the corresponding coefficient of difficulty.

3.3.14* Prove that the oracle-estimator (3.3.10) attains the lower bound of Theorem 3.2. Hint: Use the Parseval identity and the assumed Sobolev class of underlying densities.

3.3.15 Consider a projection series density estimator with cutoff J. Suppose that an underlying density belongs to a Sobolev class $S_0(\alpha, Q)$. What is the order in n of the cutoff that implies optimal rate of the MISE convergence?

3.3.16 Verify formula (3.3.12).

3.3.17* Suppose that $S^T(1) > 0$, that is the lifetime of interest supported on an interval larger than $[0, 1]$. What will be the corresponding formula (3.3.12)? What are the corresponding changes in (3.3.13)?

3.3.18 Comment on importance of the factor j in (3.3.13). Hint: Think about the variance.

3.3.19* Find the mean and variance of the estimator (3.3.14). Discuss the results and compare with the case of uncensored observations.

3.3.20 Explain all components of the density oracle-estimator (3.3.15).

3.3.21* Verify the assertion of Theorem 3.3.

3.3.22 Explain the role and meaning of the coefficients of difficulty d_u and d_c. Hint: Use Theorem 3.3.

3.3.23 Describe all steps of the E-estimation algorithm for density based on RC data.

3.3.24* Propose an unbiased estimator of the Sobolev functional $L^{-1}\sum_{j \in B}\theta_j^2$ for the density model and RC data. Here L is cardinality of the set (block) B of indexes.

3.3.25*. Explain the underlying motivation of the aggregation (3.3.21).

3.3.26* Find the bias and the variance of the Fourier coefficient (3.3.21).

3.3.27 Propose estimators of S^C and f^C.

3.3.28 Explain the diagrams in Fig. 3.2.

3.3.29 Use Fig. 3.2 and find optimal parameters of the E-estimator. Hint: Use the argument nsim.

3.3.30 Can Kaplan-Meier estimator have the same quality of estimation of the right tail of S^T as of its left tail?

3.3.31 Using Fig. 3.2, compare estimation of the density's right tail by the four estimators.

3.3.32 Based on Fig. 3.2, can a censored-data estimate be better than the uncensored-data estimate?

3.3.33 Using Fig. 3.2, explore the importance of a good estimation of the distribution of C.

3.3.34 Change the underlying densities f^T in Fig. 3.2 and analyze the outcome.

3.3.35 Explain diagrams in Fig. 3.3.

3.3.36 How does the rate of censoring affect the four estimators used in Fig. 3.3?

3.3.37 What parameters affect the rate of censoring? Use Fig. 3.3 to check your answer.

3.3.38 Present analysis of results in Table 3.1.

3.3.39 Explain the diagrams in Fig. 3.4. Does the aggregation shed a new light on the lung cancer survival time?

3.3.40 Why is the problem of density estimation over unknown support is so complicated for right-censored data?

3.3.41 Explain why the MISE criterion $\mathbb{E}\{\int_0^\infty (\check{f}(t) - f^T(t))^2 dt\}$ can be used for the case of an unknown finite support of T.

3.3.42 Verify formula (3.3.23).

3.3.43 Explain Assumption 3.2. Can it be relaxed?

3.3.44 Explain the sequences γ_n and k_γ.

3.3.45 What is the underlying idea of the estimator (3.3.29)? Hint: Think about the right tail.

3.3.46 Explain formulas (3.3.30) and (3.3.31).

3.3.47 What is main message of Theorem 3.4?

3.3.48 What is the main message of Theorem 3.5?

3.3.49 Use Fig. 3.5 and check the assertions of Theorems 3.4.and 3.5.

3.4.1 Definition of the hazard rate function.

3.4.2 What is called the force of mortality?

3.4.3* Explain the assumptions and the assertions of Theorem 3.6.

3.4.4 Why is it necessary to consider the restricted MISE over an interval $[0, r]$?

3.4.5 Does the right-censoring slow down the rate of hazard rate estimation?

3.4.6 Prove the relation (3.4.6).

3.4.7 Verify formula (3.4.7).

3.4.8* Consider (3.4.7). What will be the right side of this formula if T are C are dependent?

3.4.9 Verify (3.4.8).

3.4.10 Why is the Fourier coefficient written as the expectation in (3.4.9)?

3.4.11 Find the mean and variance of the Fourier coefficient estimator (3.4.10).

3.4.12 Find the mean and variance of the empirical survival (3.4.11).

3.4.13 Explain how the coefficient of difficulty d_r may be estimated. What may the estimator be used for?

3.4.14 Explain all diagrams in Fig. 3.6, and then comment on the used E-estimator.

3.4.15 Explain how to choose a feasible restriction r. Then check your explanation using Fig. 3.6.

3.4.16 Using Fig. 3.6, consider several different underlying distributions of T and compare the recommended restrictions r.

3.4.17* Propose a hazard rate estimator based on censored observations. Then propose aggregation of uncensored and censored observations. Hint: Use the methodology developed for density estimation.

3.5.1 Describe the regression problem with right-censored response.

3.5.2 What is the underlying idea of the Buckley-James imputation procedure?

3.5.3 Explain the Assumption 3.3. Is it a necessary one for consistent regression estimation?

3.5.4 Explain the Assumption 3.4.

3.5.5 What is the role of the anchor in the sharp-minimax asymptotic theory?

3.5.6 Why do we need the assumption (3.5.1)?

3.5.7 Explain and then discuss the assumptions and results of Theorem 3.7.

3.5.8 Does Theorem 3.7 contradict the principle of dominance of uncensored observations over censored observations?

3.5.9 Compare the coefficient of difficulty (3.5.4) for censored regression with the coefficient of difficulty for a regression based on direct observations. Hint: Recall the paragraph below Theorem 3.7.

3.5.10 Verify (3.5.7) and (3.5.8).

3.5.11 Using integration by parts, verify (3.5.10).

3.5.12 Find the mean and variance of the Fourier coefficient estimator (3.5.12).

3.5.13* Prove that the estimator (3.5.14) is sharp minimax. Hint: Calculate the MISE and compare with the lower bound of Theorem 3.7.

3.5.14* Explain the underlying idea of the Fourier coefficient estimator (3.5.15). Then find its mean and variance. Can the estimator be improved? Hint: Review the discussion of Fourier coefficient estimators in Sect. 1.2.5.

3.5.15 Explain diagrams in Fig. 3.7.

3.5.16 Using Fig. 3.7, compare performance of the four estimators.

3.5.17 Find optimal parameters of the E-estimator for simulations used in Fig. 3.7.

3.5.18* For Fig. 3.7, use several different underlying experiments and then explain how the underlying distributions affect performance of the estimators.

3.5.19 Consider the lung cancer example presented in Fig. 3.8 and explain possible conclusions from the presented estimates.

3.5.20* Use Fig. 3.8 with different parameters of the E-estimator, and then comment on how they affect the estimates.

3.5.21 Explain the model of regression with right-censored predictor.

3.5.22 Present several examples of regression with right-censored predictor.

3.5.23 Does the right-censored predictor slow down rate of the MISE convergence?

3.5.24 Verify (3.5.18).

3.5.25 Repeat Fig. 3.9 several times and describe the scattergrams and the estimates.

3.5.26* Find optimal parameters of the E-estimator for the simulation in Fig. 3.9.

3.5.27* Propose a consistent regression estimator based on censored predictors.

3.6.1 What is the definition of a natural nuisance function? Present examples. Does the natural nuisance function (3.6.1) depend on the estimand?

3.6.2 Explain how the natural nuisance function (3.6.1) can be estimated.

3.6.3 Verify formula (3.6.4).

3.6.4 Prove formula (3.6.5).

3.6.5 Explain formula (3.6.6.).

3.6.6 Show that (3.6.7) is unbiased estimator of bivariate function $\sum_{j=0}^{J_p} \kappa_j(t)\varphi_j(x)$.

3.6.7 Explain how E-estimator and the universal hard-thresholding estimator (3.6.8) are related.

3.6.8* Evaluate the mean and variance of the estimator (3.6.9).

3.6.9 Explain diagrams in Fig. 3.10.

3.6.10 Using Fig. 3.10, comment on how hard-thresholding affects the estimates.

3.6.11 Using Fig. 3.10, propose better parameters for the estimator.

3.6.12 Definition of conditional survival function.

3.6.13 What are the RC observations used for estimation of the conditional survival function?

3.6.14 Definition of the conditional cumulative hazard.

3.6.15 Verify all equalities in (3.6.10).

3.6.16 What is the underlying idea of using the Fourier coefficients (3.6.11)?

3.6.17 Is the Fourier coefficient (3.6.11) a constant or a function in t? Hint: Can the answer depend on the underlying distribution?

3.6.18 Explain the underlying idea of estimator (3.6.12).

3.6.19* Find the mean and variance of estimator (3.6.12).

3.6.20* Find the mean and variance of estimator (3.6.13).

3.6.21 Explain the estimator (3.6.14).

3.6.22 Explain the estimator (3.6.15).

3.6.23 What is the underlying idea of formula (3.6.16)?

3.6.24 Explain the diagrams in Fig. 3.11.

3.6.25 What is the simulation used in Fig. 3.11?

3.6.26* Use Fig. 3.11 and propose better parameters for the used estimator.

3.6.27 Repeat Fig. 3.11 with different sample sizes and discuss the estimates.

3.6.28 Definition of the conditional hazard rate. Present several practical examples where this characteristic may be of interest.

3.6.29 Explain formula (3.6.17).

3.6.30 What is the conditional cumulative hazard?

3.6.31 Is it possible to estimate the conditional hazard rate over the support of T?

3.6.32 What is the rectangle $\mathcal{R}$ used for? What is the r?

3.6.33 Explain the Assumption 3.5. Do you think that this assumption is necessary for consistent estimation?

3.6.34 What is the used assumption about smoothness of the conditional hazard?

3.6.35 What is the role of a conditional hazard anchor in the oracle's approach?

3.6.36 Explain the considered function class (3.6.19).

3.6.37 What is the underlying idea of the arrays A_1 and A_2 used in the series expansion (3.6.24)?

3.6.38 Explain the lower bound (3.6.26) and under what conditions it is valid.

3.6.39 Explain why the coefficient of difficulty (3.6.27) increases to infinity as $r \to \infty$.

3.6.40 Describe the low-frequency and high-frequency components of the oracle-estimator (3.6.28).

3.6.41* The oracle-estimator (3.6.28) can perform dimension reduction and adaptation to smoothness of the conditional hazard rate. Explain why it has that nice statistical properties. Hint: Think about the used structure of blocks.

3.6.42* Describe statistical properties of the natural nuisance function estimator (3.6.31).

3.6.43* Is the Fourier coefficient estimator (3.6.32) asymptotically unbiased?

3.6.44 Explain the underlying idea of the estimator (3.6.33) for the coefficient of difficulty.

3.6.45 Explain the assumptions and the results of Theorem 3.9.

3.6.46 Why does inequality (3.6.38) imply that the estimator (3.6.36) adjusts to an underlying dimensionality of the conditional hazard rate?

3.6.47* Use Fig. 3.12 to propose optimal parameters of the used estimator.

3.6.48 How can you explain the fact that the estimator (3.6.36) performs adaptation and dimension reduction without numerical optimization or hypotheses testing?

3.6.49 Describe the tensor-product basis used in Sect. 3.6.4 for the case of a mixed multivariate predictor.

3.6.50 Check that (3.6.39) is indeed the inner product. What is the corresponding norm?

3.6.51 Explain the system of indexes used by the series approximation (3.6.42).

3.6.52 How does the series approximation (3.6.42) help in solving the problem of dimension reduction?

3.6.53 Explain the function class (3.6.45). Further, what is the effective smoothness for that class?

3.6.54* Find the mean and the variance of the Fourier coefficient estimator (3.6.46).

3.6.55* Find the mean and the variance of the estimator (3.6.47).

3.6.56 Explain the underlying idea of the estimator (3.6.48).

3.6.57 Estimator (3.6.49) is the U-statistic. What is its kernel?

3.6.58 Explain the low-frequency and high-frequency parts of the estimator (3.6.50).

3.6.59 What is the main practical implication of Theorem 3.10?

3.6.60* Describe E-estimator motivated by Theorem 3.10.

3.7.1 Is there a simple generalization of the Kaplan-Meier univariate survival function estimator to the case of a bivariate survival function?

3.7.2 Explain the right-censored model for a pair of lifetimes of interest. Hint: Check (3.7.1).

3.7.3 Consider the problem of consistent estimation of the bivariate survival function. Can the Assumption 3.8 be relaxed?

3.7.4 What is the bivariate cumulative hazard?

3.7.5* Verify (3.7.3). Hint: Recall a line integral.

3.7.6 Verify formulas (3.7.4) and (3.7.5).

3.7.7 Verify (3.7.6).

3.7.8* Find the mean and variance of the estimator (3.7.8).

3.7.9 Verify (3.7.9).

3.7.10 Verify (3.7.10).

3.7.11 Verify (3.7.11).

3.7.12 Explain the two components of the estimator (3.7.14).

3.7.13* Find the mean and variance of the empirical bivariate survival function (3.7.15).

3.7.14 Explain the simulation used in Fig. 3.13.

3.7.15 Repeat Fig. 3.13 and describe outcomes in all six diagrams.

3.7.16 How can N11 be increased by changing controlled parameters of Fig. 3.13?

3.7.17 Compare errors in Fig. 3.13 with errors in Fig. 3.1. Repeat simulations 10 times and make a conclusion.

3.7.18* How can one estimate the conditional probability $\mathbb{P}(T > t | X > x)$?

3.7.19* Suggest estimators of $\mathbb{E}\{T | X > x\}$ and $\mathbb{E}\{T | T > t\}$. Hint: These are univariate functions.

3.7.20 Compare outcomes of Figs. 3.13 and 3.14.

3.7.21 What is the basis used for bivariate density estimation? Hint: Check (3.7.23).

3.7.22 What is the main result of Theorem 3.11?

3.7.23 Verify formula (3.7.29).

3.7.24* Find the mean and the variance of the estimator $\tilde{d}$ defined in (3.7.33).

3.7.25 What is the practical implication of Theorem 3.12 for the E-estimation algorithm?

3.7.26 Explain diagrams in Fig. 3.15.

3.7.27 May the N00 affect quality of estimating $f^{T,X}$?

3.8.1 What is the MRL?

3.8.2 Verify (3.8.2).

3.8.3* Prove (3.8.3). Hint: The proof can be found in the references mentioned in the Notes.

3.8.4 Verify validity of formula (3.8.5) for the empirical MRL.

3.8.5 Present several practical examples where the BMRL may be of interest.

3.8.6 Explain the underlying idea of the empirical BMRL (3.8.8).

3.8.7 Consider a right-censored lifetime and propose an empirical MRL using the method of moments estimator of Sect. 3.2.

3.8.8* Consider the univariate right-censoring and suggest an estimator of the BMRL.

3.8.9* Explain the proposed methodology of estimating the BMRL based on the empirical CRS. Hint: Begin with definitions.

3.8.10 Describe the underlying experiments and diagrams in Fig. 3.16.

3.8.11 Repeat Fig. 3.16 with different argument *tset*. For what parameters t the estimation is the best and worst?

3.8.12 Repeat Fig. 3.16 ten times and explain the effect of the CRS on the BMRL.

3.8.13 Use Fig. 3.16 to describe possible shapes of the CRS and the BMRL.

3.8.14 Use Fig. 3.16 with different sample sizes n and describe the outcomes.

3.8.15* The formula $\mathbb{E}\{T\} = \int_0^\infty S^T(t)dt$ allows us to introduce the so-called equilibrium distribution with density $f_e(t) = S^T(t)/\mathbb{E}\{T\}$. Find the corresponding equilibrium survival function S_e^T and hazard rate h_e^T.

3.8.16 Let S_e^T be the equilibrium survival function defined in the previous exercise. Verify or disprove the relation $\int_t^\infty x f^T(x)dx = t S^T(t) + \mathbb{E}\{T\}S_e^T(t)$.

3.8.17* In actuarial science the conditional expectation $\mathbb{E}\{T|T > p_\alpha\}$, where p_α is the 100αth percentile, is called the tail-value-at-risk. Propose its estimators based on a sample of size n from T and a corresponding right-censored sample.

3.10   Notes

3.1 There are many excellent books and reviews devoted to survival analysis and right-censored data, ranging from practical guides to mathematically rigorous monographs. Let us mention Kalbfleisch and Prentice (2002), Klein and Moeschberger (2003), Martinussen and Scheike (2006), Aalen et al. (2008), Hosmer et al. (2008), Kosorok (2008), Allison (2014), Gijbels (2010), Guo (2010), Fleming and Harrington (2011), Mills (2011), Royston and Lambert (2011), van Houwelingen and Putter (2011), Wienke (2011), Chen et al. (2012), Crowder (2012), Kleinbaum and Klein (2012), Klugman et al. (2012), Liu (2012), Lee and Wang (2013), Li and Ma (2013), Allison (2014), Klein et al. (2014), Harrell (2015), Moore (2016), Tutz and Schmid (2016), Ghosal and van der Vaart (2017), Efromovich (2018), Emura and Chen (2018), Karim and Islam (2019), Zhou (2019), Broström (2021), Lokhnygina (2021), O'Quigley (2021), Collett (2023), Yu and Guo (2024). See also the reviews and interesting applications in Oakes (1981), Luo and Tsai (2009), Ning et al. (2010), Peng and Taylor (2014), Wang et al. (2019), Turkson et al. (2021), Dey et al. (2022), Deltuvaite-Thomas et al. (2023), Lin and Wang (2023), Lyu et al. (2023), Beis et al. (2024), Jiang and Guterman (2024), Luo et al. (2024), Wiegrebe et al. (2024).

Description of R packages devoted to survival analysis can be found in Allignol and Latouche (2023), Augugliaro et al. (2023), Perez-Jaume et al. (2023).

Apart of right-censoring, survival analysis also deals with left-censoring. Under left-censoring, the lifetime of interest T is censored on the left by a censoring variable C if available observations are from the pair $(V, \Delta) := (\max(T, C), I(T \geq C))$. For instance, when a physician asks a patient about the onset of a particular disease, the answer may be either a specific date or that the onset occurred prior to some specific date. In this case the variable of interest is left-censored. It is important to know that left-censoring may be "translated" into a right-censoring. To do this, choose a value A that is not less than all available left-censored observations, and then consider new observations that are A minus left-censored observations. Then the new observations become right-censored. The latter explains how to use all presented results for the left-censored data.

3.2 The modern survival analysis, and distribution estimation in particular, are based on the pathbreaking product-limit methodology of Kaplan and Meier (1958) for nonparametric estimation of survival function by a stepwise function with steps at uncensored lifetimes. The product-limit methodology is based on the understanding that censored observations are dominated by uncensored. Moreover, in that seminal paper censored observations are referred to as "losses" and "incomplete". And sure enough, rigorous proof of the dominance was done later in Efron (1967), Breslow and Crowley (1974), Meier (1975), Aalen (1978) and optimality in Wellner (1982). Other popular and closely related survival function estimators are the Nelson–Aalen and Nelson-Aalen-Breslow estimators, see Efromovich (2018). Asymptotically these estimators have the same statistical properties.

Proofs of the formulated results can be found in Efromovich (2022a).

Let us mention several open research topics. (1) The known empirical survival function estimators are stepwise. It is of interest to propose a continuous estimator that matches and, may be, dominates the stepwise estimators. For the case of directly observed lifetimes it is known that a continuous series estimator may outperform the empirical survival function in terms of the second order efficiency, see Efromovich (2001b). There are no similar results for right-censored data. (2) If the sample size is small and the rate of censoring is high, it may be of interest to develop an estimator that aggregates uncensored and censored observations. As we know, the aggregation should be done in the frequency domain. Accordingly, the aggregation is related to the previously formulated topic of a continuous series survival function estimator.

3.3 For density estimation, the dominance of uncensored observations over censored ones is established in Efromovich (2001a). Namely, it is shown that the oracle, who knows data and distribution of the censoring variable, can attain the sharp constant and rate of the MISE (mean integrated squared error) convergence using only uncensored observations. In other words, the oracle does not need censored observations for efficient estimation of the density.

Interesting discussion and many examples can be found in Brunel and Comte (2005), Gill (2006), Brunel et al. (2009), Talamakrouni et al. (2016), Cui and Hanning (2019), Ewnety et al. (2023), Janssen and Veraverbeke (2023), Wang et al. (2023).

Ill-posed problems is a familiar topic in mathematics and statistics, see Tichonov (1998) and Efromovich (1999a).

The principle of dominance of uncensored observations serves well when the sample size is large and there is a light censoring. In this case considering censored observations as "losses" is feasible. At the same time, there are well documented practical examples with high rate of censoring. Buon et al. (2008) reports censoring rates up to 50% in pharmacokinetic studies, Yildiray (2008, 2013) reports heavy censoring up to 60% in loan defaults data, Chu et al. (2008) reports up to 50% censored data in cancer survival data. For testing new medication Cummings (2013) reports censoring rates up to 92.1% for patients with liver transfer while referring to "normal" censoring rates at the level between 30% and 50%. Similarly high rates have been discussed in Zhou (2019), Bayawa et al. (2022), Yu et al. (2022), Yu et al. (2023), and Cheng et al. (2023b). The issue becomes even more acute if T is a vector. See interesting discussion and examples in Enders (2010), Moore (2016), Ross et al. (2019), Efromovich (2022a), Czado and Van Keilegom (2023), Janssen and Veraverbeke (2023), Salerno and Li (2023).

Necessity of the independence between T and C for consistent estimation is due to Tsiatis (1975).

Estimation under shape restrictions is an important part of survival analysis and special procedures are suggested for taking the restrictions into consideration, see a discussion in Groeneboom and Jongbloed (2014) and Srivastava and Klassen (2016). For the E-estimation, there is no need to make any adjustment. Indeed, the algorithm includes Step 3 that involves a projection on the class of assumed functions. Theoretical justification of this approach can be found in Efromovich (2001a).

The small cell lung cancer (SCLC) clinical study data is from Ying et al. (1995). The data contains right-censored survival lifetimes in days and age in years. The right-censoring is caused by administrative end of the study, and according to the paper it is independent of the survival lifetime and the age. Let us note that about 15% of all lung cancer cases are the SCLC and this is the most aggressive type of lung cancer with extremely short survival times after the cancer diagnosis. A plausible explanation of the two strata in Fig. 3.4 can be found in the publication Maksymiuk et al. (1994) in the *Journal of Clinical Oncology* devoted to the lung cancer study. According to the publication, the survival of participants was primarily defined by the binary stage of cancer, limited or extensive. This is what the density estimates tell us about. The conclusion about two strata is different from the unimodal on $[0, \infty)$ Bayesian density estimate in Poynor and Kottas (2019).

For the problem of estimation with unknown support and the developed methodology, let us mention Giné and Guillou (2002) where a kernel convolution of Kaplan-Meier estimator is studied and it is shown that, under some mild assumptions, the density estimator is uniformly consistent over an increasing interval with right end point being the mth order statistic. This result is based on a series of mathematically beautiful assertions about expected values of the supremum of empirical processes and exponential inequalities. The interested reader can find the exciting history of the research and further references in that paper. Review of other interesting results on consistent estimation and estimation over fixed intervals can be found in Efromovich (2018), Legrand (2021), Shi et al. (2024).

The theory and methodology of the discussed aggregation are developed in Efromovich (2024d) where the proofs can be found. Proofs of theorems presented in Sect. 3.3.5 can be found in Efromovich (2024c).

The theory of sharp-minimax blockwise-shrinkage and corresponding proofs can be found in Efromovich (1985, 1999a, 2000b).

Let us mention several interesting and open research topics. (1) Analysis of Figs. 3.3 and 3.4, and the discussion at the end of Sect. 3.3.4, indicates the necessity of developing the methodology of frequency-spatial aggregation that assigns larger weight to the right tail of censored observations. One of the possible solutions is to divide the support into subintervals and perform the developed aggregation in the frequency domain for each subinterval. This approach requires using special bases that can sew estimates for aperiodic functions on each subinterval. A feasible choice is the polynomial-trigonometric basis proposed and used in Efromovich (2019c, 2021a) and special wavelet and multiwavelet bases described in Efromovich (1999a, b, 2000c, 2001c, 2004e, 2009b), Efromovich et al. (2004), Efromovich and Valdez-Jasso (2010), Efromovich and Smirnova (2014a, b), Tymes, Pereyra and Efromovich (2010). (2) It is a challenging problem to develop the theory of optimal aggregation. The main challenge is that the Kaplan-Meier dominance of uncensored observations holds for large samples, and the new theory should bypass it and propose an approach for small sizes and heavily right-censoring cases. The methodology of Bunea et al. (2007) and Cai (2008) may be of interest. (3) Left-truncation is another data modification that often occurs in survival analysis, see Tsai et al. (1987), Efromovich (2018). It occurs when smaller lifetimes may be skipped (not included in the sample). Developing aggregation for left truncated and right-censored data is a challenging problem. Here the methodology of Efromovich and Chu (2018a, b) may be instrumental. (4) Missing data is another important topic, see the discussion and methodology in Efromovich (2018), Ben Elouefi and Saadaoui (2023).

3.4 Estimation of the hazard rate, based on right-censored observations, is a classical topic in survival analysis. Its excellent overview can be found in books Cox and Oakes (1984), Cohen (1991), Andersen et al. (1993), Klein and Moeschberger (2003), Fleming and Harrington (2011), Lee and Wang (2013), Collet (2014), Harrell (2015), Efromovich (2018), as well as in papers Uzunogullari and Wang (1992), Cao et al. (2005), Brunel and Comte (2008), Qian and Betensky (2014), Hagar and Dukic (2015), Shi et al. (2015), Bremhorsta and Lamberta (2016), Dai et al. (2016), Talamakrouni et al. (2016), and Wang et al. (2017). Bayesian approach is discussed in Ghosal and van der Vaart (2017).

The sharp-minimax approach and the presented theoretical results are based on Efromovich and Chu (2018a, b) where a numerical study and practical examples of the analysis of cancer data and the longevity in a retirement community may be found. More discussion can be found in Efromovich (2018).

Let us present several open research topics. (1) Developing the theory and methodology of estimation based on censored observations. (2) Efficient aggregation of uncensored and censored observations. There is no currently known solution of this problem. (3) Missing right-censored observations including the MNAR. (4) Considering left-truncated and right-censored lifetimes. (5) Efficient estimation in the presence of right-censoring and measurement errors, see Comte et al. (2017). (6) Using wavelet bases, see a discussion in Antoniadis, Gregoire and Nason (1999). (7) Change point is another interesting topic, see Efromovich (1999a), Rabhi and Asgharian (2017).

3.5 It is well documented in the survival analysis literature that the Kaplan-Meier's understanding of the dominance of uncensored observations was also pathbreaking in regression estimation. The dominance is at the core of the seminal papers Cox (1972, 1975) where the methodology of partial likelihood is proposed. Later, using information calculations, it was established in Efron (1977) and Oakes (1977) that the Cox's estimator is nearly fully efficient. Using the dominance principle, Buckley and James (1979) suggested to replace censored responses by their conditional expectations calculated using uncensored observation. This novel imputation approach and the methodology got their name, and the estimator was rigorously studied in Ritov (1990), Jin et al. (2006), and also see a discussion of the missingness and imputation in Rubin (1987), van Buuren (2018), da Silva (2023). Discussion of several other related ideas, based on the dominance principle, can be found in Miller and Halpern (1982), Kohler, Mathe and Pinter (2002), Chen and Cheng (2006), Jin et al. (2006), Aalen et al. (2008), Delecroix et al. (2008), Su and Wang (2012), Shirazi et al. (2013), Lopuhaa and Musta (2018), Salah and Yousri (2019), Zhong et al. (2022), Chen et al. (2023a), Collett (2023), Fernandez et al. (2023).

There is also a large literature specifically devoted to nonparametric regression with censored responses. The pioneering result is Fan and Gijbels (1994) where uncensored observations are used to construct nonparametric imputation of censored responses, and then a nonparametric estimator is based on the transformed responses. This is an interesting and technically challenging nonparametric development of the Buckley-James imputation method. Interesting and sophisticated regression estimators, based on the dominance of uncensored observations, Kaplan-Meier methodology and Buckley-James imputation, can be found in Li and Doss (1995), Gross and Lai (1996), Kim and Truong (1998), Lewbel and Oliver (2002), Brunel and Comte (2006), Delecroix, Lopez and Patilea (2008), Li et al. (2008), Huang (2010), Shirazi et al. (2013), Efromovich (2018), Zhao et al. (2020b), Hao et al. (2021), Beyhum and Van Keilegom (2023), Candès et al. (2023), James et al. (2023), Lautier et al. (2023), Liu et al. (2023), Valeriano et al. (2023).

In the classical nonparametric curve estimation theory for directly observed data, there is the renown principle of equivalence between density estimation and regression, see a discussion in Efromovich (1999a). The principle states that the same asymptotic results should hold for the both classical problems. Accordingly, it could be expected from Sect. 3.3 that regression estimation based on censored responses is ill-posed with respect to uncensored responses. Surprisingly, this us not the case, and censored responses yield the same rate of the MISE convergence as uncensored ones. The latter is an interesting and important example when the principle of equivalence does not hold.

There is an excellent recent review Lotspeich et al. (2024) devoted to practical examples and methodology, including a number of interesting ad hoc methods, for regression with right-censored predictor. Nonparametric approaches are discussed in Efromovich (2018).

Nonparametric regression with measurement errors in predictor is discussed in Efromovich (1999a). The first rate optimal result is due to Fan and Truong (1993), the first sharp-minimax result (the best constant) is due to Efromovich (2023a).

Proofs of the presented results can be found in Efromovich (2024d).

Now let us formulate several interesting topics for future research. (1) Aggregation for regression with mixed multivariate predictor. It may be expected that here, similarly to the univariate predictor, censored observations are not ill-posed and the aggregation is feasible both asymptotically and for small samples. The main complication is in estimation of the natural nuisance function which is an interesting problem on its own. (2) Regression with a categorical variable which may affect smoothness of the regression function in continuous covariates and possibly the dimensionality. A classical setting in multivariate nonparametric analysis is when the function of interest $f(x_1, \ldots, x_k, u)$ depends on k continuous variables $(x_1, \ldots, x_k)$ and a categorical u. Then it is assumed that the function belongs to an anisotropic (having different smoothness in the continuous variables) Sobolev class, see Hoffmann and Lepski (2002) and Efromovich (2018). At the same time, there are plenty of practical examples where the categorical covariate u affects the smoothness. For instance, stage of disease, diabetes, smoking, homelessness may dramatically affect both smoothness and even the support of longevity. A possible approach is to: (i) Introduce corresponding families of anisotropic Sobolev classes where smoothness is defined by categorical variable. Then, similarly to Hoffman and Lepski (2002) and Efromovich (2018), develop sharp-minimax theory; (ii) Use special blocks and thresholding procedures, motivated by Efromovich (2022c, 2024a,b), to develop adaptive sharp-minimax estimators; (iii) Test the estimators via simulations and practical examples. The dimension reduction part is even more challenging. Here for each u the function of interest f may depend on a subvector $\mathbf{x} := (x_{k_1}, \ldots, x_{k_\nu})$, $(k_1, \ldots, k_\nu) \in \mathcal{K}(u)$, $\nu \leq k$ and the task is to recognize that and attain the corresponding rate of the MISE convergence. (3) Time-varying covariate regression. This is a classical and also hot topic in survival analysis. Interesting reviews can be found in Cai et al. (2007), Enders (2010), Molenberghs et al. (2014), Lv et al. (2018), Yang et al. (2018), Zhang et al. (2018), Moreno-Betancur et al. (2018), Zhang et al. (2018), Jiang et al. (2020), Lee et al. (2021), Legrand (2021), Qian et al. (2021), Austin, Fang and Lee (2022), Spreafico et al. (2022), Liao et al. (2023), Ruth et al. (2023), Xue (2024). The main used methodology is to employ classical models with time-varying estimated parameters. A different methodology can be explained via the following example. Consider estimation of $\mathbb{E}\{T | \{x(t), 0 \leq t \leq r\}\}$ where $x(t)$ is observed with continuous in time measurement error being a stochastic process, that is we observe $\mathcal{Y}(t) = x(t) + \mathcal{E}(t)$, $0 \leq t \leq r$. In functional regression literature it is often assumed that $x(t)$ is observed in a classical nonparametric regression model with independent regression errors, see excellent reviews and discussions in Müller and Yao (2008), Morris (2015), Aneiros et al. (2022). It is shown in Efromovich (2023a) that the filtering model for the covariate yields a new curse in ill-posedness of functional regression and the necessity to employ a new methodology of functional regression with measurement errors in predictor. Important references are Carroll et al. (1991), Carroll et al. (2006), Chiou and Müller (2009), Lee and Song (2022). (4) Presented examples indicate an interesting opportunity to improve right tail of an estimate by aggregating uncensored-data and censored-data estimates in time domain whenever uncensored observations are sparse in right tail. The classical theory and methodology, discussed in Nemirovski (2000), Rigollet and Tsybakov (2007), Dai et al. (2014), would explore aggregation $\lambda \tilde{f}_u(t) + (1 - \lambda)\tilde{f}_c(t)$, $t \in [0, r]$. For the right-censored data using a function $\lambda(t)$, $t \in [0, r]$ in place of a constant

λ looks promising following the methodology of Efromovich (2022a, b). (5) A technically involved problem is to aggregate censored predictors. (6) Regression with dependent censored data, see a discussion in El Ghouch and Van Keilegom (2008, 2009). (7) Regression with missing data, see a discussion in McKeague and Subramanian (2001), Zou and Liang (2017), Efromovich (2018). (8) Efficient estimation with left truncated and right-censored data, see Zhang and Zhou (2013) and Efromovich (2018).

3.6 There is a rich literature devoted to conditional distributions, including interesting ad hoc methods. At the same time, little is known about methodology and theory of sharp minimax (efficient) estimation.

The Cox proportional hazards model is a classical approach that performs well in cases where an underlying model fits the Cox model and may be too restrictive otherwise, see Cox and Oakes (1984). Partially linear hazard models with varying coefficients are considered in Cai et al. (2007), additive Cox models in Lu et al. (2018). Nonparametric estimation of the conditional hazard is a more flexible and technically challenging alternative, see a discussion in McKeague and Utikal (1990), Li and Doss (1995), Van Keilegom and Veraverbeke (2001) where martingale and counting process techniques are employed. Using splines to estimate the conditional log-hazard rate is discussed in Kooperberg et al. (1995). Survival analysis often involves a vector of continuous and ordinal/nominal categorical covariates (predictors), see the above-presented literature as well as Kang et al. (2018) where additive models are used, Cui and Hannig (2019) and companion discussions are devoted to nonparametric generalized fiducial inference, the book Prentice and Zhao (2018) presents a comprehensive overview of multivariate models. Interesting discussion of applications can be found in Li et al. (2023), Zhao and Feng (2020) study deep neural networks, Balan and Putter (2020) explores frailty methodology, Huang and Su (2021) use penalized splines in concave extended linear models and study optimal rates, deep extended hazard (DeepEH) and deep learning for partially linear Cox models are suggested in Zhong et al. (2022). Theoretically beautiful results about consistency of the kernel conditional Kaplan-Meier estimator can be found in Dabrowska (1989).

Using multivariate Sobolev function classes (ellipsoids) for filtering and regression problems and the case of direct observations was first suggested in Hoffmann and Lepski (2002). Mathematical discussion of these classes can be found in Nikolskii (1975).

Let us comment on the notion of natural nuisance function. The oracle defines a function (or a vector of functions) as the natural nuisance function if using it together with data yields a sharp minimax estimation. Let us present several particular examples that shed light on the notion. For nonparametric regression model $Y = m(X) + \sigma(X)\xi$ and available sample from pair (X, Y), the natural nuisance function is the design density f^X of the predictor X. The oracle uses the design density f^X and a sample to construct a sharp minimax estimator of the regression. Note that in this example the natural nuisance function is the nuisance function for both the oracle and the statistician because f^X is not related to the estimated regression function. Our second example is estimation of the hazard rate function h^T based on a sample from T considered in Efromovich (2016a). For this problem the natural nuisance function is the survival function S^T. This is an interesting example because knowledge of S^T is equivalent to knowing the estimand h^T , and hence S^T is not a nuisance function for the statistician. Further, for this problem the oracle uses data and the natural nuisance function to construct a sharp minimax oracle-estimator of h^T that can be mimicked by a data-driven estimator. Our final example is the considered estimation of $h^{T|X}$, when the natural nuisance function is $p(t, x) = f^X(x) S^{C|X}(t|x) S^{T|X}(t|x)$. For the statistician the only nuisance functions are the design density f^X and the conditional survival $S^{C|X}$ of the censoring variable C, but the presence of factor $S^{T|X}$ in the natural nuisance function p allows the statistician to directly estimate p instead of estimation of the factors. The latter is the interesting specific of the studied problem. Another remark is that the bivariate nuisance function can be estimated with the univariate rate defined by its smoothness in x. Further, the minimal required smoothness of $p(t, x)$ in (t, x) is untied from smoothness of the estimand $h^{T|X}(t|x)$.

Excellent and thorough discussion of hard-thresholding can be found in the monograph Johnstone (2024).

Proofs of the presented results can be found in Efromovich (2022a, 2024a,b).

Now let us formulate several open topics. (1) An interesting setting, specifically in biostatistical and actuarial applications, is when a categorical variable affects smoothness and dimensionality of conditional distribution in continuous covariates. A specific example is the effect of smoking or zip code on longevity. (2) Missing data are typical in survival data. For a nonparametric regression different estimators are optimal for missing responses and predictors. It is of interest to explore these settings for conditional hazard. (3) Sequential estimation with assigned risk is an interesting and important practical problem due to unknown smoothness of conditional hazard and unknown censoring mechanisms. (4) Measurement errors in covariates is another familiar problem in survival data. It will be of interest to explore optimal nonparametric estimation for this setting. (5) Theory and methodology of aggregation of censored and uncensored lifetimes. (6) Time-varying covariate. (7) There are many interesting topics with truncated, censored and dependent data, see Liang and de Una-Álvarez (2011), Wang and Chan (2018). (8) Estimation under shape restrictions is an important part of survival analysis and special procedures are suggested for taking the restrictions into consideration, see a discussion in Groeneboom and Jongbloed (2014) and Srivastava and Klassen (2016). For the E-estimation, there is no need to make any adjustment, instead, after calculating an E-estimate, it is sufficient to take a projection on the class of assumed functions and make the estimate bona fide. Theoretical justification of this approach can be found in Efromovich (2001a).

3.7 If we are interested in recovery of the survival function S^T of a right-censored T, the fundamental underlying idea of the empirical cumulative distribution function is no longer applicable, and instead a product-limit methodology is used. See the original paper Kaplan and Meier (1958) and a discussion in books Moore (2016) and Efromovich (2018). What is even more surprising, that there is no straightforward extension of the univariate product-limit methodology to the bivariate case. Instead, a number of sophisticated and mathematically involved procedures, ranging from EM, hazard gradient, partial differential equations and copula to nonparametric MLE and solving an inhomogeneous Volterra equation via the Peano series, have been proposed. The interested reader can find an insightful discussion of this topic in Campbell (1981), Tsai et al. (1986), Dabrowska (1988), Oakes (1989), Pruitt (1993), Frees et al. (1995), Wang and Wells (1997), Hougaard (2000), Akritas and Van Kellogom (2003), Crowder (2012), Lopez (2012), Li and Ma (2013), Prentice (2016), Prentice and Zhao (2018), Collett (2023). Nonparametric estimation of the joint probability density $f^{X,Y}$ in the presence of censoring is less explored. The nonparametric literature is primarily devoted to differentiation of smoothed estimators of the joint cumulative distribution function. An interesting discussion of the bivariate density estimation can be found in Wells and Yeo (1996), Dabrowska et al. (1998), Kooperberg (1998), Crowder (2012), Seok et al. (2014) and Ghosal and van der Vaart (2017).

Proofs of the presented results can be found in Efromovich (2022a).

Open research topics: (1) Develop a smooth estimator of the bivariate survival function. Further, similarly to the univariate case and direct observations, there is a possibility to develop a second-order efficient estimator of the survival function. Here the approach of Efromovich (2001b, 2004c) may be instrumental. (2) Multivariate survival function is another important topic for research. Here approaches of Efromovich (1999a, 2000b, 2010c, 2011c) may be useful. (3) Aggregation of censored and uncensored observaions. (4) Developing a Bayesian approach, see Ghosal and van der Vaart (2017).

3.8 Mean residual life (MRL) is used in different fields ranging from medical study and biostatistics to actuarial science and reliability research. Then the corresponding terminology is used. For instance, in the actuarial theory of loss models, the lifetime is called the loss, the MRL is called the mean excess loss function, and the residual survival is the excess survival. Review of properties of the MRL, including proofs and applications, can be found in Guess and Proschan (1988), Klein and Moeschberger (2003), Sun et al. (2012), Khmaladze (2013), Mansourvar et al. (2016), Hall and Wellner (2020), Macdonald et al. (2018), Hu et al. (2022). Nice discussion of the multivariate and conditional mean residual life function can be found in Nair and Nair (1989), Kulkarni and Rattihalli (2002), Rojo and Ghebremichael (2006), McLain and Ghosh (2011), Wang et al. (2015), and Ahmadi (2022). Bayesian perspective is discussed in Poynor and Kottas (2019). Restricted MRL is studied in Andersen, Hansen and Klien (2004), Cortese et al. (2017). BMRL is studied in Efromovich (2025).

The MRL is often used to shed light on the right tail. If the MRL $m^T(t)$ is increasing in t, the distribution is considered to have a heavy tail. If it is decreasing in t, the distribution is considered to have a light tail. Accordingly, estimated MRLs may be used to compare tails of several populations.

Important open research topic is the conditional MRL and BMRL given covariates. For example, the times from cancer surgery to its relapse and then to death may be related to the individual's gender, age, and weight.

References

1. Aalen, O. (1978). Nonparametric inference for a family of counting processes. *The Annals of Statistics* **6** 701–726.
2. Aalen, O, Borgan, O., and Gjessing, H. (2008). *Survival and Event History Analysis: A Process Point of View (Statistics for Biology and Health)*. New York: Springer.
3. Ahmadi, R. (2022). A bivariate process-based mean residual lifetime model for maintenance and inspection planning. *Computers and Industrial Engineering* **163** 107792.
4. Akritas, M. and Van Kellogom, I. (2003). Estimation of bivariate and marginal distribution with censored data. *Journal of Royal Statistical Society Ser. B* **65** 457–471.
5. Allignol, A. and Latouche, A. (2023). CRAN task view: Survival analysis.
6. Allison, P. (2014). *Event History and Survival Analysis*. 2nd ed. Thousand Oaks: SAGE Publications.
7. Amico, M. and Van Keilegom, I. (2018). Cure models in survival analysis. *Annual Review of Statistics and Its Application* **5** 311–342.
8. Andersen P., Borgan, O., Gill, R.D., and Keiding, N. (1993). *Statistical Models Based on Counting Processes*. New York: Springer.
9. Andersen, P., Hansen, M. and Klein, J. (2004). Regression analysis of restricted mean survival time based on pseudo-observations. *Lifetime Data Analysis* **10** 335–350.
10. Anderson-Bergman, C. (2017). icenReg: Regression models for interval censored data in R. *Journal of Statistical Software* **81** 1–23.
11. Anderson-Bergman, C. and Yu, Y. (2016). Computing the log concave npmle for interval censored data. *Statistics and Computing* **26** 813–826.
12. Aneiros, G., Novo, S. and Vieu, P. (2022). Variable selection in functional regression: A review. *Journal of Multivariate Analysis* **188** 1–13.
13. Antoniadis, A., Gregoire, G., and Nason, G. (1999). Density and hazard rate estimation for right-censored data by using wavelets methods. *Journal of the Royal Statistical Society: Series B (Statistical Methodology)* **61** 63–84.
14. Augugliaro, L., Sottile, G., Wit, E.C. and Vinciotti, V. (2023). cglasso: An R Package for Conditional Graphical Lasso Inference with Censored and Missing Values. *Journal of Statistical Software* **105** 1–58.

© The Editor(s) (if applicable) and The Author(s), under exclusive license to Springer Nature Switzerland AG 2026

S. Efromovich, *Survival Analysis*, Synthesis Lectures on Mathematics & Statistics, https://doi.org/10.1007/978-3-031-82814-0

209

15. Balan, T. and Putter, H. (2020). A tutorial on frailty models. *Statistical methods in medical research* **29** 3424–3454.

16. Bary, N.K. (1964). *A Treatise on Trigonometric Series.* Oxford: Pergamon Press.

17. Bayowa, B., Ebenezer, O., Osafu, E. and Ayantunji, A. (2022). A simulated based comparative study of some tests for checking homogeneity of non crossing survival curves under high censoring rates. *Journal of Applied Probability and Statistics* **17** 87–99.

18. Becker, D., Braun, W. and White, B. (2017). Interval-censored unimodal kernel density estimation via data sharpening. *Journal of Statistical Computation and Simulation* **87** 2023–2037.

19. Beis, G., Iliopoulos, A. and Papasotiriou, I. (2024). An overview of introductory and advanced survival analysis methods in clinical applications: where have we come so far? *Anticancer Research* **44** 471–487.

20. Bellman, R.E. (1961). *Adaptive Control Processes.* Princeton: Princeton University Press.

21. Ben Elouefi, R. and Saadaoui, F. (2023). Inverse-probability-weighted logrank test for stratified survival data with missing measurements. *Statistica Neerlandica* **77** 113–129.

22. Beyhum, J. and Van Keilegom, I. (2023). Robust censored regression with l_1-norm regularization. *Test* **32** 146–162.

23. Bickel, P. and Doksum, K. (2007). *Mathematical Statistics.* 2nd ed. London: Prentice Hall.

24. Birge, L. (1999). Interval censoring: a nonasymptotic point of view. *Mathematical Methods of Statistics* **8** 285–298.

25. Bouaziz, O., Brunel, E. and Comte, F. (2019). Nonparametric survival function estimation for data subject to interval censoring, case 2. *Journal of Nonparametric Statistics* **31** 952–987.

26. Braun, J., Duchesne, T. and Stafford, J. (2005). Local likelihood density estimation for interval censored data. *The Canadian Journal of Statistics* **33** 39–60.

27. Bremhorsta, V. and Lamberta, F. (2016). Flexible estimation in cure survival models using Bayesian p-splines. *Computational Statistics and Data Analysis* **93** 270–284.

28. Breslow, N. and Crowley, J. (1974). A large sample study of the life-table and product-limit estimates under random censorship. *The Annals of Statistics* **3** 437–453.

29. Breslow, N. and Hu, K. (2018). *Survival Analysis of Case-Control Data: A Sample Survey Approach*, in "Handbook of Statistical Methods for Case-Control Studies". Boca Raton: CRC Press.

30. Broström, G. (2021). *Event History Analysis with R.* Boca Raton: Chapman & Hall.

31. Brunel, E. and Comte, F. (2005). Penalized contrast estimation of density and hazard rate with censored data. *Sankhya* **67** 441–475.

32. Brunel, E. and Comte, F. (2006). Nonparametric adaptive regression estimation in presence of censoring. *Mathematical Methods of Statistics* **15** 233–255.

33. Brunel, E. and Comte, F. (2008). Adaptive estimation of hazard rate with censored data. *Communications in Statistics - Theory and Methods* **37** 1284–1305.

34. Brunel, E. and Comte F. (2009). Cumulative distribution function estimation under interval censoring case. *Electronic Journal of Statistics* **3** 1–24.

35. Brunel, E., Comte, F., and Guilloux, A. (2009). Nonparametric density estimation in presence of bias and censoring. *Test* **18** 166–194.

36. Buckley, J. and James, I. (1979). Linear regression with censored data. *Biometrika* **66** 429–436.

37. Bunea, F., Tsybakov, A. and Wegkamp, M. (2007). Aggregation for Gaussian regression. *The Annals of Statistics* **35** 1674–1697.

38. Buon, W., Fletcher, C. and Brundage, R. (2008). Impact of censoring data below an arbitrary quantification limit on structural model misspecification *Journal of Pharmacokinet Pharmacodyn* **35** 101–116.

39. Butzer, P. and Nessel, R. (1971). *Fourier Analysis and Approximations.* New York: Academic Press.

40. Cai, J., Fan, J., H. Zhou and Y. Zhou (2007). Hazard models with varying coefficients for multivariate failure time data. *The Annals of Statistics* **35** 324–354.

41. Cai, T. and Low, M. (2006). Adaptive confidence balls. *The Annals of Statistics* **34** 202–228.

42. Cai, T. (2008). On information pooling, adaptability and superefficiency in nonparametric function estimation. *Journal of Multivariate Analysis* **99** 421–436.

43. Cai, T. and Betensky, R. (2003). Hazard regression for interval-censored data with penalized spline. *Biometrics* **59** 570–579.

44. Campbell, G. (1981). Nonparametric Bivariate Estimation with Randomly Censored Data. *Biometrika* **68** 417–422.

45. Candès, E., Lei, L. and Ren, Z. (2023). Conformalized survival analysis. *Journal of the Royal Statistical Society Series B: Statistical Methodology* **85** 24–45.

46. Cao, R., Janssen, P., and Veraverbeke, N. (2005). Relative hazard rate estimation for right censored and left truncated data. *Test* **14** 257–280.

47. Carroll, R. and Ruppert, D. (1988). *Transformation and Weighting in Regression.* Boca Raton: Chapman & Hall.

48. Carroll, R., Van Rooij, A. and Ruymgaart, F. (1991). Theoretical aspects of ill-posed problems in statistics. *Acta Appl Math* **24** 113–140.

49. Carroll, R., Ruppert, D., Stefanski, L., and Crainceanu, C. (2006). *Measurement Error in Nonlinear Models: A Modern Perspective.* 2nd ed. Boca Raton: Champan & Hall.

50. Casella, G. and Berger, R. (2002). *Statistical Inference.* 2nd ed. New York: Duxbury.

51. Chagny, G. (2015). Adaptive warped kernel estimators. *Scandinavian Journal of Statistics* **42** 336–360.

52. Chan, K. (2013). Survival analysis without survival data: connecting length-biased and case-control data. *Biometrika* **100** 764–770.

53. Chen, D., Sun, J., and Peace, K. (2012). *Interval-Censored Time-to-Event Data: Methods and Applications.* Boca Raton: Chapman & Hall.

54. Chen, C., Chen, H. and Peng, Y. (2023). Mean residual life cure models for right-censored data with and without length-biased sampling. *Biometrical Journal* 2100368.

55. Chen, Y. and Cheng, S. (2006). Linear life expectancy regression with censored data. *Biometrika* **93** 303–313.

56. Cheng, X., Wang, S., Wang, H. and Ng., S. (2023). Deep survival forests for extremely high censored data. *Applied Intelligence* **53** 7041–7055.

57. Chiou, J. and Müller, H. (2009). Modeling hazard rates as functional data for the analysis of cohort lifetables and mortality forecasting. *Journal of American Statistical Association* **104** 572–585.

58. Choi, T., Park, S., Cho, H. and Choi, S. (2024). Interval-censored linear quantile regression. *Journal of Computational and Graphical Statistics*, (just-accepted) 1–23.

59. Chu, P., Wang, J., Hwang, J. and Chang, Y. (2008). Estimation of life expectancy and the expected years of life lost in patients with major cancers: extrapolation of survival curves under high-censored rates. *Value in Health* **11** 1102–1109.

60. Cohen, A.C. (1991). *Truncated and Censored Samples: Theory and Applications.* New York: Marcel Dekker.

61. Collett, D. (2023). *Modeling Survival Data in Medical Research.* 4th ed. Boca Raton: Chapman & Hall.

62. Comte, F., Mabon, G., and Samson, A. (2017). Spline regression for hazard rate estimation when data are censored and measured with error. *Statistica Neerlandica* **71** 115–140.

63. Cortese, G., Holmboe, S., and Scheike, T. (2017). Regression models for the restricted residual mean life for right-censored and left-truncated data. *Statistics in Medicine* **36** 1803–1822.

64. Cosslett, S. (1983). Distribution-free maximum likelihood estimator of the binary choice model. *Econometrics* **51** 765–782.

65. Cox, D.R. (1972). Regression models and life-tables. *Journal of the Royal Statistical Society: Series B (Methodological)* **34** 187–202.
66. Cox, D.R. (1975). Partial likelihood. *Biometrika* **62** 269–276.
67. Cox, D.R. and Oakes, D. (1984). *Analysis of Survival Data.* London: Chapman & Hall.
68. Crowder, M. (2012). *Multivariate Survival Analysis and Competing Risks.* Boca Raton: Chapman & Hall.
69. Cui, Y. and Hannig, J. (2019). Nonparametric generalized fiducial inference for survival functions under censoring, *Biometrika* **106**, 501–518.
70. Cui, Y., Kosorok, M., Sverdrup, E., Wager, S. and Zhu, R. (2023). Estimating heterogeneous treatment effects with right-censored data via causal survival forests, *Journal of the Royal Statistical Society Series B: Statistical Methodology* **85** 179–211.
71. Cui, Y., Hannig, J. and Kosorok, M. R. (2024). A unified nonparametric fiducial approach to interval-censored data. *Journal of the American Statistical Association* **119** 2230–2241.
72. Cummings, S. (2013). *The Use of Survival Analysis Techniques Among Highly Censored Data Sets.* www.witenberg.edu, 1–38.
73. Czado, C. and Van Keilegom, I. (2023). Dependent censoring based on parametric copulas. *Biometrika* **110** 721–738.
74. Dabrowska, D. (1988). Kaplan-Meier estimate on the plane. *The Annals of Statistics* **16** 1475–1489.
75. Dabrowska, D. (1989). Uniform consistency of the kernel conditional Kaplan-Meier estimate. *The Annals of Statistics* **17** 1157–1167.
76. Dabrowska, D., Duffy, D. and Zhang, Z. (1998). Hazard and Density Estimation from Bivariate Censored Data, *Journal of Nonparametric Statistics* **10** 67–93.
77. Dai, D., Rigollet, P., Xia, L. and Zhang, T. (2014). Aggregation of affine estimators. *Electronic Journal of Statistics* **8** 302–327.
78. Dai, H., Restaino, M., and Wang, H. (2016). A class of nonparametric bivariate survival function estimators for randomly censored and truncated data. *Journal of Nonparametric Statistics* **28** 736–751.
79. da Silva, J. L. P. (2023). A comparison of multiple imputation methods for the analysis of survival data with outcome related missing covariate values. *Sigmae* **12** 76–89.
80. Deltuvaite-Thomas, V., Verbeeck, J., Burzykowski, T., Buyse, M., Tournigand, C., Molenberghs, G. and Thas, O. (2023). Generalized pairwise comparisons for censored data: an overview. *Biometrical Journal* **65** 2100354.
81. DeVore, R.A. and Lorentz, G.G. (1993). *Constructive Approximation.* New York: Springer-Verlag.
82. Dey, T., Lipsitz, S.R., Cooper, Z., Trinh, Q.D., Krzywinski, M. and Altman, N. (2022). Survival analysis – time-to-event data and censoring. *Nature Methods* **19** 906–908.
83. Diao, G. and Yuan, A. (2019). A class of semiparametric cure models with current status data. *Lifetime Data Analysis* **25** 26–51.
84. Du, M., Hu, T. and Sun, J. (2019). Semiparametric probit model for informative current status data. *Statistics in Medicine* **38** 2219–2227.
85. Dym, H. and McKean, H.P. (1972). *Fourier Series and Integrals.* London: Academic Press.
86. Efromovich, S. (1980a). Information contained in a sequence of observations. *Problems of Information Transmission* **15** 178–189.
87. Efromovich, S. (1980b). On sequential estimation under conditions of local asymptotic normality. *Theory of Probability and its Applications* **25** 27–40.
88. Efromovich, S. (1984). Estimation of a spectral density of a Gaussian time series in the presence of additive noise. *Problems of Information Transmission* **20** 183–195.
89. Efromovich, S. (1985). Nonparametric estimation of a density with unknown smoothness. *Theory of Probability and its Applications* **30** 557–568.

90. Efromovich, S. (1986). Adaptive algorithm of nonparametric regression. *Proc. of Second IFAC symposium on Stochastic Control*. Vilnuis: Science, 112–114.
91. Efromovich, S. (1989). On sequential nonparametric estimation of a density. *Theory of Probability and its Applications* **34** 228–239.
92. Efromovich, S. (1992). On orthogonal series estimators for random design nonparametric regression. *Computing Science and Statistics* **24** 375–379.
93. Efromovich, S. (1994a). On adaptive estimation of nonlinear functionals. *Statistics and Probability Letters* **19** 57–63.
94. Efromovich, S. (1994b). On nonparametric curve estimation: multivariate case, sharp-optimality, adaptation, efficiency. *CORE Discussion Papers* **9418** 1–35.
95. Efromovich, S. (1994c). Nonparametric curve estimation from indirect observations. *Computing Science and Statistics* **26** 196–200.
96. Efromovich, S. (1995a). Thresholding as an adaptive method (discussion). *Journal of Royal Statistical Society ser. B* **57** 343.
97. Efromovich, S. (1995b). On sequential nonparametric estimation with guaranteed precision. *The Annals of Statistics* **23** 1376–1392.
98. Efromovich, S. (1996a). On nonparametric regression for iid observations in general setting. *The Annals of Statistics* **24** 1126–1144.
99. Efromovich, S. (1996b). Adaptive orthogonal series density estimation for small samples. *Computational Statistics and Data Analysis* **22** 599–617.
100. Efromovich, S. (1997a). Density estimation for the case of supersmooth measurement error. *Journal of the American Statistical Association* **92** 526–535.
101. Efromovich, S. (1997b). Robust and efficient recovery of a signal passed through a filter and then contaminated by non-Gaussian noise. *IEEE Transactions on Information Theory* **43** 1184–1191.
102. Efromovich, S. (1997c). Quasi-linear wavelet estimation involving time series. *Computing Science and Statistics* **29** 127–131.
103. Efromovich, S. (1998a). On global and pointwise adaptive estimation. *Bernoulli* **4** 273–278.
104. Efromovich, S. (1998b). Data-driven efficient estimation of the spectral density. *Journal of the American Statistical Association* **93** 762–770.
105. Efromovich, S. (1998c). Simultaneous sharp estimation of functions and their derivatives. *The Annals of Statistics* **26** 273–278.
106. Efromovich, S. (1999a). *Nonparametric Curve Estimation: Methods, Theory, and Applications*. New York: Springer.
107. Efromovich, S. (1999b). Quasi-linear wavelet estimation. *The Journal of the American Statistical Association* **94** 189–204.
108. Efromovich, S. (1999c). How to overcome the curse of long-memory errors. *IEEE Transactions on Information Theory* **45** 1735–1741.
109. Efromovich, S. (1999d). On rate and sharp optimal estimation. *Probability Theory and Related Fields* **113** 415–419.
110. Efromovich, S. (2000a). Can adaptive estimators for Fourier series be of interest to wavelets? *Bernoulli* **6** 699–708.
111. Efromovich, S. (2000b). On sharp adaptive estimation of multivariate curves. *Mathematical Methods of Statistics* **9** 117–139.
112. Efromovich, S. (2000c). Sharp linear and block shrinkage wavelet estimation. *Statistics and Probability Letters* **49** 323–329.
113. Efromovich, S. (2001a). Density estimation under random censorship and order restrictions: from asymptotic to small samples. *The Journal of the American Statistical Association* **96** 667–685.

114. Efromovich, S. (2001b). Second order efficient estimating a smooth distribution function and its applications. *Methodology and Computing in Applied Probability* **3** 179–198.
115. Efromovich, S. (2001c). Multiwavelets and signal denoising. *Sankhya ser. A* **63** 367–393.
116. Efromovich, S. (2002). Discussion on random rates in anisotropic regression. *The Annals of Statistics* **30** 370–374.
117. Efromovich, S. (2003a). On the limit in the equivalence between heteroscedastic regression and filtering model. *Statistics and Probability Letters* **63** 239–242.
118. Efromovich, S. (2004a). Density estimation for biased data. *The Annals of Statistics* **32** 1137–1161.
119. Efromovich, S. (2004b). Financial applications of sequential nonparametric curve estimation. In *Applied Sequential Methodologies*, eds. N.Mukhopadhyay, S.Datta, and S.Chattopadhyay. 171–192.
120. Efromovich, S. (2004c). Distribution estimation for biased data. *Journal of Statistical Planning and Inference* **124** 1–43.
121. Efromovich, S. (2004d). On sequential data-driven density estimation. *Sequential Analysis Journal* **23** 603–624.
122. Efromovich, S. (2004e). Analysis of blockwise shrinkage wavelet estimates via lower bounds for no-signal setting. *Annals of the Institute of Statistical Mathematics* **56** 205–223.
123. Efromovich, S. (2004f). Oracle inequalities for Efromovich–Pinsker blockwise estimates. *Methodology and Computing in Applied Probability* **6** 303–322.
124. Efromovich, S. (2004g). Discussion on "Likelihood ratio identities and their applications to sequential analysis" by Tze L. Lai. *Sequential Analysis Journal* **23** 517–520.
125. Efromovich, S. (2004h). Adaptive estimation of error density in heteroscedastic nonparametric regression. In: Proceedings of the 2nd International workshop in Applied Probability IWAP 2004, Univ. of Piraeus, Greece, 132–135.
126. Efromovich, S. (2005a). Univariate nonparametric regression in the presence of auxiliary covariates. *Journal of the American Statistical Association* **100** 1185–1201.
127. Efromovich, S. (2005b). Estimation of the density of regression errors. *The Annals of Statistics* **33** 2194–2227.
128. Efromovich, S. (2007a). A lower-bound oracle inequality for a blockwise-shrinkage estimate. *Journal of Statistical Planning and Inference* **137** 176–183.
129. Efromovich, S. (2007b). Universal lower bounds for blockwise-shrinkage wavelet estimation of a spike. *Journal of Applied Functional Analysis* **2** 317–338.
130. Efromovich, S. (2007c). Adaptive estimation of error density in nonparametric regression with small sample size. *Journal of Statistical Inference and Planning* **137** 363–378.
131. Efromovich, S. (2007d). Sequential design and estimation in heteroscedastic nonparametric regression. Invited paper with discussion. *Sequential Analysis* **26** 3–25.
132. Efromovich, S. (2007e). Response on sequential design and estimation in heteroscedastic nonparametric regression. *Sequential Analysis* **26** 57–62.
133. Efromovich, S. (2007f). Optimal nonparametric estimation of the density of regression errors with finite support. *Annals of the Institute of Statistical Mathematics* **59** 617–654.
134. Efromovich, S. (2007g). Conditional density estimation. *The Annals of Statistics* **35** 2504–2535.
135. Efromovich, S. (2007h). Comments on nonparametric inference with generalized likelihood ratio tests. *Test* **16** 465–467.
136. Efromovich, S. (2007i). Applications in finance, engineering and health sciences: Plenary Lecture. *Abstracts of IWSM-2007*, Auburn University, 20–21.
137. Efromovich, S. (2008a). Optimal sequential design in a controlled nonparametric regression. *Scandinavian Journal of Statistics* **35** 266–285.
138. Efromovich, S. (2008b). Adaptive estimation of and oracle inequalities for probability densities and characteristic functions. *The Annals of Statistics* **36** 1127–1155.

139. Efromovich, S. (2008c). Nonparametric regression estimation with assigned risk. *Statistics and Probability Letters* **78** 1748–1756.
140. Efromovich, S. (2009a). Lower bound for estimation of Sobolev densities of order less 1/2. *Journal of Statistical Planning and Inference* **139** 2261–2268.
141. Efromovich, S. (2009b). Multiwavelets: theory and bioinformatic applications. *Communications in Statistics – Theory and Methods* **38** 2829–2842
142. Efromovich, S. (2009c). Optimal sequential surveillance for finance, public health, and other areas: discussion. *Sequential Analysis* **28** 342–346.
143. Efromovich, S. (2010a). Sharp minimax lower bound for nonparametric estimation of Sobolev densities of order 1/2. *Statistics and Probability Letters* **80** 77–81.
144. Efromovich, S. (2010b). Oracle inequality for conditional density estimation and an actuarial example. *Annals of the Institute of Mathematical Statistics* **62** 249–275.
145. Efromovich, S. (2010c). Orthogonal series density estimation. *WIREs Computational Statistics* **2** 467–476.
146. Efromovich, S. (2010d). Dimension reduction and oracle optimality in conditional density estimation. *Journal of the American Statistical Association* **105** 761–774.
147. Efromovich, S. (2011a). Nonparametric regression with predictors missing at random. *Journal of the American Statistical Association* **106** 306–319.
148. Efromovich, S. (2011b). Nonparametric regression with responses missing at random. *Journal of Statistical Planning and Inference* **141** 3744–3752.
149. Efromovich, S. (2011c). Nonparametric estimation of the anisotropic probability density of mixed variables. *Journal of Multivariate Analysis* **102** 468–481.
150. Efromovich, S. (2012a). Nonparametric regression with missing data: theory and applications. *Actuarial Research Clearing House* **1** 1–15.
151. Efromovich, S. (2012b). Sequential analysis of nonparametric heteroscedastic regression with missing responses. *Sequential Analysis* **31** 351–367.
152. Efromovich, S. (2013a). Nonparametric regression with the scale depending on auxiliary variable. *The Annals of Statistics* **41** 1542–1568.
153. Efromovich, S. (2013b). Notes and proofs for nonparametric regression with the scale depending on auxiliary variable. *The Annals of Statistics* **41**, 1–29.
154. Efromovich, S. (2013c). Adaptive nonparametric density estimation with missing observations. *Journal of Statistical Planning and Inference* **143** 637–650.
155. Efromovich, S. (2014a). On shrinking minimax convergence in nonparametric statistics. *Journal of Nonparametric Statistics* **26** 555–573.
156. Efromovich, S. (2014b). Efficient nonparametric estimation of the spectral density in the presence of missing observations. *Journal of Time Series Analysis* **35** 407–427.
157. Efromovich, S. (2014c). Nonparametric regression with missing data. *Computational Statistics* **6** 265–275.
158. Efromovich, S. (2014d). Nonparametric estimation of the spectral density of amplitude-modulated time series with missing observations, *Statistics and Probability Letters* **93** 7–13.
159. Efromovich, S. (2014e). Nonparametric curve estimation with incomplete data, *Actuarial Research Clearing House* **15** 31–47.
160. Efromovich, S. (2015). Two-stage nonparametric sequential estimation of the directional density with complete and missing observations. *Sequential Analysis* **34** 425–440.
161. Efromovich, S. (2016a). Minimax theory of nonparametric hazard rate estimation: efficiency and adaptation. *Annals of the Institute of Statistical Mathematics* **68** 25–75.
162. Efromovich, S. (2016b). Estimation of the spectral density with assigned risk. *Scandinavian Journal of Statistics* **43** 70–82.
163. Efromovich, S. (2016c). What an actuary should know about nonparametric regression with missing data. *Variance* **10** 145–165.

164. Efromovich, S. (2017). Missing, modified and large-p-small-n data in nonparametric curve estimation. *Calcutta Statistical Association Bulletin* **69** 1–34.

165. Efromovich, S. (2018). *Missing and Modified Data in Nonparametric Estimation with R Examples*, Chapman & Hall: Boca Raton.

166. Efromovich, S. (2019a). On two-stage estimation of the spectral density with assigned risk in presence of missing data. *Journal of Time Series Analysis* **40** 203–224.

167. Efromovich, S. (2019b). Statistical analysis of fMRI using wavelets: big data, denoising, large-p-small-n matrices. *WIREs Statistical Computing* **11** 1–14

168. Efromovich, S. (2019c). On sharp nonparametric estimation of differentiable functions *Statistics and Probability Letters* **152** 9–14.

169. Efromovich, S. (2019d). On sequential spectral analysis of amplitude-modulated time series. *Sequential Analysis* **38** 259–278.

170. Efromovich, S. (2020a). Missing not at random and the nonparametric estimation of the spectral density. *Journal of Time Series Analysis* **41** 652–675.

171. Efromovich, S. (2021a). Sharp minimax distribution estimation for current status censoring with or without missing. *The Annals of Statistics* **49** 568–589.

172. Efromovich, S. (2021b). Current status censoring and asymptotic theory: Supplementary Materials. *The Annals of Statistics* 1–65, https://doi.org/10.1214/20-AOS1970SUPP

173. Efromovich, S. (2021c). Nonparametric curve estimation for truncated and censored data without product limit. *Variance* **14** 1–17.

174. Efromovich, S. (2022a). Nonparametric Bivariate Density Estimation for Censored Lifetimes. *The Annals of Statistics* **50** 2767–2792.

175. Efromovich, S. (2022b). Efficient nonparametric estimation of distribution for current status censoring. *Electronic Journal of Statistics* **16** 998–1057.

176. Efromovich, S. (2022c). Sequential estimation of controlled multivariate regression. *Sequential Analysis* **41** 492–511.

177. Efromovich, S. (2022d). Nonparametric multivariate regression for mesophilic and thermophilic anaerobic digestion decreasing greenhouse gas emission: Supplementary Materials. *Sequential Analysis*, https://doi.org/10.1080/07474946.2022.2129690

178. Efromovich, S. (2023a). Sharp lower bound for regression with measurement errors and its implication for ill-posedness of functional regression. *Mathematical Methods of Statistics* **32** 209–221.

179. Efromovich, S. (2023b). Functional regression, anaerobic digestion and greenhouse gas reduction: Supplementary Materials. *Mathematical Methods of Statistics* **32** https://doi.org/10.3103/S1066530723030031

180. Efromovich, S. (2024a). Conditional hazard rate estimation for right censored data. *Bernoulli* **30** 2423–2449.

181. Efromovich, S. (2024b). Conditional hazard rate estimation for right censored data: Supplementary Materials. *Bernoulli* **30** https://doi.org/10.3150/23-BEJ1679SUPP

182. Efromovich, S. (2024c). Nonparametric density estimation over its unknown support for right censored data. *Statistics and Probability Letters* **209** 1–7.

183. Efromovich, S (2024d). On aggregation of uncensored and censored observations, *Mathematical Methods of Statistics* **33** 154–181.

184. Efromovich, S. (2024e). Nonparametric regression for current status censoring. *Electronic Journal of Statistics* **18** 4916–4991.

185. Efromovich, S. (2025). Nonparametric estimation of bivariate mean residual life for right-censored observations. *STAT* **14** 1–6.

186. Efromovich, S. and Baron, M. (2010). Discussion on quickest detection problems: fifty years later" by Albert N. Shiryaev. *Sequential Analysis* **29** 398–403.

187. Efromovich, S. and Chu, J. (2018a). Hazard rate estimation for left truncated and right censored data. *Annals of the Institute of Mathematical Statistics* **70** 889–917.

188. Efromovich, S. and Chu, J. (2018b). Small LTRC samples and lower bounds in hazard rate estimation. *Annals of the Institute of Mathematical Statistics* 1–46. https://doi.org/10.1007/s10463-017-0617-x

189. Efromovich, S. and Fuksman, L. (2024a). Study of imputation procedures for nonparametric density estimation based on missing censored lifetimes. *Computational Statistics and Data Analysis*, **198** 107994.

190. Efromovich, S. and Fuksman, L. (2024b). Numerical study of missing survival data: Supplementary Materials. *Computational Statistics and Data Analysis* 1–34, https://doi.org/10.1016/j.csda.2024.107994

191. Efromovich, S. and Fuksman, L. (2024c). Missing in Survival Analysis. In: Ansari, J., et al. Combining, Modelling and Analyzing Imprecision, Randomness and Dependence. SMPS 2024. *Advances in Intelligent Systems and Computing* **1458** 134–141, Springer, Cham.

192. Efromovich, S. and Ganzburg, M. (1999). Best Fourier approximation and application in efficient blurred signal reconstruction. *Computational Analysis and Applications* **1** 43–62.

193. Efromovich, S., Grainger, D., Bodenmiller, D. and Spiro, S. (2008). Genome-wide identification of binding sites for the nitric oxide sensitive transcriptional regulator NsrR. *Methods in Enzymology* **437** 211–233.

194. Efromovich, S. and Koltchinskii, V. (2001). On inverse problems with unknown operators. *IEEE Transactions on Information Theory* **47** 2876–2894.

195. Efromovich, S., Lakey, J., Pereyra, M.C., and Tymes, N. (2004). Data-driven and optimal denoting of a signal and recovery of its derivative using multiwavelets. *IEEE Transactions on Signal Processing* **52** 628–635.

196. Efromovich, S. and Low, M. (1994). Adaptive estimates of linear functionals. *Probability Theory and Related Fields* **98** 261–275.

197. Efromovich, S. and Low, M. (1996a). On Bickel and Ritov's conjecture about adaptive estimation of some quadratic functionals. *The Annals of Statistics* **24** 682–686.

198. Efromovich, S. and Low, M. (1996b). Adaptive estimation of a quadratic functional. *The Annals of Statistics* **24** 1106–1125.

199. Efromovich, S. and Pinsker M.S. (1981). Estimation of a square integrable spectral density for a time series. *Problems of Information Transmission* **17** 50–68.

200. Efromovich, S. and Pinsker M.S. (1982). Estimation of a square integrable probability density of a random variable. *Problems of Information Transmission* **18** 19–38.

201. Efromovich, S. and Pinsker M.S. (1984). An adaptive algorithm of nonparametric filtering. *Automation and Remote Control* **11** 58–65.

202. Efromovich, S. and Pinsker M.S. (1986). Adaptive algorithm of minimax nonparametric estimating spectral density. *Problems of Information Transmission* **22** 62–76.

203. Efromovich, S. and Pinsker, M.S. (1989). Detecting a signal with an assigned risk. *Automation and Remote Control* **10** 1383–1390.

204. Efromovich, S. and Pinsker, M. (1996). Sharp-optimal and adaptive estimation for heteroscedastic nonparametric regression. *Statistica Sinica* **6** 925–945.

205. Efromovich, S. and Salter-Kubatko, L. (2008). Coalescent time distributions in trees of arbitrary size. *Statistical Applications in Genetics and Molecular Biology* **7** 1–21.

206. Efromovich, S. and Samarov, A. (1996). Asymptotic equivalence of nonparametric regression and white noise model has its limits. *Statistics and Probability Letters* **28** 143–145.

207. Efromovich, S. and Samarov, A. (2000). Adaptive estimation of the integral of squared regression derivatives. *Scandinavian Journal of Statistics* **27** 335–352.

208. Efromovich, S. and Smirnova, E. (2014a). Wavelet estimation: minimax theory and applications. *Sri Lankan Journal of Applied Statistics* **15** 17–31.

209. Efromovich, S. and Smirnova, E. (2014b). Statistical analysis of large cross-covariance and cross-correlation matrices produced by fMRI Images. *Journal of Biometrics and Biostatistics* **5** 1–8.

210. Efromovich, S. and Thomas, E. (1996). Application of nonparametric binary regression to evaluate the sensitivity of explosives. *Technometrics* **38** 50–58.

211. Efromovich, S. and Valdez-Jasso, Z.A. (2010). Aggregated wavelet estimation and its application to ultra-fast fMRI. *Journal of Nonparametric Statistics* **22** 841–857.

212. Efromovich, S. and Wu, J. (2017). Dynamic nonparametric analysis of nonstationary portfolio returns and its application to VaR and forecasting. *Actuarial Research Clearing House* 1–25.

213. Efromovich, S. and Wu, J. (2018a). Wavelet analysis of big data contaminated by large noise in an fMRI study of neuroplasticity. *Methodology and Computing in Applied Probability* **20** 1381–1402.

214. Efromovich, S. and Wu, J. (2018b). Proofs and complementary materials for wavelet analysis of Big Data contaminated by large noise in an fMRI study of neuroplasticity. Supplementary Materials. *Methodology and Computing in Applied Probability* 1–47. https://doi.org/10.1007/s11009-018-9626-3.

215. Efromovich, S. and Wu, J. (2023a). Efficient Nonparametric Spectral Density Estimation with Censored Observations. *Communications in Statistics - Theory and Methods* **53** 6671–6694.

216. Efromovich, S. and Wu, J. (2023b). Numerical study of adaptive spectral density estimator for censored data. Supplementary Materials. *Communications in Statistics - Theory and Methods* **53** 1–11. https://doi.org/10.1080/03610926.2023.2250484

217. Efron, B. (1967). The two sample problem with censored data. *Procedures of 5th Berkeley Symposium* **4** 831–854.

218. Efron, B. (1977). The efficiency of Cox's likelihood function for censored data. *Journal of the American Statistical Association* **72** 557–565.

219. El Ghouch, A. and Van Keilegom, I. (2008). Nonparametric regression with dependent censored data. *Scandinavian Journal of Statistics* **35** 228–247.

220. El Ghouch, A. and Van Keilegom, I. (2009). Local linear quantile regression with dependent censored data. *Statistica Sinica* **19** 1621–1640.

221. Emura, T. and Chen, Y. (2018). *Analysis of Survival Data with Dependent Censoring: Copula-Based Approaches*. Singapore: Springer.

222. Enders, C. (2010). *Applied Missing Data Analysis*. New York: The Guilford Press.

223. Eubank, R.L. (1988). *Spline Smoothing and Nonparametric Regression*. New York: Marcel and Dekker.

224. Ewnetu, W., Gijbels, I. and Verhasselt, A. (2023). Flexible two-piece distributions for right censored survival data. *Lifetime Data Analysis* **29** 34–65.

225. Fan, J. and Truong, Y.K. (1993). Nonparametric regression with errors in variables, *The Annals of Statistics* **21** 1900–1925.

226. Fan, J. and Gijbels, I. (1994). Censored regression: local linear approximations and their applications. *Journal of the American Statistical Association* **89** 560–570.

227. Feng, Y. and Chen, Y. (2018). Regression analysis of current status data with auxiliary covariates and informative observation times. *Lifetime Data Analysis* **24** 293–309.

228. Fernandez, T., Gretton, A., Rindt, D. and Sejdinovic, D. (2023). A Kernel log-rank test of independence for right-censored data, *Journal of the American Statistical Association* **118** 925–936.

229. Fleming, T.R. and Harrington, D.P. (2011). *Counting Processes and Survival Analysis*. New York: Wiley.

230. Frees, E., Carriere, J. and Valdez, E. (1995). Annuity Valuation with Dependent Mortality. *Actuarial Research Clearing House* **2** 31–80.

231. Gardiner, J. (2021). Restricted mean survival time estimation: nonparametric and regression methods. *Journal of Statistical Theory and Practice* **15** 1–15.

232. Ghosal, S. and van der Vaart, A. (2017). *Fundamentals of Nonparametric Bayesian Inference*. Cambridge: Cambridge University Press.

233. Gijbels, I. (2010). Censored data. *Wiley Interdisciplinary Reviews: Computational Statistics* **2** 178–188.

234. Gill, R. (2006). *Lectures on Survival Analysis*. New York: Springer.

235. Gill, R. and Levit, B. (1995). Applications of the van Trees inequality: A Bayesian Cramér-Rao bound. *Bernoulli* **1** 59–79.

236. Giné, E. and Nickl, R. (2010). Confidence bands in density estimation. *The Annals of Statistics* **38** 1122–1170.

237. Glad, I., Hjort, N., and Ushakov, N. (2003). Correction of density estimators that are not densities. *The Scandinavian Journal of Statistics* **30** 415–427.

238. Gomez, G., Calle, M.L., Oller, R. and Langohr, K. (2009). Tutorial on methods for interval-censored data and their implementation in R. *Statistical Modeling* **9** 259–297.

239. Green, P. and Silverman, B. (1994). *Nonparametric Regression and Generalized Linear Models: a Roughness Penalty Approach*. London: Chapman & Hall.

240. Groeneboom, P., Jongbloed, G. and Witte, B. (2010). Maximum smoothed likelihood estimation and smoothed maximum likelihood estimation in the current status model. *The Annals of Statistics* **38** 352–387.

241. Groeneboom, P. and Ketelaars, T. (2011). Estimators for the interval censoring problems. *Electronic Journal of Statistics* **5** 1797–1845.

242. Groeneboom, P. and Jongbloed, G. (2014). *Nonparametric Estimation under Shape Constraints: Estimators, Algorithms and Asymptotics*. Cambridge: Cambridge University Press.

243. Groeneboom, P. and Hendrickx, K. (2018). Current status linear regression. *The Annals of Statistics* **46** 1415–1444.

244. Gross, T. and Lai, T. (1996). Nonparametric estimation and regression analysis with left-truncated and right-censored data. *Journal of the American Statistical Association* **91** 1166–1180.

245. Grummer-Strawn, L. (1993). Regression analysis of current status data: An application to breast feeding. *Journal of the American Statistical Association* **88** 758–765.

246. Guess, F. and Proschan, F. (1988). Mean residual life: theory and applications. *Handbook of Statistics* **7** 215–224.

247. Guo, S. (2010). *Survival Analysis*. Oxford: Oxford University Press.

248. Hagar, Y. and Dukic, V. (2015). Comparison of hazard rate estimation in R. arXiv: 1509.03253v1

249. Hall, W. and Wellner, J.A. (2020). *Estimation of Mean Residual Life*. In: Almudevar, A., Oakes, D., Hall, J. (eds) *Statistical Modeling for Biological Systems*. Cham: Springer.

250. Han, K. and Jung, I. (2022). Restricted mean survival time for survival analysis: a quick guide for clinical researchers. *Korean Journal of Radiology* **23** 495–499.

251. Hao, M., Kin-yat Liu, K., Xu, W. and Zhao, X. (2021). Semiparametric Inference for the Functional Cox Model, *Journal of the American Statistical Association* **116** 1319–1329.

252. Harrell, F. (2015). *Regression Modeling Strategies: With Applications to Linear Models, Logistic and Ordinal Regression, and Survival Analysis*. 2nd ed. London: Springer.

253. Hastie, T.J. and Tibshirani, R. (1990). *Generalized Additive Models*. London: Chapman & Hall.

254. Helsel, D. (2011). *Statistics for Censored Environmental Data Using Minitab and R*. 2nd. ed. New York: Wiley.

255. Hoffmann, M. and Lepski, O. (2002). Random rates in anisotropic regression, with discussion. *The Annals of Statistics* **30** 325–396.

256. Hoffmann, M. and Nickl, R. (2011). On adaptive inference and confidence bands. *The Annals of Statistics* **39** 2383–2409.

257. Honda, T. (2004). Nonparametric regression with current status data. *Annals of the Institute of Statistical Mathematics* **56** 49–72.
258. Horowitz, J. and Savin, N. (2000). Binary response models: logits, probits and semiparametrics. *Journal of Economic Perspectives* **15** 43–56.
259. Horowitz, J. (2009). *Semiparametric and Nonparametric Methods in Econometrics*. New York: Springer.
260. Horowitz, J. and Lee, S. (2017). Nonparametric estimation and inference under shape restrictions. *Journal of Econometrics* **201** 108–126.
261. Hosmer D., Lemeshow, S., and May, S. (2008). *Applied Survival Analysis: Regression Modeling of Time-to-Event Data*. 2nd ed. New York: Wiley.
262. Hougaard, P. (2000). *Analysis of Multivariate Survival Data*. New York: Springer.
263. Hsu, L., Gorfine, M. and Zucker, D. (2018). On Estimation of the hazard function from population-based case-control studies. *Journal of the American Statistical Association* **113** 560–570.
264. Hu, C., Fan, H. and Wang, Z. (2022). *Residual Life Prediction and Optimal Maintenance Decision for a Piece of Equipment*. Singapore: Springer.
265. Huang, J. and Wellner, J. (1997). Interval censored survival data: a review of recent progress. In: Lin, D.Y., Fleming, T.R. (eds) *Proceedings of the First Seattle Symposium in Biostatistics. Lecture Notes in Statistics* **123**. New York: Springer.
266. Huang, J. (1996). Efficient estimation for the proportional hazards model with interval censoring. *The Annals of Statistics* **24** 540–566.
267. Huang, Y. (2010). Quantile calculus and censored regression. *The Annals of Statistics* **38** 1607–1637.
268. Ingster, Yu. and Suslina, I. (2003). *Nonparametric Goodness-of-Fit Testing Under Gaussian Models*. New York: Springer
269. Izenman, A. (2008). *Modern Multivariate Statistical Techniques: Regression, Classification, and Manifold Learning*. New York: Springer.
270. James, G., Witten, D., Hastie, T., Tibshirani, R. and Taylor, J. (2023). Survival analysis and censored data. In *An Introduction to Statistical Learning: with Applications in Python*. 469–502). Cham: Springer International Publishing.
271. Jankowski, H. and Wellner, J. (2009). Nonparametric estimation of a convex bathtub-shaped hazard function. *Bernoulli* **15** 1010–1035.
272. Janssen, P. and Veraverbeke, N. (2023). Nonparametric estimation of univariate and bivariate survival functions under right censoring: a survey. *Metrika* **87** 211–245.
273. Jewell, N. and van der Laan, M. (2004a). Current status data: review, recent developments and open problems. In: *Advances in Survival Analysis*, eds. N. Balakrishnan and C.R. Rao. 625–643. Amsterdam: Elsevier.
274. Jewell, N. and van der Laan, M. (2004b). Case-control current status data. *Biometrika* **91** 529–541.
275. Jiang, T. (2022). Nonparametric regression with the scale depending on auxiliary covariates and missing data. *Journal of Nonparametric Statistics* **35** 302–322.
276. Jiang, F., Cheng, Q., Yin, G. and Shen, H. (2020). Functional censored quantile regression. *Journal of the American Statistical Association* **115** 931–944.
277. Jiang, F. and Guterman, E. (2024). Survival analysis. *Statistical Methods in Epilepsy* 124–142.
278. Jin, Z., Lin, D. and Ying, Z. (2006). On least-squares regression with censored data. *Biometrika* **93** 147–161.
279. Johnstone, I. (2024). *Gaussian Estimation: Sequence and Wavelet Models*. Manuscript. Stanford: University of Stanford.
280. Kalbfleisch, J. and Prentice, R. (2002). *The Statistical Analysis of Failure Time Data*. 2nd ed. New York: Springer.

281. Kang, S., Lu, W. and Zhang, J. (2018). On estimation of the optimal treatment regime with the additive hazards model. *Statistica Sinica* **28** 1539.

282. Kaplan, E. and Meier, P. (1958). Nonparametric estimation with incomplete observations. *JASA* **53** 457–481.

283. Karim, M.R. and Islam, M. (2019). *Reliability and Survival Snalysis*. Singapure: Springer.

284. Karrison, T. (1987). Restricted mean life with adjustment for covariates, *Journal of the American Statistical Association* **82** 1169–1176.

285. Keogh, R. and Cox, D. (2014). *Case-Control Studies*. Cambridge: Cambridge Univ. Press.

286. Khmaladze, E. (2013). *Statistical Methods with Applications to Demography and Life Insurance*. Boca Raton: CRC Press.

287. Kim, H. and Truong, Y. (1998). Nonparametric regression estimates with censored data: local linear smoothers and their applications. *Biometrics* **54** 1434–1444.

288. Klein, J. and Moeschberger, M. (2003). *Survival Analysis: Techniques for Censored and Truncated Data*. New York: Springer.

289. Klein, J., van Houwelingen, H., Ibrahim, J., and Scheike, T. (2014). *Handbook of Survival Analysis*. Boca Raton: Chapman & Hall.

290. Kleinbaum, D. and Klein, M. (2012). *Survival Analysis*. 3rd ed. New York: Springer.

291. Klugman, S., Panjer, H., and Willmot, G. (2012). *Loss Models: From Data to Decisions*. 4th ed. New York: Wiley.

292. Kohler, M., Mathe, K. and Pinter, M. (2022). Prediction from randomly right censored data, *Jounal of Multivariate Analysis* **80** 73–100.

293. Koley, T. and Dewanji, A. (2024). Use of additional information for current status data with two competing risks and missing failure types. *Sankhya B* 1–29.

294. Kolmogorov, A.N. and Fomin, S.V. (1957). *Elements of the Theory of Functions and Functional Analysis*. Rochester: Graylock Press.

295. Kooperberg, C. and Stone, C. (1992). Logspline density estimation for censored data. *Journal of Computational and Graphical Statistics* **1** 301–328.

296. Kooperberg, C., Stone, C. and Truong, Y. (1995). Hazard regression. *Journal of the American Statistical Association* **90** 78–94.

297. Kooperberg, C. (1998). Bivariate density estimation with an application to survival analysis. *Journal of Computational and Graphical Statistics*, 7, 322–341.

298. Kosorok, M. (2008). *Introduction to Empirical Processes and Semiparametric Inference*. New York: Springer.

299. Kulkarni, H. and Rattihalli, R. (2002). Nonparametric estimation of a bivariate mean residual life function. *Journal of the American Statistical Association* **97** 907–917.

300. Lautier, J., Pozdnyakov, V. and Yan, J. (2023). Pricing time-to-event contingent cash flows: A discrete-time survival analysis approach. *Insurance: Mathematics and Economics* **110** 53–71.

301. Lee, D., Chen, N. and Ishwaran, H. (2021). Boosted nonparametric hazards with time-dependent covariates. *The Annals of Statistics* **49** 2101–2128.

302. Lee, E. and Wang, J. (2013). *Statistical Methods for Survival Analysis*. 4th ed. New York: Wiley.

303. Legrand, C. (2021). *Advanced Survival Models*. Boca Raton: Chapman&Hall.

304. Lepski, O. (2022). Theory of adaptive estimation. *Procedures of International Congress of Mathematics* **7** 5478–5498.

305. Lewbel, A. and Oliver, L. (2002). Nonparametric censored and truncated regression. *Econometrika* **70** 765–779.

306. Li, B. and Song, J. (2022). Dimension reduction for functional data based on weak conditional moments. *The Annals of Statistics* **50** 107–128.

307. Li, G. and Doss, H. (1995). An approach to nonparametric regression for life history data using local linear fitting. *The Annals of Statistics* **23** 787–823.

308. Li, G. and Zhang, C.-H. (1998). Linear regression with interval censored data. *Annals of Statistics* **26** 1306–1327.
309. Li, J. and Ma, S. (2013). *Survival Analysis in Medicine and Genetics*. Boca Raton: Chapman & Hall.
310. Li, L., MacGibbon, B. and Valenta, C. (2008). On the optimality of wavelet-based nonparametric regression with censored data. Journal of Applied Probability and Statistics **3** 243–261.
311. Li, Q. and Racine, J. (2007). *Nonparametric Econometrics: Theory and Practice*. Princeton: Princeton University Press.
312. Li, S., Hu, T., Zhao, X. and Sun, J. (2019). A class of semiparametric transformation cure models for interval-censored failure time data. *Computational Statistics and Data Analysis* **133** 153–165.
313. Li, S., Sun, J., Tian, T. and Cu, X. (2020). Semiparametric regression analysis of doubly censored failure time data from cohort studies. *Lifetime Data Analysis* **26** 315–338.
314. Li, S., Wang, P. and Sun, J. (2017). Regression analysis of current status data in the presence of depended censoring with applications to tumorigenicity experiments. *Computational Statistics and Data Analysis* **110** 75–86.
315. Li, W., Ma, H., Faraggi, D. and Dinse, G. (2023). Generalized mean residual life models for survival data with missing censoring indicators. *Statistics in Medicine* **42** 264–280.
316. Liang H. and de Una-Álvarez J. (2011). Wavelet estimation of conditional density with truncated, censored and dependent data. *Journal of Multivariate Analysis* **102** 448–467
317. Liao, C., Su, C., Huang, H. and Lin, C. (2023). Improved survival analyses based on characterized time-dependent covariates to predict individual chronic kidney disease progression. *Biomedicines* **11**, 1664–1689.
318. Lin, D., Oakes, D, and Ying, Z. (1998). Additive hazard regression with current status data. *Biometrika* **85** 289–298.
319. Lin, T. and Wang, W. (2023) Flexible modeling of multiple nonlinear longitudinal trajectories with censored and non-ignorable missing outcomes. *Statistical Methods in Medical Research* **32** 593–608.
320. Little, R. and Rubin, D. (2002). *Statistical Analysis with Missing Data*. New York: Wiley.
321. Liu, H. and Qin, J. (2018). Semiparametric probit models with univariate and bivariate current status data. *Biometrics* **74** 68–76.
322. Liu, X. (2012). *Survival Analysis: Models and Applications*. New York: Wiley
323. Liu, X., Dong, X., Zhang, L., Chen, J. and Wang, C. (2023). Least squares support vector regression for complex censored data. *Artificial Intelligence in Medicine*, **136** 102497.
324. Lokhnygina, Y. (2021). Nonparametric Survival Analysis. In: Piantadosi, S., Meinert, C.L. (eds) Principles and Practice of Clinical Trials, 1717–1742. Cham: Springer.
325. Lopez, O. (2012). A generalization of Kaplan-Meier estimator for analyzing bivariate mortality under right-censoring and left-truncation with applications to model-checking for survival copula models. *Insurance: Mathematics and Economics* **51** 505–516.
326. Lopuhaa, H. and Musta, E. (2018). Smoothed isotonic estimators of a monotone baseline hazard in the Cox model. *Scandinavian Journal of Statistics* **45** 753–791.
327. Lotspeich, S., Ashner, M., Vazquez, J., Richardson, B., Grosser, K., Bodek, B. and Garcia, T. (2024). Making sense of censored covariates: statistical methods for studies of huntington's disease. *Annual Review of Statistics and Its Application* **11** 5.1–5.23.
328. Lu, M., Lu, T. and Li, C. (2018). Efficient estimation of partially linear additive Cox model under monotonicity constraint. *Journal of Statistical Planning and Inference* **192** 18–34.
329. Lu, S., Wu, J. and Lu, X. (2019). Efficient estimation of the varying-coefficient partially linear proportional odds model with current status data. *Metrika* **82** 173–194.
330. Lu, T., Li, H., Li, S. and Sun, L. (2023). Efficient estimation for the proportional hazards model with left-truncated and interval-censored data. *Stat* **12** e628.

331. Luo, J., Xie, L., Yang, H., Yin, X. and Zhang, Y. (2024). Machine learning for time-to-event prediction and survival clustering: a review from statistics to deep neural networks. In: Cruz, C., Zhang, Y., Gao, W. (eds) Intelligent Computers, Algorithms, and Applications. IC 2023. *Communications in Computer and Information Science* Singapore: Springer.

332. Luo, X. and Tsai, W. (2009). Nonparametric estimation for right-censored length-biased data: a pseudo-partial likelihood approach. *Biometrika* **96** 873–886.

333. Lv, S., Jiang, J, Zhou, S. and Huang, J. (2018). Estimating high-dimensional additive Cox model with time-dependent covariate processes. *Scandinavian Journal of Statistics* **45** 900–922.

334. Lyu, L., Cheng, Y. and Wahed, A. S. (2023). Imputation-based Q-learning for optimizing dynamic treatment regimes with right-censored survival outcome. *Biometrics* **79** 3676–3689.

335. Ma, L., Hu, T. and Sun, J. (2015). Maximum likelihood regression analysis of dependent current status data. *Biometrika* **102** 731–738.

336. Ma, S. and Kosorok, M. (2005). Penalized log-likelihood estimation for partly linear transformation models with current status data. *The Annals of Statistics* **33** 2256–2290.

337. Macdonald, A., Richards, S. and Currie, I. (2018). *Modeling Mortality with Actuarial Applications*. Cambridge: Cambridge University Press.

338. Maksymiuk, A, Jett, J., Earle, J., Su, J., Diegert, F., Mailliard, J., Kardinal, C., Krook, J., Veeder, M., Wiesenfeld, M., Tschetter, L. and Levitt, R. (1994). Sequencing and schedule effects of cisplatin plus etoposide in small cell lung cancer results of a north central cancer treatment group randomized clinical trial. *J. Clinical Oncology* **12** 70–76.

339. Malov, S. (2019). Nonparametric estimation for a current status right-censored data model. *Statistica Neerlandica* **73** 475–495.

340. Manski, C. and McFadden, D. (1981). *Structural Analysis of Discrete Data with Econometric Applications*. Cambridge: The MIT Press.

341. Manski, C. (1988). Identification of binary response models. *Journal of American Statistical Association* **83** 729–738.

342. Mansourvar, Z., Martinussen, T. and Scheike, T. (2016). An additive–multiplicative restricted mean residual life model. *Scandinavian Journal of Statistics* **43** 487–504.

343. Martinussen, T. and Scheike, T. (2006). *Dynamic Regression Models for Survival Data*. New York: Springer.

344. McKeague, I. and Utikal, K. (1990). Stochastic calculus as a tool in survival analysis: a review. *Applied Mathematics and Computation* **38** 23–49.

345. McKeague, I. and Subramanian, S. (2001). Product-limit estimators and Cox regression with missing censoring information. *Scandinavian Journal of Statistics* **25** 589–601.

346. McLain, A. and Ghosh, S. (2011). Nonparametric estimation of the conditional mean residual life function with censored data. *Lifetime Data Analysis* **17** 514–532.

347. McMahan, C., Wang, L. and Tebbs, J. (2013). Regression analysis for current status data using the EM algorithm. *Statistics in Medicine* **32** 4452–4466.

348. Meier, P. (1975). Estimation of a distribution function with incomplete observations. *Journal of Applied Probability* **12** 67–87.

349. Meister, A. (2009). *Deconvolution Problems in Nonparametric Statistics*. New York: Springer.

350. Miller, R. (1981). *Survival Analysis*. New York: Wiley.

351. Miller, R. and Halpern, J. (1982). Regression with censored data. *Biometrika* **69** 521–531.

352. Mills, M. (2011). *Introducing Survival and Event History Analysis*. Thousand Oaks: Sage.

353. Molenberghs, G., Fitzmaurice G., Kenward, M., Tsiatis A., and Verbeke G. (Eds.) (2014). *Handbook of Missing Data Methodology*. Boca Raton: Chapman & Hall.

354. Moore, D. (2016). *Applied Survival Analysis Using R*. New York: Springer.

355. Moreno-Betancur, M., Carlin J., Brilleman, S., Tanamas, S., Peeters, A. and Wolfe, R. (2018). Survival analysis with time-dependent covariates subject to missing data or measurement error: Multiple Imputation for Joint Modeling (MIJM). *Biostatistics* **19** 479–496.

356. Morris, J. (2015). Functional regression. *Annual Review of Statistics and Its Application* **2** 321–359.
357. Müller, H.-G. and Yao, F. (2008). Functional additive models. *Journal of American Statistical Association* **103** 1534–1544.
358. Murphy, S., van der Vaart, A. and Wellner, J. (1999). Current status regression. *Mathematical Methods of Statistics* **8** 407–425.
359. Nair, K. and Nair, N. (1989). Bivariate mean residual life. *IEEE Transactions on Reliability*, **38**, 362–364.
360. Nemirovskii, A.S. (2000). *Topics in Non-Parametric Statistics*. New York: Springer.
361. Nikolskii, S.M. (1975). *Approximation of Functions of Several Variables and Embedding Theorems*. New York: Springer-Verlag.
362. Ning, J. Qin, J., and Shen, Y. (2010). Nonparametric tests for right-censored data with biased sampling. *Journal of Royal Statistical Society, B* **72** 609–630.
363. Oakes, D. (1977). The asymptotic information in censored survival data. *Biometrika* **64** 441–448.
364. Oakes, D. (1981). Survival times: aspects of partial likelihood. *International Statistical Review* **49** 235–264.
365. Oakes, D. (1989). Bivariate survival models induced by frailties. *Journal of American Statistical Association* **84** 487–493.
366. O'Quigley, J. (2021). *Survival Analysis*. Cham: Springer International Publishing.
367. Peng, Y. and Taylor, J. (2014). Cure models. *Handbook of survival analysis* **34** 113–134.
368. Perez-Jaume, S., Carrasco, J. L. and Perez-Jaume, M. S. (2023). Package "ThresholdROCsurvival".
369. Petrov, V. (1975). *Sums of Independent Random Variables*. New York: Springer.
370. Plancade, S. (2013). Adaptive estimation of the conditional cumulative distribution function from current status data. *Journal of Statistical Planning and Inference* **143** 1466–1485.
371. Poynor, V. and Kottas, A. (2019). Nonparametric bayesian inference for mean residual life functions in survival analysis. *Biostatistics* **20** 240–255.
372. Prakasa Rao, B.L.S. (1983). *Nonparametric Functional Estimation*. New York: Academic Press.
373. Prentice R. (2016). Higher dimensional Clayton–Oakes models for multivariate failure time data. *Biometrika* **103** 231–236
374. Prentice, R. and Zhao, S. (2018). Nonparametric estimation of the multivariate survivor function: the multivariate Kaplan–Meier estimator. *Lifetime Data Analysis* **24** 3–27.
375. Pruitt, R. (1993). Identifiability of bivariate survival curves from censored data. *Journal of American Statistical Association* **88** 573–579.
376. Tymes, N., Pereyra, M.C. and Efromovich S. (2000). The Application of Multiwavelets to Recovery of Signals. *Computing Science and Statistics* **33** 234–241.
377. Qian, J. and Betensky, R. (2014). Assumptions regarding right censoring in the presence of left truncation. *Statistics and Probability Letters* **87** 12–17.
378. Qian, T., Yoo, H., Klasnja, P., Almirall, D. and Murphy, S. (2021). Estimating time-varying causal excursion effects in mobile health with binary outcomes (with Discussion). *Biometrika* **108** 507–527.
379. Qiu, J., Gu, E., Zhou, D., Lawrence, J., Bai, S. and H. Hung (2019). Estimation of conditional restricted mean survival time with counting process. *Journal of Biopharmaceutical Statistics* **29** 800–809.
380. Quick, C., Dey, R. and Lin, X. (2021). Regression models for understanding COVID-19 epidemic dynamics with incomplete data. *Journal of the American Statistical Association* **116** 1561–1577.
381. Rabhi, Y. and Asgharian, M. (2017). Inference under biased sampling and right censoring for a change point in the hazard function. *Bernoulli* **23** 2720–2745.

382. Rabinowitz, D., Tsiatis, A. and Aragon, J. (1995). Regression with interval-censored data. *Biometrika* **82** 501–5013.

383. Rabinowitz, D. and Jewell, N. (1996). Regression with doubly censored current status data. *Journal of Royal Statistical Society* **58** 541–550.

384. Rigollet, P. and Tsybakov, A. (2007). Linear and convex aggregation of density estimators. *Mathematical Methods of Statistics* **16** 260–280.

385. Ritov, Y. (1990). Estimation in a linear regression model with censored data. *The Annals of Statistics* **18** 303–328.

386. Rojo, J. and Ghebremichael, M. (2006). Estimation of two ordered bivariate mean residual life functions. *Journal of Multivariate Analysis* **97** 431–454.

387. Ross, L., Prentice, R. and Zhao, S. (2019). *The Statistical Analysis of Multivariate Failure Time Data: A Marginal Modeling Approach*. Boca Raton: CRC Press.

388. Ross, S. (2023). *A First Course in Probability*. 10th ed. Upper Saddle River: Prentice Hall.

389. Rossini, A. and Tsiatis, A. (1996). A semiparametric proportional odds regression model for the analysis of current status data. *Journal of American Statistical Association* **91** 713–721.

390. Royston, P. and Lambert, P. (2011). *Flexible Parametric Survival Analysis Using Stata: Beyond the Cox Model*. College Station: Stata Press.

391. Rubin, D.B. (1987). *Multiple Imputation for Nonresponse in Surveys*. New York: Wiley.

392. Ruth, D.M., Wood, N.L. and VanDerwerken, D.N. (2023). Fully nonparametric survival analysis in the presence of time-dependent covariates and dependent censoring. *Journal of Applied Statistics* **50** 1215–1229.

393. Sakhanenko, L. (2015). Asymptotics of suprema of weighted Gaussian fields with applications to kernel density estimators. *Theory of Probability and its Applications* **59** 415–451.

394. Sakhanenko, L. (2017). In search of an optimal kernel for a bias correction method for density estimators. *Statistics and Probability Letters* **122** 42–50.

395. Salah, K. and Yousri, S. (2019). Nonparametric relative regression under random censorship model. *Statistics and Probability Letters* **151** 116–122.

396. Salerno, S. and Li, Y. (2023). High-dimensional survival analysis: Methods and applications. *Annual Review of Statistics and its Application* **10** 25–49.

397. Samarov, A. and Tsybakov, A. (2007). Aggregation of density estimators and dimension reduction'. In *Advances in Statistical Modeling and Inference*. Essays in Honor of Kjell A. Doksum, V. Nair, ed., 233–251.

398. Scott, D. (2015). *Multivariate Density Estimation: Theory, Practice, and Visualization*. 2nd ed. New York: Wiley.

399. Seok, J., Tian, L. and Wong, W. (2014). Density estimation on multivariate censored data with optional Polya tree. *Biostatistics* 15, 182–195.

400. Serfling, R. (1980). *Approximation Theorems of Mathematical Statistics*. New York: Wiley.

401. Shao, J. and Wang, L. (2016). Semiparametric inverse propensity weighting for nonignorable missing data. *Biometrika* **103** 175–187.

402. Shen, X. (2000). Linear regression with current status data. *Journal of American Statistical Association* **95** 842–852.

403. Shi, J., Chen, X. and Zhou, Y. (2015). The strong representation for the nonparametric estimation of length-biased and right-censored data. *Statistics and Probability Letters* **104** 49–57.

404. Shi, J., Xu, J. and Xu, J. (2024). Strong asymptotic properties of kernel smoothing estimation for NA random variables with right censoring. *Communications in Statistics - Theory and Methods* **53** 4531–4541.

405. Shirazi, E., Doosti , H., Niroumand, H. and Hosseinioun, N. (2013). Nonparametric regression estimates with censored data based on block thresholding method *Journal of Statistical Planning and Inference* **143** 1150–1165.

406. Silverman, B.W. (1986). *Density Estimation for Statistics and Data Analysis*. London: Chapman & Hall.
407. Spreafico, M., Ieva, F. and Fiocco, M. (2023). Modeling time-varying covariates effect on survival via functional data analysis: application to the MRC BO06 trial in osteosarcoma. *Statistical Methods and Applications* **32** 271–298.
408. Srivastava, A. and Klassen, E. (2016). *Functional and Shape Data Analysis*. New York: Springer.
409. Su, Y. and Wang, J. (2012). Modeling left-truncated and right censored survival data with longitudinal covariates. *The Annals of Statistics* **40** 1465–1488.
410. Sullivan, T., Lee, K., Ryan, P., and Salter, A. (2017). Treatment of missing data in follow-up studies of randomized controlled trials: a systematic review of the literature. *Clinical Trials* **14** 387–395.
411. Sun, J. (2007). *The Statistical Analysis of Interval-Censored Failure Time Data*. New York: Springer.
412. Sun, J. and Zhao, X. (2013). *The Statistical Analysis of Interval-Censored Failure Time Data*. New York: Springer.
413. Sun, L., Song, X. and Zhang, Z. (2012). Mean residual life models with time-dependent coefficients under right censoring. *Biometrika* **99** 185–197.
414. Susarla, V. and Van Ryzin, J. (1980). Large sample theory for an estimator of the mean survival time from censored samples. *The Annals of Statistics* **8** 1002–1016.
415. Talamakrouni, M., Van Keilegom, I., and El Ghouch, A. (2016). Parametrically guided nonparametric density and hazard estimation with censored data. *Computational Statistics and Data Analysis* **93** 308–323.
416. Tian, L. and Cai, T. (2006). On the accelerated failure time model for current status and interval censored data. *Biometrika* **93** 329–342.
417. Tikhonov, A.N. (1998). *Nonlinear Ill-Posed Problems*. New York: Springer.
418. Train, K. (2009). *Discrete Choice Methods with Simulations*. Cambridge: Cambridge University Press.
419. Tsai, W., Jewell, N. and Wang, M. (1987) A note on the product-limit estimator under right censoring and left truncation. *Biometrika* **74**, 883–886.
420. Tsai, W., Leurgans, S. and Crowley, J. (1986). Nonparametric Estimation of a Bivariate Survival Function in the Presence of Censoring. *The Annals of Statististics* **14** 1351–1365.
421. Tsiatis, A. (1975). A nonidentifiability aspect of the problem of competing risks. *Proceedings of the National Academy of Sciences* **72** 20–22.
422. Tsybakov, A. (2009). *Introduction to Nonparametric Estimation*. New York: Springer.
423. Turkson, A., Ayiah-Mensah, F. and Nimoh, V. (2021). Handling censoring and censored data in survival analysis: a standalone systematic literature review. *International Journal of Mathematics and Mathematical Sciences* 9307475.
424. Tutz, G. and Schmid, M. (2016). *Modeling Discrete Time-to-Event Data*. Cham: Springer.
425. Uzunogullari, U. and Wang, J. (1992). A comparison of hazard rate estimators for left truncated and right censored data. *Biometrika* **79** 297–310.
426. Valeriano, K. A., Galarza, C. E., Matos, L. A. and Lachos, V. H. (2023). Likelihood-based inference for the multivariate skew-t regression with censored or missing responses. *Journal of Multivariate Analysis* **196** 105–134.
427. van Buuren, S. (2018). *Flexible Imputation of Missing Data, 2nd ed*. Boca Raton: Chapman & Hall.
428. van de Geer, S. (1993). Hellinger-consistency of certain nonparametric likelihood estimators. *The Annals of Statistics* **21** 14–44.
429. Vandenbroucke, J. and Pearce, N. (2014). Case-control studies: basic concepts. *International Journal of Epidemiology* **41** 1480–1489.

430. van der Laan, M., Bickel, P. and Jewell, N. (1997). Singly and doubly censored current status data: estimation, asymptotics and regression. *Scandinavian Journal of Statistics* **24** 289–307.

431. van der Vaart, A. (1991). On differentiable functionals. *The Annals of Statistics* **19** 178–204.

432. van Es, B. and Graafland, C. (2017). *Nonparametric Kernel Density Estimation for Univariate Current Status Data*. Online Manuscript. arXiv: 1707.00544v1

433. van Houwelingen, H. and Putter, H. (2011). *Dynamic Prediction in Clinical Survival Analysis*. Boca Raton: Chapman & Hall.

434. Van Keilegom, I. and Veraverbeke, N. (2001). Hazard rate estimation in nonparametric regression with censored data. *Annals of the Institute of Statistical Mathematics* **53** 730–745.

435. Walter, G. (1994). *Wavelets and Other Orthogonal Systems with Applications*. London: CRC Press.

436. Wang, C. and Chan, K. (2018). Quasi-likelihood estimation of a censored autoregressive model with exogenous variables. *Journal of the American Statistical Association* **113** 1135–1145.

437. Wang, P., Li, Y. and Reddy, C. (2019). Machine learning for survival analysis: a survey. *ACM Computing Surveys* **51** 1–36.

438. Wang, P., Tong, X. and Sun, J. (2018). A semiparametric regression cure model for doubly censored data. *Lifetime Data Analysis* **24** 492–508.

439. Wang, W. and Wells, M. (1997). Nonparametric estimators of the bivariate survival function under simplified censoring conditions *Biometrika* **84** 863–880.

440. Wang, Y., Zhou, Z., Zhou, X., and Zhou, Y. (2017). Nonparametric and semiparametric estimation of quantile residual lifetime for length-biased and right-censored data. *The Canadian Journal of Statistics* **45** 220–250.

441. Wang, X., Balakrishnan, N., Guo, B. and Jiang, P. (2015). Residual life estimation based on bivariate non-stationary gamma degradation process. *Journal of Statistical Computation and Simulation* **85** 405–421.

442. Wang, Y., Yuan, X. and Wang, C. (2023). Estimation and variable selection for single-index models with non ignorable missing data. *Communications in Statistics-Theory and Methods* **53** 42–64.

443. Wasserman, L. (2006). *All of Nonparametric Statistics*. New York: Springer.

444. Wellner, J. (1982). Asymptotic optimality of the product limit estimator. *The Annals of Statistics* **10** 595–602.

445. Wellner, J. (1995). Interval censoring, case 2: alternative hypotheses. Analysis of censored data. *IMS Lecture Notes Monogr. Ser.* **27** 271–291.

446. Wells, M. and Yeo, K. (1996). Density estimation with bivariate censored data. *Journal of American Statistical Association* **91** 1566–1574.

447. Wiegrebe, S., Kopper, P., Sonabend, R. and Bender, A. (2024). Deep learning for survival analysis: a review. *Artificial Intelligence Reviews* **57** 1–34.

448. Wienke, A. (2011). *Frailty Models in Survival Analysis*. Boca Raton: Chapman & Hall.

449. Windmeijer, F. (1995). Goodness-of-fit measures in binary choice models. *Econometric Reviews* **14** 101–116.

450. Wu, Y. and Zhang, Y. (2012). Partially monotone tensor spline estimation of the joint distribution function with bivariate current status data. *The Annals of Statistics* **40** 1609–1636.

451. Xu, D., Zhao. S., Hu, T. and Sun, J. (2019). Regression analysis of informatively interval-censored failure time data with semiparametric linear transformation model. *Journal of Nonparametric Statistics* **31** 663–679.

452. Xue, H., Lam, K. and Li, G. (2004). Sieve maximum likelihood estimator for semiparametric regression models with current status data. *Journal of American Statistical Association* **99** 346–356.

453. Xue, L. (2024). Empirical likelihood and estimation in varying coefficient models with right censored data. *Computational Statistics* **39** 1683–1707.

454. Yang, S. (2000). Functional estimation under interval censoring case 1. *Journal of Statistical Planning and Inference* **89** 135–144.
455. Yang, S., Tsiatis, A. and Blazing, M. (2018). Modeling survival distribution as a function of time to treatment discontinuation: A dynamic treatment regime approach. *Biometrics* **74** 900–909.
456. Yildiray, Y. (2008). Estimating default probabilities of CMBS loans with clustering and heavy censoring. *The Journal of Real Estate Finance and Economics* **37** 93–111.
457. Yildiray, Y. (2013). Estimating default probabilities of CMBS loans with clustering and heavy censoring. *Journal of Real Estate and Financial Economics* **37** 93–111.
458. Ying, Z., Jung, S. and Wei, L. (1995). Survival analysis with median regression models, *Journal of the American Statistical Association* **90** 178–184.
459. Yu, H. and Guo, Y. (2024). *Survival Analysis*. In *Textbook of Medical Statistics: For Medical Students* 191-208. Singapore: Springer Nature.
460. Yu, M., Zhao, W., Zhou, Y. and Wu, C. (2023). Robust online detection on highly censored data using a semi-parametric EWMA chart, *Journal of Statistical Computation and Simulation* **93** 1403–1419.
461. Zhang, C., Wu, Y. and Yin, G. (2020). Restricted mean survival time for interval-censored data. *Statistics in Medicine* **39** 3879–3895.
462. Zhang, F. and Zhou, Y. (2013). Analyzing left-truncated and right-censored data under Cox model with long-term survivors. *Acta Mathematicae Applicatae Sinica* **29** 241–252.
463. Zhang, Z., Reinikainen, J., Adeleke K., Pieterse ME, Groothuis-Oudshoorn C. (2018). Time-varying covariates and coefficients in Cox regression models. *Annals of Translation Medicine* **121** 1–10.
464. Zhao H., Wu Q., Li G., and Sun J. (2020). Simultaneous estimation and variable selection for interval-censored data with broken adaptive ridge regression. *Journal of the American Statistical Association* **115** 204–216.
465. Zhao, L. and Feng, D. (2020). Deep neural networks for survival analysis using pseudo values. *IEEE Journal of Biomedical and Health Informatics* **24** 3308–3314.
466. Zhong, Q., Mueller, J. and Wang, J. (2022). Deep learning for the partially linear Cox model, *Annals of Statistics* **50** 1348–1375.
467. Zhou, Q. and Wong, K. (2024). Improving estimation efficiency of case-cohort studies with interval-censored failure time data. *Statistical Methods in Medical Research* 09622802241268601.
468. Zhong, Y. and Schaubel, D. (2022). Restricted mean survival time as a function of restricted time. *Biometrika* **78** 192–201.
469. Zhou, M. (2019). *Empirical Likelihood Method in Survival Analysis*. Boca Raton: Chapman & Hall.
470. Zhou, Q., Hu, T. and Sun, J. (2017). A sieve semiparametric maximum likelihood approach for regression analysis of bivariate interval-censored failure time data. *Journal of the American Statistical Association* **112** 664–672.
471. Zou, Y. and Liang, H. (2017). Wavelet estimation of density for censored data with censoring indicator missing at random. *A Journal of Theoretical and Applied Statistics* **51** 1214–1237.

Index